Satellites for Atmospheric Sciences 2

SCIENCES

Space Exploration and Technology,
Field Director – Jean-Luc Lefebvre

Space and Earth, Subject Head – Jean-Luc Lefebvre

Satellites for Atmospheric Sciences 2

Meteorology, Climate and Atmospheric Composition

Coordinated by
Thierry Phulpin
Didier Renaut
Hervé Roquet
Claude Camy-Peyret

iSTE

WILEY

First published 2023 in Great Britain and the United States by ISTE Ltd and John Wiley & Sons, Inc.

ISTE Ltd
27-37 St George's Road
London SW19 4EU
UK

www.iste.co.uk

John Wiley & Sons, Inc.
111 River Street
Hoboken, NJ 07030
USA

www.wiley.com

Library of Congress Control Number: 2023941805

British Library Cataloguing-in-Publication Data
A CIP record for this book is available from the British Library
ISBN 978-1-78945-141-2

ERC code:
PE10 Earth System Science
PE10_1 Atmospheric chemistry, atmospheric composition, air pollution
PE10_2 Meteorology, atmospheric physics and dynamics
PE10_14 Earth observations from space/remote sensing
PE9 Universe Sciences
PE9_15 Space Sciences
PE9_17 Instrumentation - telescopes, detectors and techniques

Contents

Acknowledgments

The coordinator is very grateful to Didier Renaut, Hervé Roquet and Claude Camy-Peyret for their significant help in editing this book through their advice, proofreading, corrections to chapters and help with the organization. Their support has been invaluable in bringing together quality contributions in a book that will cover the different facets of the topic *Space and Atmosphere* in a comprehensive manner. Gratitude is also expressed to Jérome Lafeuille and Jean Pailleux for their proofreading and improvements to various chapters. Finally, special thanks to the journal *La Météorologie* for having authorized the use of the glossary established for the special issue of the journal, No. 97 in 2019, devoted to satellite missions for meteorology and climate.

List of Acronyms

(3D/4D)-Var	(3/4 Dimensional) – Variational
4A/OP	Automatized Atmospheric Absorption Atlas/OPerational
AC-VC	Atmospheric Composition Virtual Constellation
ACCP	Aerosols, Clouds, Convection and Precipitation
ACX	GeoXO Atmospheric Composition
ADM	Atmospheric Dynamics Mission
AERIS	Data and Services for the Atmosphere
AERONET	AErosol RObotic NETwork
AI	Artificial Intelligence
AMV	Atmospheric Motion Vector
AOD	Aerosol Optical Depth
APHRODITE	Asian Precipitation – Highly-Resolved Observational Data Integration Towards Evaluation

APT	Automatic Picture Transmission
ARTS	Atmospheric Radiative Transfer Simulator
ASI	Agenzia Spaziale Italiana (Italian Space Agency)
AVCS	Advanced Vidicon Camera System
BC	Black Carbon
BIPM	International Bureau of Weights and Measures
BRDF	Bidirectional Reflectance Distribution Function
C3IEL	Cluster for Cloud evolution, ClImatE and Lightning
C3S	Copernicus Climate Change Service
CAMS	Copernicus Atmosphere Monitoring Service
CAPE	Convective Available Potential Energy
CALIPSO	Cloud-Aerosol Lidar and Infrared Pathfinder Satellite Observation
CATS	Cloud-Aerosol Transport System
CCI	Climate Change Initiative
CCSDS	Consultative Committee on Space Data Systems
CDR	Climate Data Record
CDRD	Cloud Dynamics and Radiation Database
CEOS	Committee on Earth Observation Satellites
CFC	ChlorFluoroCarbon
CGMS	Coordination Group for Meteorological Satellites
CI	Convection Initiation
CIMSS	Cooperative Institute for Meteorological Satellite Studies

CLW	Cloud Liquid Water
CM	Cloud Mask
CM-SAF	Climate Monitoring – Satellite Application Facility
CMA	Chinese Meteorological Agency
CMEMS	Copernicus Marine Environment Monitoring Service
CMORPH	CPC MORPHing technique
CMV	Cloud Motion Vector
CNES	Centre national d'études spatiales France
CNRM	Centre national de la recherche météorologique France
CNRS	Centre national de la recherche scientifique France
CNSA	Chinese National Space Administration
COCCON	COllaborative Carbon Column Observing Network
CoMet	Carbon Dioxide and Methane field campaign
CONTRAIL	Comprehensive Observation Network for TRace gases by AIrLiner
COP	Conference Of the Parties
CORRA	COmbined Radar-Radiometer Algorithm
CPC	Climate Prediction Center
CRM	Cloud Resolving Model
CRTM	Community Radiative Transfer Model
CTTH	Cloud Top Temperature and Height
CWDP	Commercial Weather Data Pilot

CYGNSS	CYclone Global Navigation Satellite System
DAOD	Differential AOD
DBNet	Direct Broadcast NETwork WMO-OMM
DCS	Data Collection System
DDA	Discrete Dipole Approximation
DIAL	DIfferential Absorption Lidar
DISORT	DIScrete Ordinate Radiative Transfer
DLR	Deutsches Zentrum für Luft- und Raumfahrt Allemagne
DMSP	Defense Meteorological Satellite Program
DNB	Day and Night Band
DOFS	Degrees of Freedom for Signal
DU	Dobson Unit
DVB	Digital Video Broadcast
DVB-S	Digital Video Broadcast by Satellite
EC	European Commission
ECMWF	European Centre for Medium-term Weather Forecast
ECOSTRESS	Ecostress Spectral Library Database
ECV	Essential Climate Variable
EDR	Environmental Data Record
EE	Earth Explorer
EEA	European Environment Agency
EF	Emission Factor

ELDO	European Launcher Development Organisation
ENSO	El Niño Southern Oscillation
EOS	Earth Observing System
EOSDIS	EOS Data Information System
EPS	EUMETSAT Polar System
ERA	ECMWF ReAnalysis
ERB	Earth Radiation Budget
ESA	European Space Agency
ESE	Earth Science Enterprise
ESRO	European Space Research Organisation
ESSA	Environmental Science Services Administration
ESSP	Earth System Science Pathfinder
ESTO	Earth Science Technology Office
EUMETSAT	EUropean organisation for the exploitation of METeorological SATellites
EW	Earth Watch
EXIM	EXtrapolated IMagery products
FASTEM	FAST Emissivity Model
FCDR	Fundamental Climate Data Record
FCI	Flexible Combined Imager
FCover	Fractional Cover (of vegetation)
FDR	Fundamental Data Record

FGGE	First GARP Global Experiment
FFT	Fast Fourier Transform
FGGE	First Global GARP Experiment WMO-OMM
FIR	Far IR
FOR	Field Of Regard
FOV	Field Of View
FSOI	Forecast Sensitivity Observation Impact
FSR	Forecast Sensitivity to R (the observation error covariance matrix)
FT	Fourier Transform
FTS	Fourier Transform Spectrometer
FY	Feng Yun
GADS	Global Aerosol Data Set
GARP	Global Atmospheric Research Programme WMO-OMM
GAS	GMES Atmospheric Service
GASP	GOES Aerosol/Smoke Product
GAW	Global Atmospheric Watch
GCM	Global Climate Model
GCOS	Global Climate Observing System
GDP	Gross Domestic Product
GEO	GEostationary Orbit
GEO	Group on Earth Observation
GEOS-5	Goddard Earth Observing System model – version 5

GEOSS	Global Earth Observation System of Systems
GeoXO	GEOstationary eXtended Observations
GEWEX	Global Energy and Water Cycle Experiment
GHCN	Global Historical Climatology Network
GHG	GreenHouse Gas
GLM	Geostationary Lightning Mapper
GMES	Global Monitoring for Environment and Security
GMF	Geophysical Model Function
GNP	Gross National Product
GNSS	Global Navigation Satellite System
GOCCP	GCM-Oriented CALIPSO Cloud Product (CALIPSO-GOCCP)
GPCC	Global Precipitation Climatology Center
GPCP	Global Precipitation Climatology Project
GPI	GOES Precipitation Index
GPM-CO	GPM-Core Observatory
GPROF	Goddard PROFiling algorithm
GPS	Global Positioning System
GRUAN	Global Reference Upper-Air Network GCOS
GSFC	Goddard Space Flight Center
GSICS	Global Space-based Inter-Calibration System
GSIP	GOES Surface and Insolation Project
GSMaP	Global Satellite Mapping of Precipitation

GSOD	Global Summary Of the Day
GV	Ground Validation
H-SAF	Hydrology – Satellite Application Facility
HCFC	HydroChloroFluoroCarbon
HEO	High Eccentricty Orbit
HEPAD	High Energy Proton and Alpha Detector
HFC	HydroFluoroCarbon
HLOS	Horizontal Line Of Sight
HOAPS	Hamburg Ocean Atmosphere Parameters and fluxes from Satellite data
HRIT	High Rate Image Transmission
HRPT	High Resolution Picture Transmission
HSRL	High Spectral Resolution Lidar
IAGOS	In-service Aircraft for a Global Observing System
IBTrACS	International Best Track Archive for Climate Stewardship
ICOADS	International Comprehensive Ocean-Atmosphere Data Set
ICOS	Integrated Carbon Observation System
IEC	International Electrotechnical Commission
IF	Intermediate Frequency
IFS	Integrated Forecasting System
IIP	Instrument Incubator Program

IJSP	International Joint Polar System
ILS	Instrument Line Shape
IMD	Indian Meteorological Department
IMERG	Integrated Multi-satellitE Retrievals for GPM
IMS	Ice Mapping System
IMS	Interactive Multisensor Snow
INDOEX	INDian Ocean EXperiment
INPE	Instituto Nacional de Pesquisas Espaciais Brésil
InVEST	In-space Validation of Earth Science Technology
IOC	Intergovernmental Oceanographic Commission UNESCO
IODC	Indian Ocean Data Coverage
IPCC	Intergovernmental Panel on Climate Change
IPDA	Integrated Path Differential Absorption
IPSL	Institut Pierre-Simon Laplace France
IPWG	International Precipitation Working Group CGMS
IR	Infra-Red
IROWG	International Radio Occultation Working Group CGMS
ISA	Israel Space Agency
ISCCP	International Satellite Cloud Climatology Project
ISO	International Standardization Organization
ISRF	Instantaneous Spectral Response Function
ISRO	Indian Space Research Organization

ISS	International Space Station
ITOS	Improved TIROS Operational System
ITWG	International TOVS Working Group CGMS
IUPAC	International Union of Pure and Applied Chemistry
IWP	Ice Water Path
IWWG	International Winds Working Group CGMS
JAXA	Japan Aerospace eXploration Agency
JCSDA	Joint Center for Satellite Data Assimilation
JMA	Japan Meteorological Agency
JPL	Jet Propulsion Laboratory
JPSS	Joint Polar Satellite System
JRA	Japanese ReAnalysis
KMA	Korean Meteorological Agency
LAI	Leaf Area Index
LAMP	Laboratoire de météorologie physique France
LATMOS	Laboratoire atmosphères, milieux, observations spatiales France
LEO	Low Earth Orbit
LI	Lifted Index
LIDORT	LInearized Discrete Ordinate Radiative Transfer
LMD	Laboratoire de météorologie dynamique France
LMT	Local Mean solar Time
LMT	LowerMost Troposphere

LOA	Laboratoire d'optique atmosphérique France
LOS	Line Of Sight
LRIT	Low Rate Information Transmission
LRPT	Low Resolution Picture Transmission
LRR	Laser Retro-Reflector
LSE	Land Surface Emissivity
LSI	Low Stream Approximation
LST	Land Surface Temperature
LTE	Local Thermodynamic Equilibrium
LUT	Look-Up Table
MAG	Mission Advisory Group
MAGIC	Monitoring Atmospheric composition and Greenhouse gases through multi-Instrument Campaigns
MEO	Middle Earth Orbit
MEPED	Medium Energy Proton and Electron Detector
MERRA	Modern-Era Retrospective analysis for Research and Applications
MOP	Meteosat Operational Programme
MOZAIC	Measurement of OZone and water vapour by Airbus In-service airCraft)
MSWEP	Multi-Source Weighted-Ensemble Precipitation
MTP	Meteosat Transition Programme
MW	MicroWave

MWR	MW Radiometer
NASA	National Aeronautics and Space Administration
NCEP	National Center for Environmental Prediction
NDACC	Network for the Detection of Atmospheric Composition Change
NDVI	Normalized Difference Vegetation Index
NEDT	Noise Equivalent Differential Temperature
NESDIS	National Environmental Satellite, Data, and Information Service (NOAA)
NIR	Near InfraRed (~ 1.0 to 0.7 μm)
NMVOC	Non-Methane Volatile Organic Compounds
NOAA	National Oceanic and Atmospheric Administration
NOAA/NESDIS	National Oceanic and Atmospheric Administration Satellite and Information Service
NPP	NPOESS Preparatory Program
NRT	Near Real-Time
NSOAS	National Satellite Ocean Application Service
NSOSA	NOAA Satellite Observing System Architecture study
NSSDC	NASA Space Science Data Center
NUCAPS	NOAA Unique Combined Atmospheric Processing System
NWP	Numerical Weather Prediction
NWP-SAF	Numerical Weather Prediction – Satellite Application Facility

OceanRAIN	Ocean Rain And Ice-phase precipitation measurement Network
ODS	Ozone Depleting Substances
OI	Optimal Interpolation
OLR	Outgoing Longwave Radiation
OLS	Operational Linescan System
OPAC	Optical Properties of Aerosols and Clouds
OPD	Optical Path Difference
OSCAR	Observing Systems Capability Analysis and Review tool
OSE	Observing System Experiments
OSSA	Office of Space Science Applications
OSSE	Observing System Simulation Experiments
OSTIA	Operational Sea surface Temperature and Ice Analysis
PACRAIN	Pacific Rainfall Database
PAN	PeroxyAcetyl Nitrate
PAOB	PAid OBservation
PAR	Photosynthetically Active Radiation
PATMOS-x	Pathfinder ATMOSpheres – eXtended
PERSIANN	Precipitation Estimation from Remotely Sensed Information using Artificial Neural Networks
PERSIANN-CCS	PERSIANN-Cloud Classification System
PM	Particulate Matter
PM2.5	Particulate Matter (diameter smaller than 2.5 μm)

PMW	Passive MicroWave
PNPR	Passive microwave Neural network Precipitation Retrieval
POA	Primary Organic Aerosol
POEM	Polar-Orbit Earth-Observation Mission
PPS	Precipitation Processing System
PRPS	Precipitation Retrieval and Profiling Scheme
PSC	Polar Stratospheric Cloud
QA4EO	Quality Assurance Framework for Earth Observation
QI	Quality Index
QRNN	Quantile Residual Neural Network
R&T	Research & Technology
RDT	Rapidly Developing Thunderstorm
RF	RadioFrequency
RO	Radio Occultation
ROC	Reactive Organic Carbon
ROSCOSMOS	Russian Federal Space Agency
RR	Rain Rate
RSS	Rapid Scanning Service
RT	Radiative Transfer
RTE	Radiative Transfer Equation
RTTOV	Rapid Transmission for TOVs
RTTOV-SCAT	RTTOV with SCATtering

SAF	Satellite Application Facility
SAG	Science Advisory Group
SAGE	Stratospheric Aerosol and Gas Experiment
SAM-II	Stratospheric Aerosol Measurement II
SC	Snow Cover
SCaMPR	Self-Calibrating Multivariate Precipitation Retrieval
SEM	Space Environment Monitor
SFCG	Space Frequency Coordination Group
SI	International System
SIC	Sea Ice Concentration
SNPP	Suomi National Polar-orbiting Partnership
SNR	Signal to Noise Ratio
SOA	Secondary Organic Aerosol
SORCE	SOlar Radiation and Climate Experiment
SOS	Sum Of Squares
SPARC	Stratosphere-troposphere Processes And their Role in Climate
SR	Scattering Ratio
SRSOR	Super Rapid Scan Operations for GOES-R
SSA	Single Scattering Albedo
SSEC	Space Science and Engineering Center
SST	Sea Surface Temperature
STAR	(Center for) Satellite Applications and Research

SWE	Snow Water Equivalent
SWIR	ShortWave InfraRed (~ 2.5 to 1.0 μm)
SZA	Solar Zenithal Angle
Tb	Brightness temperature
TCCON	Total Carbon Column Observing Network
TCHP	Tropical Cyclone Heat Potential
TCWV	Total Column Water Vapor
TED	Total Energy Detector
TEMPEST-D	Temporal Experiment for Storms and Tropical Systems – Demonstration
TEMPO	Tropospheric Emissions: Monitoring of Pollution
TIR	Thermal InfraRed (~ 15 to 5 μm)
TLS	Temperature of the Lower Stratosphere
TLT	Temperature of the Lower Troposphere
TMPA	TRMM Multisatellite Precipitation Analysis
TMS	Temperature of the Mid-Stratosphere
TMT	Temperature of the Mid-Troposphere
TOAR	Tropospheric Ozone Assessment Report
TOR	Tropospheric Ozone Residual
TPW	Total Precipitable Water
TTS	Temperature of the Top Stratosphere
TUS	Temperature of the Upper Stratosphere
TUT	Temperature of the Upper Troposphere

TV	TeleVision
UMORA	Unified Microwave Ocean Retrieval Algorithm
UN	United Nations
UNESCO	United Nations Educational, Scientific and Cultural Organization
UNFCCC	United Nations Framework Convention on Climate Change
UT/LS	Upper Troposphere/Lower Stratosphere
UTH	Upper Tropospheric Humidity
UV	Ultraviolet (~ 400 to 200 nm)
UV-B	Ultraviolet-B (~ 315 to 280 nm)
UVAI	Ultraviolet Aerosol Index
UVN	Ultraviolet, Visible and Near-infrared
VAAC	Volcanic Ash Advisory Centre
VarBC	Variational Bias Correction
VHI	Vegetation Health Index
VIS	VISible
VLIDORT	V-LInearized Discrete Ordinate Radiative Transfer
VOC	Volatile Organic Compound
WCRP	World Climate Research Program
WEFAX	Weather Facsimile
WF	Weighting Function

WMO	World Meteorological Organization
WWW	World Weather Watch

Introduction

Thierry PHULPIN
Retired from CNES and Météo-France, Toulouse, France

Observation of the Earth System from space

Satellite images are now in common use during television weather forecasts and are now also accessible on smartphones. Nowadays, everyone is introduced to this vision of the Earth from space at school. Thanks to satellites and their imagers, planet Earth appears to everyone as a familiar environment and is seen as a system in which all environments are interconnected. When a storm breaks out, its intensity is marked by the depth of the gyre formed by the depression and translates into very powerful winds whose impact on the ocean is very noticeable in data from oceanographic satellites. A volcanic eruption or an earthquake leaves traces in the atmosphere (ash clouds, ionospheric signals, etc.), which can be perceived and monitored. Heavy precipitation visualized using microwave instruments and droughts measured with sounders leave surface footprints that can be detected with satellites intended for monitoring continental resources. The relationship between the nature of the surface (oceans, desert, soil, bare ground, urban environment, crops, forests, etc.) and its state at a given moment (roughness, water state), in connection with atmospheric conditions, is decisive for meteorology via the mechanisms of thermal emission, or evaporation. All of these relationships can be characterized, inferred or measured by satellites, some intended for oceanography and others for monitoring land surfaces, or even for meteorology. These links between the various compartments of the Earth System began to be clearly highlighted as soon as the human or animal populations inhabiting the Earth began to move between the continents, thanks to discoveries by explorers. These discoveries, by allowing the development of a trade economy, led to the establishment of an interdependent system and had a major impact on a planetary scale, particularly on the various ecosystems. We could therefore

Satellites for Atmospheric Sciences 2,
coordinated by Thierry PHULPIN, Didier RENAUT, Hervé ROQUET and Claude CAMY-PEYRET.

already speak of an Earth System as early as the 16th century. However, it is the links between the observed natural phenomena that have made it possible to construct the notion of the climate system. Intuitively, we have perceived over the centuries that many phenomena responsible for the evolution of measurable variables were linked. These concepts can be visualized in diagrams, for example, Figure I.1. However, thanks to satellites, it has been possible to advance in the understanding of the processes (in addition to field studies), quantify certain phenomena at scales not possible from ground measurements alone, establish parametric links between certain variables and ultimately improve the physics of the models and then calibrate or validate these models.

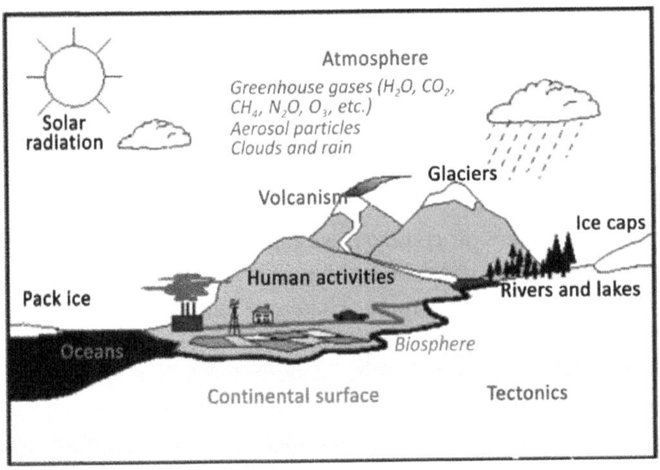

Figure I.1. *The different compartments of the Earth System, taken into account in a climate model. For a color version of this figure, see www.iste.co.uk/phulpin/satellites2.zip*

It is therefore logical to address the various subsystems of the Earth in a work devoted to space science, and to consider them in a coherent approach.

The atmosphere

Volume 1 is devoted to satellite observation of the atmosphere and the sciences it has developed: meteorology, atmospheric composition, as well as knowledge of the climate and its evolution.

According to Blum (2019)[1], it is the mastery of meteorology that would have allowed the WWII allies to land, demonstrating the major strategic importance of investing in an observation network. Space science and technology started early on, having been boosted by know-how of German engineers who developed the V2. The first photos of the Earth were taken in 1960, which highlighted how important the images were for providing an overview of the spatial distribution of cloud cover and its movement. The US then created a program of meteorological satellites, preceding the Soviets, followed by Europe and Japan. The history of these developments will be discussed in Chapter 1 of Volume 1. For meteorology, it is useful for the satellite measurements to be taken at the same solar time, to better follow the evolution of the thermodynamic parameters by comparing them with the readings of the previous day at the same time, hence the use of Sun-synchronous polar orbits. On the other hand, the ability to observe the Earth continuously from geostationary orbit is also very attractive for deducing information about the wind from the apparent movements of clouds and observing the development of storms or the movement of cyclones. The orbits used for these missions and their advantages are described in Chapter 8 of Volume 1. It quickly became clear that the long-term continuity of observations under identical conditions is necessary to integrate these observations into the international network of meteorological measurements and to facilitate their use in forecasting models. We then saw a dichotomy of missions appear; on the one hand, a series of identical operational satellites for a program spanning 20 or 30 years and, on the other hand, satellites equipped with innovative instruments for research. In the US, this dichotomy is reflected in the division of responsibilities between NOAA and NASA. These agencies pioneered meteorology from space. The history of their programs to date is presented in the first chapters of Part 1 of Volume 1. This part shows how the programs are developed to meet the needs of users, using the most advanced technologies whose developments are supported by the agencies that lead in this field. Following the American agencies, Chapters 4–6 of Volume 1 are devoted to the European programs of ESA, EUMETSAT and CNES. We see in Chapter 7 of Volume 1 the role played by the World Meteorological Organization in coordinating operational projects from the US, Europe or Asia. The chapter also discusses the issue of data acquisition anywhere on the planet for real-time use.

Part 2 of Volume 1 is devoted to the physical basis of satellite observations for atmospheric sciences. Chapter 8 of Volume 1 presents the principles of orbital mechanics to determine how to meet the need for observations by meteorologists or orbits more suited to different issues such as monitoring precipitation in intertropical

---

1. Blum, A. (2019). *The Weather Machine. How We see in the Future*. The Bodley Head, London.

zones or air quality in the most polluted regions. The first operational satellites only offered observation in a visible "channel", similar to a photograph from the sky, and another in an infrared channel giving an idea of the temperature of the objects targeted. These passive observations have allowed advances in cloud mapping and sea temperature but have proven insufficient to access other meteorological information. The range of observations was quickly extended with observations in multiple optical channels, and the observation domain was extended to the microwave domain, allowing for, due to the digitization of measurements, the transition to the quantitative use of observations. Measurement physics, developed to propose passive or active measurement techniques and better use the resulting data, is presented in Chapter 9 of Volume 1. In Chapter 10 of Volume 1, we present the problem of finding atmospheric variables of interest from the observations and the main methods used for this purpose.

Part 1 of this book is more specifically devoted to meteorology. With regard to the three-dimensional distribution of temperature, humidity and wind, the main data useful for meteorology has long been provided by radiosondes. However, these observations are rare over the oceans and are therefore supplemented with satellite measurements of vertical distributions of temperature and humidity. The technique used is based on the selective absorption of carbon dioxide or oxygen, which is scarce but evenly distributed in the atmosphere (at least to the first order), and of water vapor and other greenhouse gases, which strongly absorb infrared or microwave radiation. This method, first tested by the Americans with the Infrared Interferometer Spectrometer (IRIS*) on the Nimbus IV satellite, then gave rise to the TIROS Operational Vertical Sounder (TOVS*) radiometric sounders, before technically evolving towards much more precise spectrometric measurements with the Atmospheric Infrared Sounder (AIRS*) and Infrared Atmospheric Sounding Interferometer (IASI*). Chapters 1 to 3 present the restitution of the main variables characterizing the state of the atmosphere, or that of the surface variables directly useful or necessary for the inversion of the atmospheric characteristics. Chapters 4 to 6 focus on the major applications of these observations: numerical weather forecasting, short-term forecasting and cyclone tracking.

Atmospheric sounding missions using spectrometry are successful by analyzing minority atmospheric components that are important for the planet or for human health, such as stratospheric ozone, gases and particles that are involved in air quality or the greenhouse effect.

The use of space observations and the benefits derived from them are presented in Part 2 of this book: detection of polluting species dangerous to health (Chapters 7

*See Glossary.

and 8), observation of clouds of desert dust and elements emitted by bush or forests. Chapter 11 is devoted to stratospheric chemistry, better known and understood thanks to the many satellite observations devoted to monitoring the hole in the ozone layer.

Because they offer precise and well-targeted observations with global spatial coverage, satellites have contributed significantly to better understanding the processes at play in the Earth's climate. In addition, the continuity of long-term observations (more than 30 years of data!) makes it possible to monitor the evolution of the climate and to attribute certain phenomena to this evolution. Part 3 of this book concerns the use of satellite observations for the analysis of climatic processes and the monitoring of climatic variables. In its introduction, Chapter 12 recalls the importance of well-calibrated data, the need to reprocess the data acquired over time in a homogeneous manner and to repeat the reprocessing of the products as and when more efficient algorithms and processes become available. Chapter 13 is devoted to the restitution of the greenhouse gases CO_2 and CH_4 and to the location of the sinks and sources of these gases, which play a major role in the warming of the atmosphere. Clouds and water vapor also play an essential role. However, these variables have been monitored from the very first satellites. Analytical techniques have been refined and new data produced with more precise instruments have arrived, making it possible to more precisely estimate the spatio-temporal distribution of atmospheric water (Chapter 14). Clouds and water vapor are also involved in precipitation: the distribution and intensity of which are regularly monitored (Chapter 15).

In addition to an index, this book also includes a glossary to clarify the definition of certain concepts, appendices including a compilation of the names of instruments and satellites in question (to be updated as and when new missions occur), as well as a list of acronyms.

PART 1

Meteorology

Introduction to Part 1

Meteorology and the Contribution of Satellite Observations

Hervé ROQUET
Direction de l'Enseignement Supérieur et de la Recherche de Météo-France, Saint-Mandé, France

I1.1. From image use...

Observation is an essential activity in meteorology. Observations are essential for real-time knowledge and monitoring of the main relevant atmospheric variables (pressure, temperature, humidity, wind, cloud variables and precipitation), as well as for the detection and monitoring of hazardous weather phenomena affecting human activities. They are also used for the preparation and monitoring of weather predictions. In particular, they make it possible to determine the initial state of the atmosphere from which numerical weather prediction models can predict its evolution. Finally, observations allow for continuous progress in the knowledge of meteorological phenomena and the physical processes that govern them.

In the history of meteorological observations, the advent of satellite imagery was a major milestone, giving meteorologists real-time access for the first time to a spatialized view of cloud systems and the meteorological phenomena they represent, such as fronts, mid-latitude weather disturbances, thunderstorms and tropical

Satellites for Atmospheric Sciences 2,
coordinated by Thierry PHULPIN, Didier RENAUT, Hervé ROQUET and Claude CAMY-PEYRET.

cyclones. The direct interpretation of an image by forecasters[1] is still an important part of the use of satellite imagery in meteorology today, whether to analyze a situation or to assess predictions derived from numerical weather predicting models.

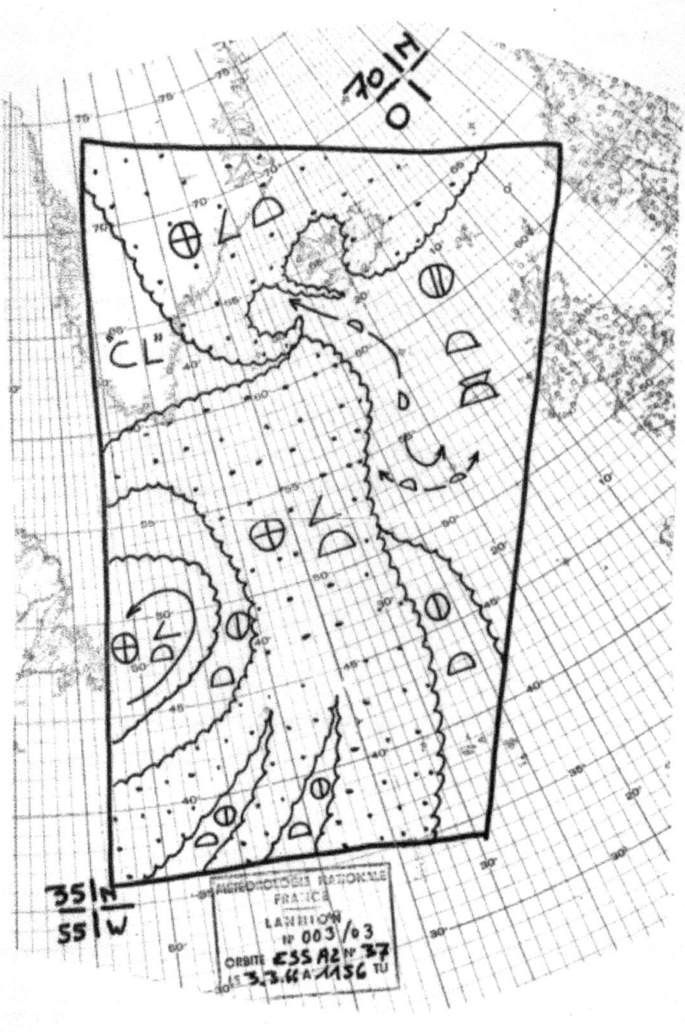

Figure I1.1. *Example of nephanalysis produced by the Météo-France Space Meteorology Center on March 3, 1966 from images taken by the Vidicon camera of the American satellite ESSA-2 (source: Météo-France, Space Meteorology Center)*

1. Persons in charge within the meteorological services of the final development of expert forecasts.

The instrument for observing on the first meteorological satellites was only a camera, which provided analog black and white photographs of the Earth and clouds of the regions they flew over. Their meteorological use was greatly limited not only by the image quality, but also by the fact that the identification of cloud types from a single black and white image in the visible domain is difficult even for an experienced user. An important step was the advent of the first radiometers with simultaneously a channel in the visible and in the thermal infrared, providing not only a usable observation at night, but also a better ability to distinguish the different types of clouds, the infrared channel providing information on the temperature and therefore on the altitude of their summit. For this reason, meteorologists have become accustomed to simultaneously exploiting images in the visible range in black and white with a natural dynamic (clouds appearing in white, because they are brighter than the surface), and images in the infrared range with an inverted dynamic (clouds also appearing in white, because they are cooler than the surface).

At the beginning of the operational use of satellite imagery in meteorology, the means of reception, processing and dissemination being essentially analogous, an operational production of nephanalyses was set up in some meteorological services (see Chapter 1). These maps, developed manually by experts in the meteorological interpretation of images (referred to as "nephanalysts"), made it possible to delineate the main cloud zones and the associated cloud cover, as well as the main types of clouds identified, and were distributed by facsimile to various weather prediction centers. This activity disappeared when computerized means of receiving, processing and displaying data became widespread, which made it possible not only to automate processing operations, such as the application of different dynamics to improve image display and the change of geographical projection, but also to transmit images via computer networks and direct display on forecasters' workstations.

The advent of geostationary meteorological satellites was a crucial advance in providing access to visible and thermal infrared images with high temporal repetitiveness, enabling meteorologists to animate them to visualize the movements and evolution of cloud systems, and thus gain access to information on the dynamics of the atmosphere. In addition, the presence of an infrared channel in the water vapor absorption band (referred to by meteorologists as the "water vapor" channel) provided additional information on humidity in the mid-troposphere. The structure and evolution of the water vapor field is an important element in anticipating or monitoring the evolution of meteorological phenomena such as weather disturbances and convective systems, and indirectly provides information on the vertical movements of the atmosphere on a large scale; drying zones are associated with subsidence movements[2] and vice versa.

Figure I1.2. *Example of an image in multispectral colored composition (March 4, 2004, 10 am UT), obtained by combining the channels of the SEVIRI imager of the Meteosat Second Generation satellite. This image processing makes it possible to visually distinguish low clouds (ochre shades) from high clouds (white shades), as well as in the absence of clouds to detect the presence of snow on the ground (shiny white) (source: Météo-France, Space Meteorology Center). For a color version of this figure, see www.iste.co.uk/phulpin/satellites2.zip*

With the advent of recent generation meteorological satellites, carrying imaging radiometers with a large number of channels (e.g. the SEVIRI imager** of Meteosat Second Generation satellites, with 12 channels whose central wavelengths are distributed between 0.6 and 13.4 microns), the direct use by forecasters of all the individual images corresponding to each channel has become difficult, if not impossible, in an operational context. However, all of these channels and their combination can provide important additional information, for example, on the microphysical properties of clouds, on the presence of volcanic dust or ash or on the type of air mass. For this reason, false-color imaging products have been developed (often called multispectral colored compositions or "RGB" images), each red, green and blue color plane using the following as input: the data of a channel or a combination of channels, with the choice of channel depending on the type of information that it is desired to highlight through the color variations. In addition to

** See Appendix 3.

2. In meteorology, subsidence is the vertical downward movement of air, which then compresses, and therefore warms and dries.

basic images of a limited number of channels, RGB products dedicated to the general recognition of clouds, the identification of the air mass, the detection of low clouds and fog, and the detection of aerosols are nowadays often used operationally by forecasters.

I1.2. ...to measurement use

But meteorologists do not just use satellite imagery. As will be shown in this section, they also use data from the various satellite instruments to access measurements of a large number of geophysical variables crucial for operational meteorology: vertical profiles of temperature and atmospheric humidity, physical and microphysical properties of clouds, wind, surface variables, etc. For this reason, satellite observations nowadays largely dominate in terms of data quantity of the other sources, and are particularly used intensively for Nowcasting and numerical weather prediction. However, despite their ever-increasing quality and diversity, they do not cover all the needs for meteorological observations, since measurements by remote sensing from space have limitations intrinsically linked to the physics of measurement, or to the current state of technology. Thus, at present, certain meteorological variables (such as atmospheric pressure) cannot be measured with sufficient accuracy, or certain regions (such as the atmospheric boundary layer) cannot be sampled with sufficient spatial or temporal resolution by satellite.

Chapter 1 presents the atmospheric temperature and humidity profiles that are produced operationally from measurements of infrared and microwave sounders, and their main uses in meteorology.

Wind at altitude is a very important meteorological variable, for which direct observations are provided either by radio soundings, the spatial coverage of which is very inhomogeneous and limited essentially to continental areas, or by airliners, the spatial coverage of which is limited to the flight path and level of the aircraft. Chapter 2 shows the long-standing contribution of satellite imagery for estimating winds at altitude, which is obtained from the temporal tracking of targets such as clouds or water vapor field structures, as well as that of measurements by Doppler lidar, the operational demonstration of which has just been made thanks to the ADM-Aeolus satellite launched in 2018 by the European Space Agency.

Weather prediction requires not only knowledge of the state of the atmosphere at a given moment, but also knowledge of certain properties of the surface, on land or at sea, which will influence the exchanges of energy, matter and quantity of movement between the two environments. Chapter 3 provides an overview of the various surface variables that are important for weather forecasting, and to which different types of satellite measurements give access.

Chapter 4 shows the essential contribution of satellite observations for data assimilation into numerical weather prediction models. Data assimilation is a set of techniques which, using available observations, are intended to determine the initial state of numerical models, from which the evolution of the atmosphere is calculated to provide weather predictions. Much progress has been made in recent decades in this area, making it possible to take full advantage of satellite observations. These techniques also make it possible to produce re-analyses of past and present climate, thanks to the series of satellite observations which have now been available for more than 40 years.

In meteorology, Nowcasting refers to all methods aimed at predicting weather conditions in the next few hours, usually for alerting purposes with a few hours of anticipation. It requires the production of predictions as quickly as possible, and updated as frequently as possible to take into account the latest observations available, which represents a strong constraint with regard to numerical weather prediction systems. For this reason, the direct use of observations, often combined with temporal extrapolation techniques, also holds an important place. Chapter 5 presents the state of the art with regard to satellite observations and the methods used for Nowcasting.

Tropical cyclones are one of the most destructive meteorological phenomena, whose direct observation remains very difficult because of the extreme conditions prevailing there. Since the advent of the first geostationary meteorological satellites, images provided have enabled forecasters to identify and estimate the characteristics of cyclones, as well as to track their trajectory. As shown in Chapter 6, although geostationary imagery remains the main source of information today for the operational monitoring of tropical cyclones, other satellite measurement systems also provide valuable information on certain meteorological variables within them.

1

Operational Sounding of Thermodynamic Variables in the Atmosphere

Thomas AUGUST
European Organisation for the Exploitation of Meteorological Satellites (EUMETSAT), Darmstadt, Germany

1.1. Introduction

Global knowledge of the vertical structure in temperature and humidity of the atmosphere is essential for weather forecasting and climate monitoring. These thermodynamic variables determine, for example, regional and large-scale air-mass circulation, as discussed in detail in Chapter 2. More locally, vertical thermal and humidity gradients are key ingredients of atmospheric stability. Knowledge of these gradients is critical to be able to alert populations and civil security components of the development of devastating weather events (massive hail, lightning, squall, flash floods, even tornadoes) or to monitor thermal inversions conducive to air quality degradation. It is also quite significant that these variables condition the storage capacity of water vapor (which is a powerful greenhouse gas) in the atmosphere as well as possible changes in the phase of the water, therefore microphysical processes associated with the formation of clouds (which are important in the global albedo of the Earth), fog layers, etc. More generally, these thermodynamic variables control the transfer of radiative energy between ground and atmosphere. They are therefore necessary to model radiative transfer, for example, for numerical weather prediction

For a color version of all the figures in this chapter, see www.iste.co.uk/phulpin/satellites2.zip.

Satellites for Atmospheric Sciences 2,
coordinated by Thierry PHULPIN, Didier RENAUT, Hervé ROQUET and Claude CAMY-PEYRET.

(Chapter 4), for modeling Earth's radiation budget (Chapter 14) or finally to be able to detect and quantify other atmospheric constituents such as aerosols, greenhouse gases and molecules involved in air quality, etc.

This information was traditionally provided mainly by ground-based measurements or sounding balloon radiosondes, with consequently very limited spatio-temporal sampling and concentrated coverage on continental surfaces of the northern hemisphere. The development of satellite observation with TOVS instruments since the 1970s has made it possible to provide initial thermodynamic information on a synoptic scale thanks to radiometers operating in infrared (TIR) and microwaves (MW). But it was with the advent of hyperspectral sounders in polar orbit in the 2000s that a decisive step was taken to access temperature and humidity profiles with real vertical resolution.

In this chapter, we first present operational strategies at NOAA and EUMETSAT to operate passive vertical sounders and meet the needs of different users. They put into practice the theoretical principles of remote sensing explained in the two previous chapters with products of temperature, humidity, clouds and also ozone which is a tracer of the dynamics in the upper atmosphere. The sounders operating at nadir have a footprint ranging from about ten to a few tens of kilometers. They have maximum sensitivity and resolution in the free troposphere and also provide information in the stratosphere (Figure 1.1) and in a more limited manner in the planetary boundary layer. Atmospheric sampling of temperature and humidity using limb radio occultation (RO) signals from geo-positioning satellite systems (GNSS) has been described in Chapter 2. The RO technique makes it possible in particular to complete the knowledge of vertical structure in the upper layers of the atmosphere. It also offers a finer vertical resolution than the passive IR and MW nadir sounders, but on generally larger horizontal scales.

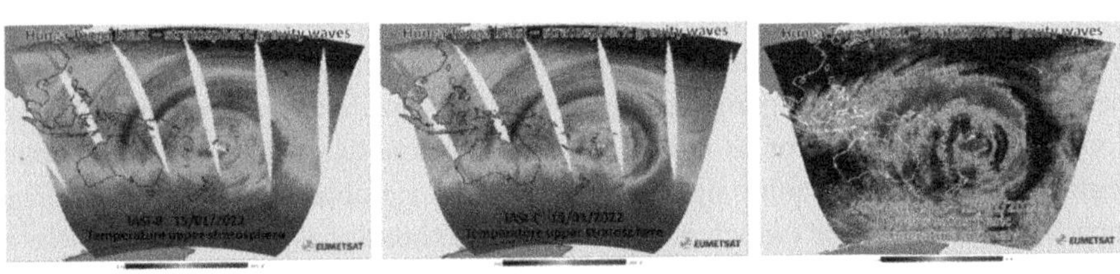

Figure 1.1. *Upper stratospheric temperature sounding by IASI-B (left) and IASI-C (middle) after the explosion of the Hunga Tonga-Hunga Ha'apai volcano on January 14, 2022. On the right: temperature anomaly noted by IASI-B and IASI-C compared to previous days. Concentric circles reveal the propagation of gravity waves generated by the explosion to an altitude of 40 km*

1.2. Operational use of TIR and MW sounders

1.2.1. *Satisfying ever-more demanding users*

The constant progress made in numerical weather prediction requires thermodynamic profiles that are increasingly resolved vertically. The same applies to very short-range prediction or nowcasting, as well as monitoring of climate, atmospheric composition and air quality.

As we have seen in Chapters 9 and 10 of Volume 1, the greater the spectral resolution of the measurements made by satellite sounders, the finer the vertical information restored. Deployed since the 1980s in polar orbit (Appendix 3), MW passive sounders operated in conjunction with broadband imagers and radiometers provided necessary vertical information until the late 1990s. Due to limited information content, however, the restitution of atmospheric profiles usually required the use of a priori information from numerical models.

The advent of hyperspectral sounders in TIR in the early 2000s made it possible to reach a milestone with vertical information content that allowed for the restitution of atmospheric profiles in temperature and humidity in a model-independent manner. They are operated in synergy with MW sounders. The respective information content of TIR and MW sounders have been discussed in Chapter 9 of Volume 1. The first hyperspectral instruments were launched in 2002 by the US with AIRS (Chapter 3 of Volume 1) on board Aqua (Aumann 2003), and by Europeans in 2006 with IASI within the EPS program (Klaes 2007). AIRS is a 2,378-channel dispersion spectrometer. A unique model was designed and is still operated by NASA. IASI (Infrared Atmospheric Sounding Interferometer) is a Fourier transform interferometer produced by the CNES in three copies, launched in 2006, 2012 and 2018, and operated by EUMETSAT. The IASI spectral information is distributed in 8,461 channels between 645 and 2,670 cm^{-1} with a 0.25 cm^{-1} sampling and a resolution of 0.5 cm^{-1}. IASI has been meeting the user requirements, with temperature profiles of an accuracy better than 1 K over layers of 1 km in the free troposphere and humidity profiles of an accuracy of 10% (relative humidity) over layers of 1–2 km. NOAA, in partnership with NASA, then developed the Cross-track Infrared Sounder (CrIS) instrument, which is also an interferometer operating in the same spectral domain as IASI. With a slightly lower resolution and less complete spectral coverage, it also aims to restitute vertical profiles with a mileage resolution. The first two CrIS models were launched in 2011 and 2017. In response to users' needs for information at different points in the daytime cycle, the partnership established in the Joint Polar Satellite System (Chapter 7 of Volume 1) enables Europeans and Americans to conduct soundings in so-called mid-morning orbits (provided by EUMETSAT) and afternoon orbits (provided by NOAA) operationally available to each other.

Across all applications, the needs over the coming period require even more resolved thermodynamic profiles, especially in the bottom layers, as well as better spatio-temporal sampling. In Europe, the response to these needs is based on two new programs. In low orbit, the second-generation polar program (EPS-SG) will deploy microwave (MWS) and hyperspectral (IASI-Next Generation or IASI-NG, developed by the CNES) sounders with increased spectral and radiometric performance. At the same time, EUMETSAT plans to operate the first operational hyperspectral sounder in geostationary orbit: IRS (InfraRed Sounder) on board MTG (Meteosat Third Generation) (Holmlund 2021). MTG-IRS will provide observations every 30 minutes over Europe with full coverage and a resolution of about 7 km (4 km at nadir, at the equator). Its spectral and radiometric characteristics, although slightly lower than IASI and a fortiori IASI-NG, will nevertheless make it possible to partially resolve the temperature and humidity profiles and to provide observations independent of numerical models. Figure 1.2 illustrates the relative spectral coverage and resolutions between IASI, IASI-NG and MTG-IRS.

Figure 1.3 provides an overview of the difference in vertical information content between imagers and hyperspectral sounders. Operating in the visible and infrared, however, they offer excellent imaging capabilities. But it is only once convective clouds are developed that satellite observation can support the very short-range prediction (Chapter 5) of associated fierce events. Providing thermodynamic information of the atmosphere resolved in time, horizontally and vertically, MTG-IRS promises to be a new revolution in support of regional applications. Figure 1.4 illustrates the respective spatio-temporal resolutions of the IASI and MTG-IRS missions. Pre-convective areas can be detected and monitored before cloud formation (Figure 1.5), thus saving time and accuracy in the management of alert bulletins.

The Russian space agency Roscosmos has developed the IKFS series of polar hyperspectral sounders flying on Meteor-M platforms. The Chinese Meteorological Administration (CMA) has developed HIRAS instruments for low Earth orbits, the first copy of which is being flown as an experimental mission on board the FY-3D satellite. The second HIRAS instrument, launched on FY-3E in July 2021 (Zang et al. 2021), has an operational purpose and is the first in early morning orbit. The

CMA also placed the first hyperspectral sounders in geostationary orbit with the first two copies of the GIIRS mission, whose operational maturity is not expected until 2025 with the third instrument. Appendix 4 summarizes the main characteristics of active and future hyperspectral sounders. It is also important to note that applications exploiting products in real time as well as the operation of climatological series have requirements on the stability over time of product performance, in addition to requirements on instantaneous accuracy. In this regard, IASI reports remarkable stability on all of its three instruments in flight, both from the radiometric and spectral calibration points of view. This is what established it as a reference in the GSICS international intercalibration standards (section 7.3.2.2, Chapter 7 of Volume 1).

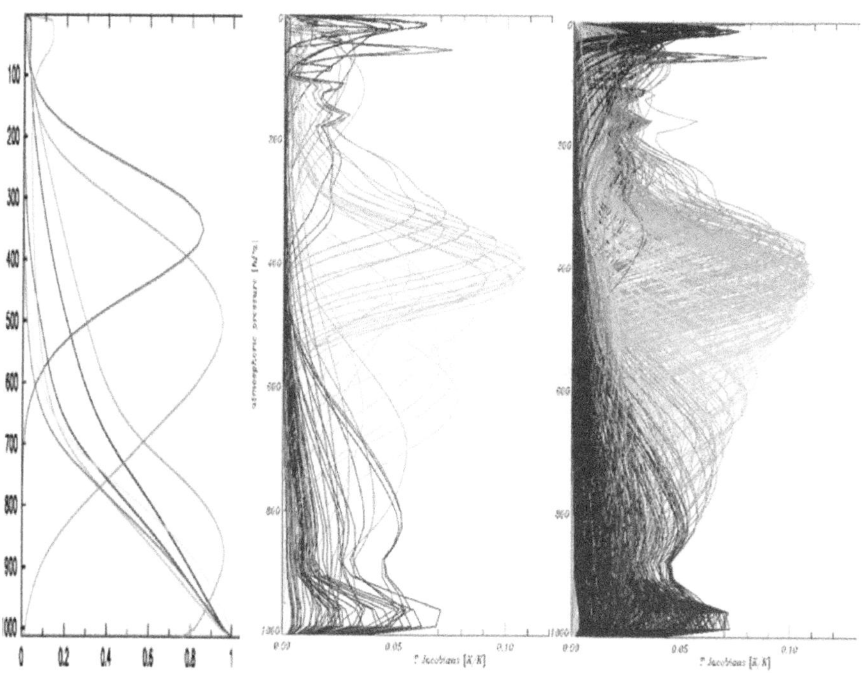

Figure 1.2. *Temperature weight functions of SEVIRI channels (left). Those of IASI are represented for a small sample of 85 channels (in the center) and 850 channels (on the right). In total, IASI contains 8,461 channels, IASI-NG will contain 16,921*

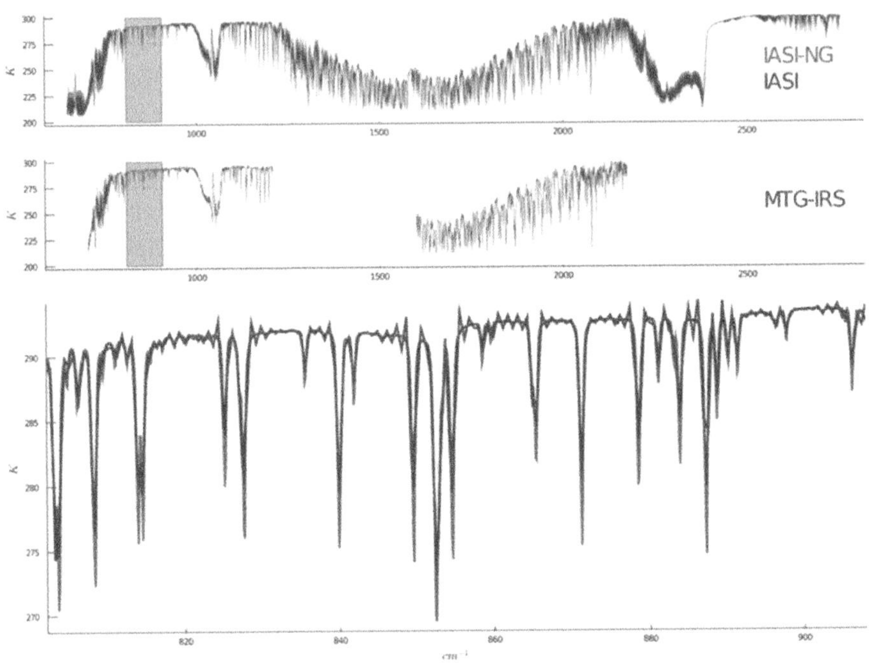

Figure 1.3. *Simulated spectra illustrating the coverage and spectral resolution of IASI, IASI-NG and MTG-IRS. The spectral extract at the bottom corresponds to a zoom on the grayed-out area in the complete spectra of the three instruments*

Figure 1.6 shows the theoretical vertical resolution of the information reproduced in temperature, humidity and ozone with the three European hyperspectral sounders for a given layer.

Figure 1.7 complements this information by presenting the number of cumulative degrees of freedom, that is, the number of independent pieces of information in a profile expected for future European sounders and with IASI.

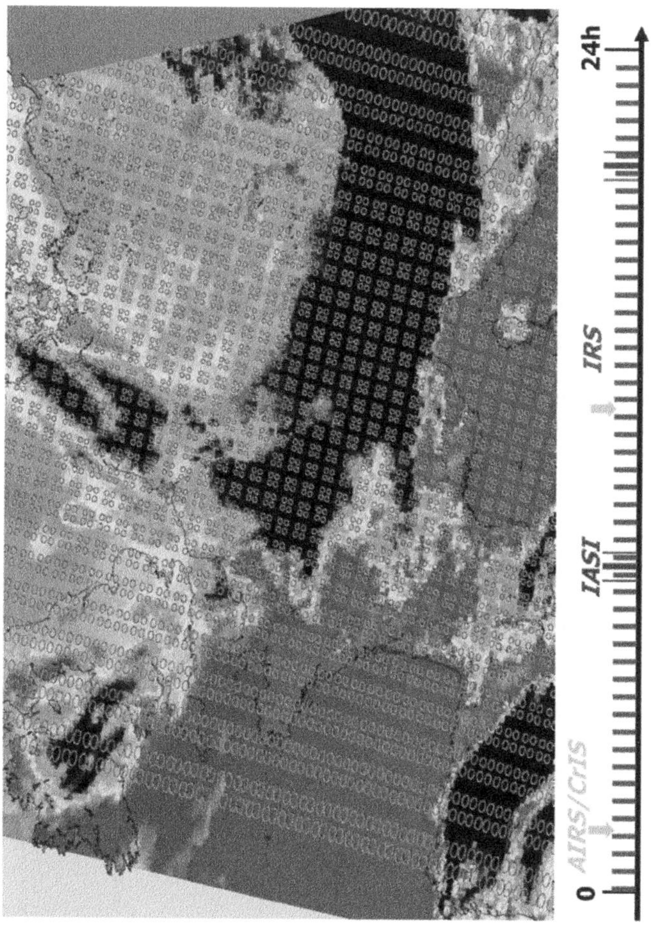

Figure 1.4. *Simulation of an MTG-IRS observation in a window channel (cold clouds in blue-black, hot surfaces in red-violet). The footprint of the IASI pixels extracted from a real orbit is superimposed. The 24-hour chronological frieze summarizes the temporal sampling allowed by MTG-IRS (every 1/2 hour over Europe) in comparison with US (AIRS and CrIS) and European (IASI-A, -B and -C) polar sounders revisiting. With contiguous pixels of about 7 km in Western Europe, MTG-IRS will also offer much better spatial coverage than polar sounders, whose disjoint pixels vary from a dozen km at nadir to a fortieth at the edge of the swath*

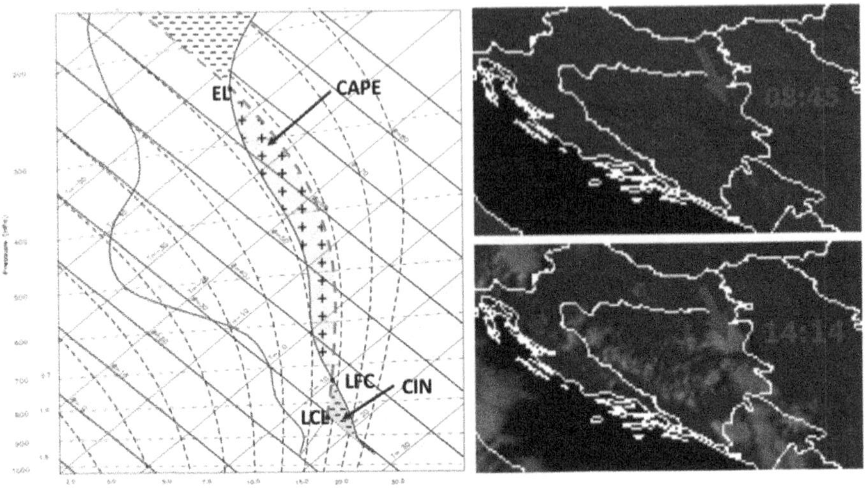

Figure 1.5. *On the left, conditional atmospheric instability highlighted by IASI-B in the Tuzla region (Bosnia and Herzegovina) on August 10, 2018 at 08:45 UTC. Humidity in red; temperature in dark blue; CAPE: potential convective energy, surface in cyan with +; CIN: inhibition on convection, surface in yellow with -. On the right, the red arrow indicates the sounding point, presented in a SEVIRI composite RGB image of the observed area. Geostationary imagery at the time of the passage of IASI (at the top) shows a clear sky. A few hours later (bottom right), the area sounded is actually covered with convective cells*

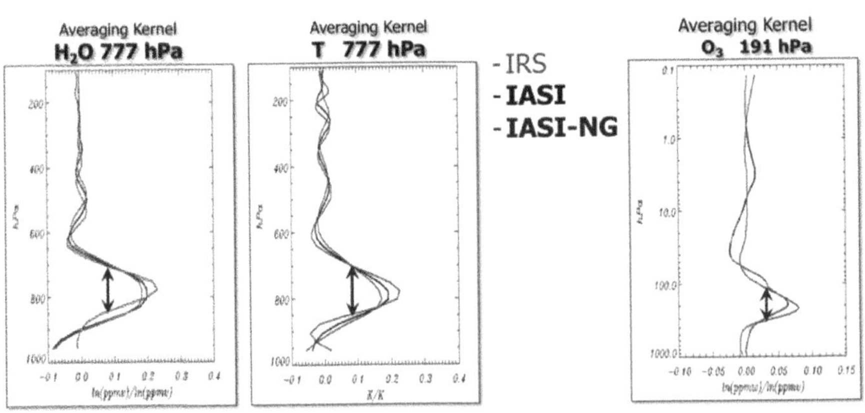

Figure 1.6. *Vertical resolution of the three sounders IASI, IASI-NG and MTG-IRS illustrated with the averaging kernels (section 10.3.2.3 of Chapter 10, Volume 1) of the moisture (left), temperature (center) and ozone (right) restitution simulated with the US Standard 1976 atmosphere (Appendix 1)*

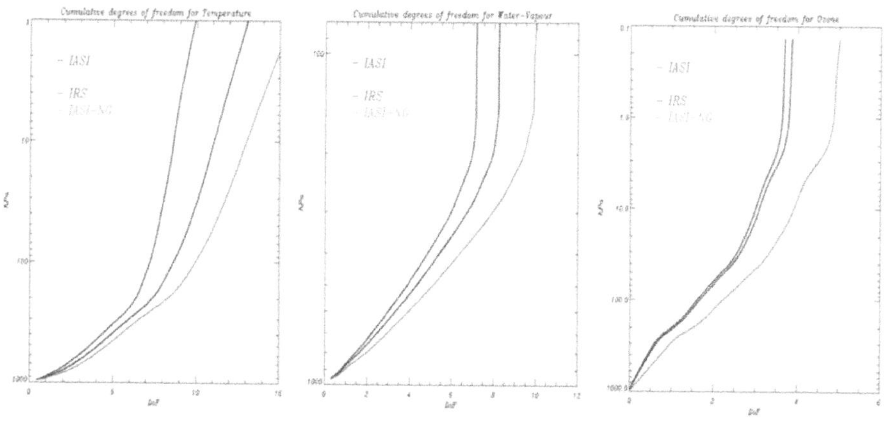

Figure 1.7. *Cumulative degrees of freedom in temperature (left), humidity (center) and ozone (right), simulated for IASI, IASI-NG and IRS instruments with the US standard atmosphere. These figures show the number of independent layers (x-axis) in the profiles rendered by the three European sounders (IASI: blue, IRS: red, IASI-NG: green) between the surface and the top of the profile (varying on the y-axis). In this configuration, IASI-NG will make it possible to distinguish nine vertical pieces of information in temperature in the troposphere (between 1,000 and 200 hPa). We can also note that above 200 hPa, none of the three sounders provides significant moisture information, while temperature information continues to increase at different rates*

1.2.2. *Clouds: an obstacle to sounding and a very useful geophysical product*

Clouds are a twofold problem for infrared atmospheric sounding. During a total and opaque cover of the pixel, they block the radiation rising from the surface and thus prevent the sounding under the top of the cloud. In the case of pixels partially covered or observed through semi-transparent clouds, information on the complete profile up to the surface is indeed present in the measurement. However, the modeling of infrared radiative transfer in clouds is still an active subject of research, which limits the operational use of these observations. MW radiation, on the other hand, has the property of being able to pass through most clouds (Figure 1.8). Sounders operating at these wavelengths therefore make it possible to reproduce some information on the vertical structure of the atmosphere in temperature and humidity, including in overcast skies, with the exception of precipitating dense clouds or very high convection. As we have seen previously, the vertical resolution obtained in general with MW is lower than in clear skies with TIR.

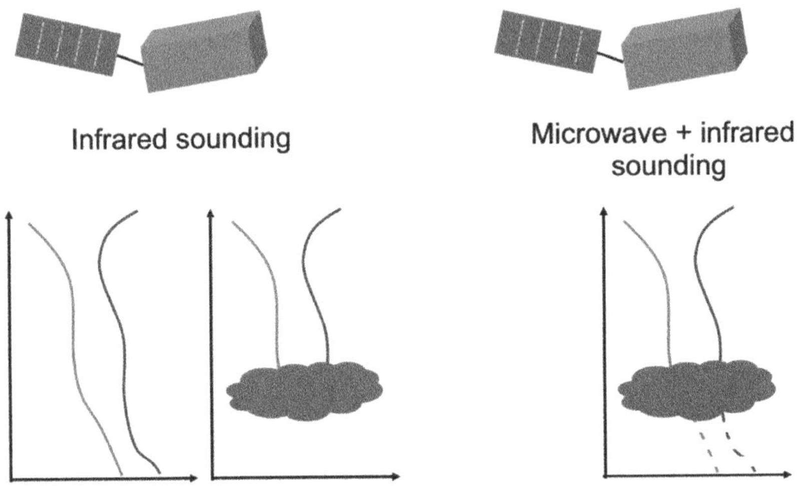

Figure 1.8. *Limitation of atmospheric sounding in the infrared in the presence of a cloud. The blue and red profiles schematize the humidity and temperature sampled, respectively*

The detection of clouds in the pixels of sounders is facilitated by the use of associated imaging missions. It is also possible to deduce the presence of clouds from the infrared spectra, by comparing the radiances observed in window channels with radiances simulated for a clear sky. A large difference between observed (OBS) and calculated (CALC) radiances gives away the presence of a cloud (McNally and Watts 2003). This method requires a description of the state of the atmosphere to model radiative transfer and is often established with numerical weather prediction models. However, its sensitivity is limited when the OBS-CALC difference is of the order of uncertainties related to knowledge of the atmosphere or to the modeling of radiation transfer (uncertainties on spectroscopy, presence of low and hot clouds, etc.).

Some characteristics of the cloud can be inferred from observations by hyperspectral sounders. By considering the cloud in a simplified way as a gray body on a layer, it is possible (Strabala 1994) to restore the effective coverage of the pixel (a combination of the geometric fraction covered and the opacity of the cloud) as well as to locate the top of the cloud. This latter information is important for the characterization of winds deduced from the products of satellite imagery (Chapter 2). Indeed, although imagers provide sufficient horizontal information to determine the direction and intensity of wind vectors, they do not intrinsically have the vertical information to assign their height. The determination of liquid water and ice content has long been studied with MW sounders. The added value of hyperspectral sounders, especially of the new generation, to find these water contents and even more to determine microphysical parameters, such as the size of droplets at the top or crystals in ice clouds, is a current topic (Léonarski 2021; Mastro 2022). Remote sensing of clouds with various sensors and associated applications are described further in Chapter 14.

1.2.3. *Demonstrating and maintaining product quality*

To demonstrate that the quality of products reaches the levels required by users and to ensure their stability over time, space agencies have developed validation and monitoring capabilities. With regard to thermodynamic variables, NOAA operates the NPROVS project (NOAA PROducts Validation System[3] which compares operational satellite soundings on a daily basis with in situ measurements provided by the meteorological radiosonde network coordinated by WMO and the GRUAN (GCOS Reference Upper-Air Network). EUMETSAT has developed a similar capability, MONA LiSA (MONitoring of Atmospheric Level 2 SAtellite products), to ensure the quality of IASI products and future sounders (EUMETSAT 2022a). Examples of product quality diagnosis and monitoring are provided in Figure 1.9. In addition to radiosondes for temperature and humidity assessment, the European network EuBrewNet routinely helps to assess the total ozone column. The quality of the total humidity column restituted by satellite can be evaluated with ground station data that continuously determine the integrated water vapor content by using the signal from navigation satellites. This method is easier and less expensive to implement than sounding balloons but only measures the total column. It has become widespread with dense networks coordinated in the United States (SuomiNet) and in Europe (EUMETNet). Figure 1.10 illustrates the validation and quality of total water vapor column operational products with IASI against European ground observations.

3. https://www.star.nesdis.noaa.gov/smcd/opdb/nprovs/.

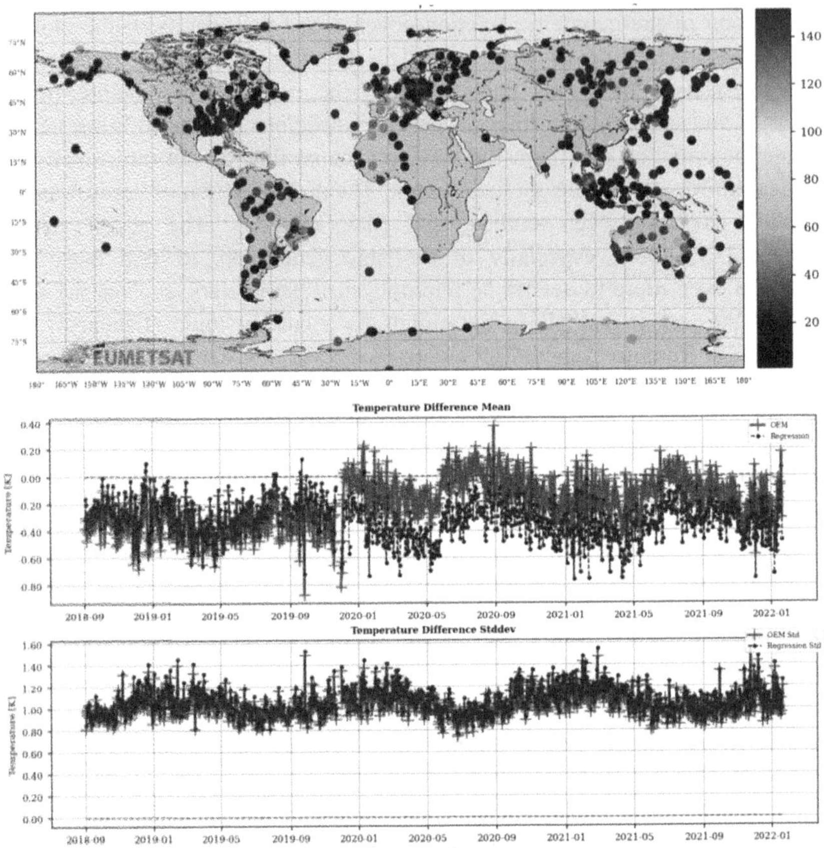

Figure 1.9. *At the top, coincidence of the in situ (radiosonde) and satellite (MetOp-B) soundings in January 2022. Time series of biases (middle) and standard deviation (bottom) between IASI products (blue: statistical restitution, red: physical method) and radiosonde at 500 hPa. We note the small improvement in biases allowed by the processor update in December 2019*

As part of the World Climate Research Programme (WCRP), the GEWEX (Global Energy and Water Exchanges) project is dedicated to the study of the water and energy cycle in the Earth system. In this context, the Data Assessment Panel has been commissioning a working group since 2011, the GEWEX Water Vapor Assessment (G-VAP[4]), to identify, assess and document the performance of the most relevant satellite products. This group's mission focuses on three essential climate variables (ECVs) established by the GCOS:

4. http://gewex-vap.org.

– total column water vapor (TCWV);

– upper tropospheric humidity (UTH);

– water vapor profiles and associated temperature (WVP).

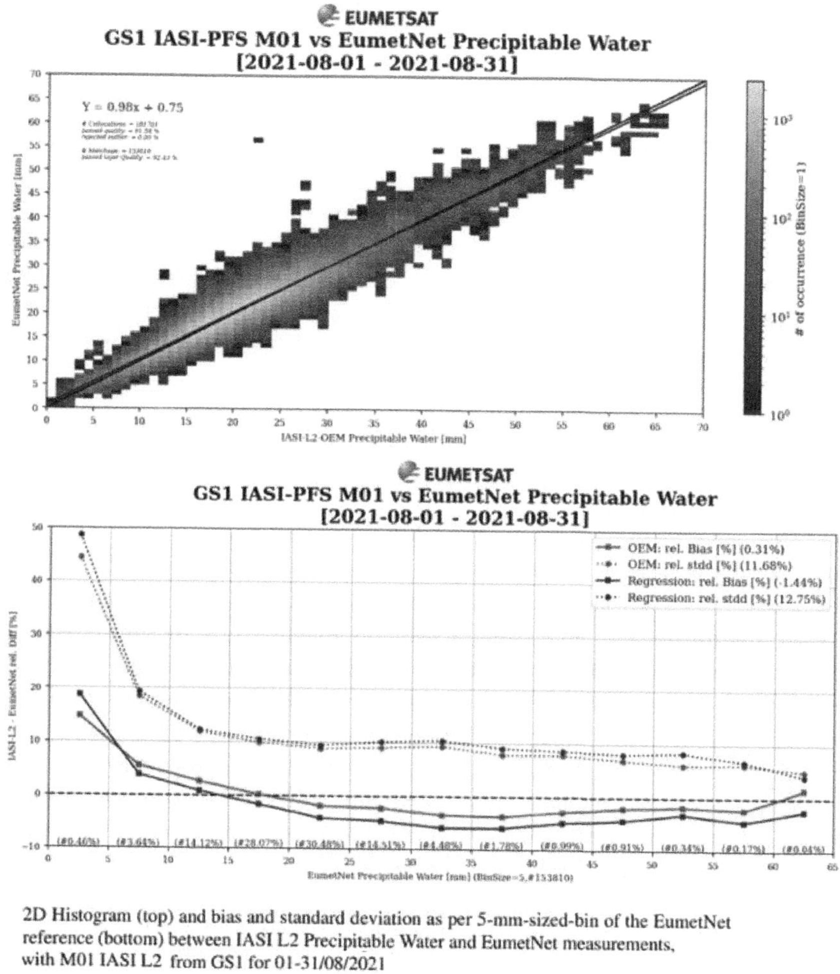

Figure 1.10. *Evaluation of the quality of the operational products (IASI+AMSU+MHS) of the total water vapor column with the EUMETNet E-GVap network (section 1.3.5). Top: 2D histogram and correlation (IASI vs. ground-GPS). Bottom: relative bias and accuracy (IASI vs. ground-GPS) depending on the water column*

It is important to remember that profiles from radiosounding have considerably better vertical resolution than that of satellite products. The evaluation of their performance by intercomparison must strictly take into account these different vertical resolutions by applying the averaging kernels to reference profiles. Another problem in the evaluation of space soundings is that radiosondes are most often launched at coordinated synoptic times (00:00, 06:00, 12:00 and 18:00 UTC), which do not correspond to the overflight by polar satellites. This difference in time can be several hours and can induce significant differences between the atmospheres observed by satellites and by sounders, in particular in bottom layers. Similarly, the radiosonde is a point measurement whose reading lasts about two hours during which it can drift over a hundred km. Depending on the homogeneity of the zone, these measurements are not necessarily representative of a pixel covering from 12 to 40 km. These questions of uncertainties related to representativeness in evaluating the performance of satellite products have been addressed by Immler (2010). The same considerations apply to the evaluation of ozone profiles with the WOUDC network coordinated by the WMO in support of the Global Atmosphere Watch (GAW) program.

1.2.4. *Different operational algorithmic strategies*

As we have seen, NOAA and EUMETSAT have developed similar and complementary atmospheric sampling polar programs within the Joint Polar System agreements. They each include MW and hyperspectral sounders operating with a multi-spectral imager with high spatial resolution. Figure 1.11 shows the dimensions and geometry of the footprints of onboard sounders in the US (Suomi-NPP, JPSS) and European (EPS) programs. Some similarities can also be found in the algorithms exploiting these missions to retrieve thermodynamic profiles. The processing chains operated by each of the agencies to generate the level 2 products first of all implement a statistical method, more commonly referred to today as "machine learning" or artificial intelligence. Nominally, this first step exploits, in synergy, observations made in the MW and in the thermal infrared. The second stage implements a variational method (optimal estimation, Rodgers (2000)) modeling the physics of radiative transfer into the atmosphere (Chapter 10 of Volume 1), and which is initialized with the result of the statistical method. It generally makes it possible to further improve the precision of profiles previously obtained with the statistical method.

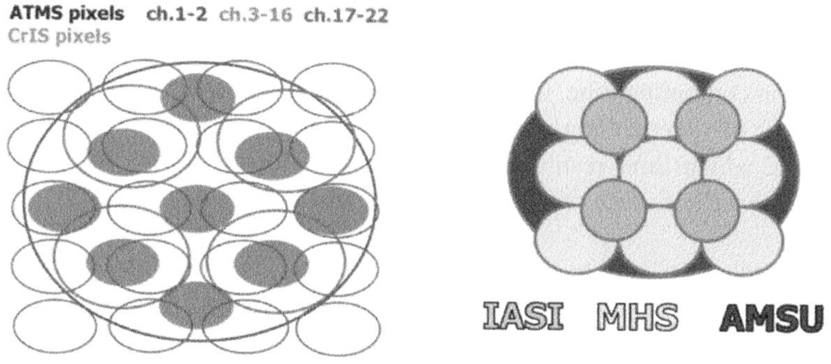

Figure 1.11. *Schematic view of the dimensions and geometry of the footprints of the American (CrIS, left) and European (IASI, right) hyperspectral polar sounders relative to their microwave companions*

However, a big difference distinguishes the US and European algorithms and responds to different strategies. As mentioned above, clouds make it difficult or impossible to carry out atmospheric soundings in infrared. In the NOAA approach (Gambarcota 2013) and in the historical continuity of NASA's AIRS products (Susskind 1998), the radiances in CrIS pixels are "decontaminated" from cloud spectral signals. The underlying assumption is that the nine (3×3) pixels of CrIS observe the same atmosphere and that only the cloudiness possibly varies from one pixel to another. The spectrum as it would have been measured in clear sky is thus deduced by simple linear regression on the spectra recorded in various pixels, weighted by their respective cloud fractions. This method, called "cloud-clearing", is the first step in the restitution of geophysical parameters. This is done by modeling the radiative transfer in clear skies, which has the advantage of being scientifically more mature and computationally faster. However, it only allows a restitution for nine pixels, or about 75 km at the nadir (300 km at the edge of the swath).

The algorithmic strategy put in place by EUMETSAT (August et al. 2012) aims to preserve and exploit the initial measurement, and to restore independent atmospheric profiles for each pixel (i.e. 12 km at nadir and 40 km at the swath edge), thus meeting users' spatial resolution requirements. In this approach, thermodynamic variables are retrieved in three stages: (i) a statistical method providing "all sky" sounding; (ii) the detection and, where appropriate, the characterization of clouds; (iii) a physical method (optimal estimation) to refine clear sky profiles (not operational yet in cloudy pixels).

To be as representative as possible, the statistical method results from supervised learning-based training of about 150 million observed real spectra associated with the ERA-5 re-analyses of the ECMWF model. In addition to atmospheric profiles, it also makes it possible to estimate errors on inverted profiles. This method is very efficient in computing time. Compared with the optimal estimation, it also offers the advantage of overcoming the uncertainties associated with radiative transfer modeling, which limit restitution in the presence of clouds as well as on certain continental surfaces with complex emissivity characteristics.

Based solely on satellite observations, the machine learning approach provides users with information that is independent of weather models. It was selected for the operational use of future IASI-NG and MTG-IRS missions (EUMETSAT 2021). In the event of unavailability of MW measurements, the retrievals are performed only with the TIR sounder. This is also de facto the configuration for MTG-IRS, which does not have an MW counterpart. The performance established by comparison with in situ radiosonde soundings is illustrated in Figure 1.12. Statistics are stratified by quality class based on the uncertainties provided along with each retrieved profile. In classes of very high quality, essentially corresponding to clear pixels, precision is better than 1 K in the free troposphere, as expected. We can observe that the contribution of microwaves is relatively limited, both in terms of precision and in terms of the size of the restituted sample. Their added value is striking in "all sky" pixels, where considerably more good quality soundings (between 1 and 1.5 K) are available than in infrared-only pixels.

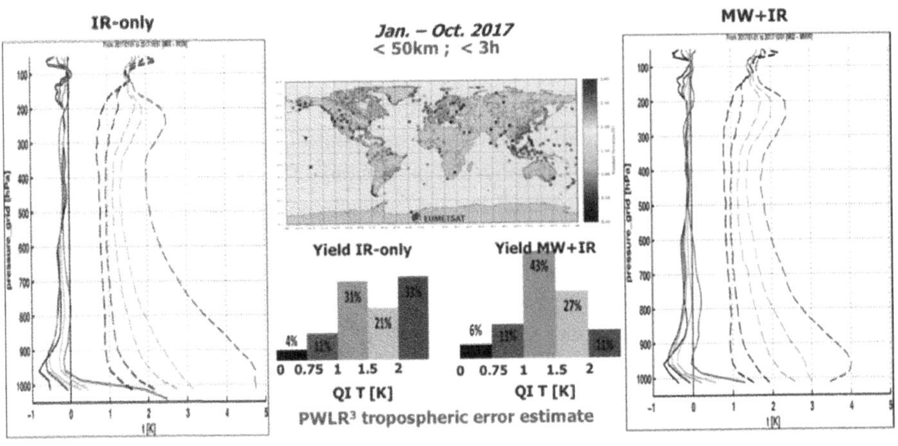

Figure 1.12. *Evaluation of operational sounding performance in the EPS program with in situ measurements by radiosounding. Co-registration criteria require a distance of better than 50 km in space and 3 h in time between the radiosonde and the IASI observation. Statistics (bias: solid lines; standard deviation: dashed lines) are shown for retrievals using IASI only (left) and with the IASI-AMSU-MHS synergy (right). Performance is evaluated for different quality classes (QI: quality indicator) expressing the uncertainty estimated by the inversion method*

The precision and speed of this method allowed for implementing a regional operational service, based on the receiving stations of the EARS network (EUMETSAT Advanced Reception System[5]). It meets the needs of regional users by providing forecast-independent descriptions of the atmosphere in less than 30 minutes after satellite observation (Figure 1.13). This service has paved the way for numerous application studies.

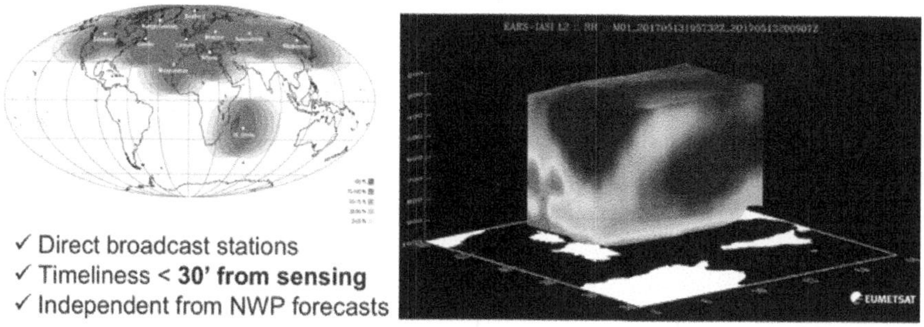

Figure 1.13. *Coverage of the EARS-IASI regional service (left) and illustration of a 3D sounding of relative humidity between Corsica and Italy in May 2017*

1.2.5. *Application perspectives*

These last generations of sounders provide users with a new type of information. Studies are underway to integrate them into weather monitoring systems, in particular with the American (NSSL[6]) and European (ESSL[7]) severe storm studies laboratories. The first results confirm the potential of these products (Hungershofer and Herold 2018; EUMETSAT 2020a, 2020b) and prepare the community for the arrival of geostationary sounders. However, the sounders have a lower sensitivity in low layers, but the convective phenomena in particular are linked to the state of the atmosphere in the first hectometers. This limitation is intrinsic to the instrument and its measurement, not to the methods of restitution. The joint use of satellite observations with data acquired on the ground (direct measurements or zenith microwave sounding, or even lidar) would constitute an operational solution and is the subject of very recent work. It could be carried out by directly combining geophysical products (Bloch 2019; Kocsis 2021) or by integrating satellite and ground measurements (luminances) into an inversion algorithm (Ebell 2013; Toporova 2021).

5. https://www.eumetsat.int/ears-iasi.
6. https://www.nssl.noaa.gov.
7. https://www.essl.org/cms/.

For numerical weather prediction, as detailed in Chapter 14, the paradigm since the late 1990s relies on the assimilation of observed radiances. Recent progress in the restitution of thermodynamic profiles has, however, re-opened the question of their assimilation in models, with very encouraging results (EUMETSAT 2019, 2022b; Coniglio 2019; Silveira 2022). These advances include performance in precision as well as the ability to characterize vertical uncertainties and sensitivities with artificial intelligence. Radiance assimilation requires the use of a radiative transfer model, which is comparatively very costly in computing time and further limits the exploitation of observations in cloudy skies or in near-surface sensitive channels. Because their use would allow much less intensive calculations, thermodynamic profiles would be of particular interest in the context of regional models where latency requirements are critical, and generally in a context of reflection on digital sobriety and associated energy expenditure.

The time series acquired with the hyperspectral sounders extend beyond 20 years already with the AIRS and IASI missions alone. They will be extended in the coming decades, with even better performance thanks in particular to IASI-NG. With higher vertical resolution than previous generations of satellite missions, they are increasingly contributing to climate modeling and monitoring (Chapters 12–14).

Access to thermodynamic information resolved vertically and over time paves the way for the characterization of atmospheric wind profiles (Chapter 2), which would provide new and very significant information for weather prediction.

Recent publications (Feltz 2017) compare sounding performance with hyperspectral instruments and radio occultation missions. These two types of missions are complementary, with radio occultation providing unique information for the study and modeling of the stratosphere.

1.3. Acknowledgments

The results on IASI Level 2 products involved the work of T. Hultberg, S. Stapelberg, M. Crapeau and A. Burini on retrieval algorithms and product validation for EUMETSAT.

1.4. References

August, T., Klaes, D., Schlüssel, P., Hultberg, T., Crapeau, M., Arriaga, A., O'Carroll, A., Coppens, D., Munro, R., Calbet, X. (2012). IASI on Metop-A: Operational level 2 retrievals after five years in orbit. *Journal of Quantitative Spectroscopy and Radiative Transfer*. doi:10.1016/j.jqsrt.2012.02.028.

Aumann, H.H., Chahine, M.T., Gautier, C., Goldberg, M.D., Kalnay, E., McMillin, L.M., Revercomb, H., Rosenkranz, P.W., Smith, W.L., Staelin, D.H. et al. (2003). AIRS/AMSU/HSB on the AQUA mission: Design, science objectives, data products and processing systems. *IEEE Transactions on Geoscience and Remote Sensing*, 41, 253–264.

Bloch, C., Knuteson, R.O., Gambacorta, A., Nalli, N.R., Gartzke, J., Zhou, L. (2019). Near-real-time surface-based CAPE from merged hyperspectral IR satellite sounder and surface meteorological station data. *Journal of Applied Meteorology and Climatology*. doi:10.1175/JAMC-D-18-0155.1.

Coniglio, M.C., Romine, G.S., Turner, D.D., Torn, R.D. (2019). Impacts of targeted AERI and doppler lidar wind retrievals on short-term forecasts of the initiation and early evolution of thunderstorms. *Monthly Weather Review*. doi: https://doi.org/10.1175/MWR-D-18-0351.1.

Ebell, K., Orlandi, E., Hünerbein, A., Löhnert, U., Crewell, S. (2013). Combining ground-based with satellite-based measurements in the atmospheric state retrieval: Assessment of the information content. *Journal of Geophysical Research Atmospheres*. doi:10.1002/jgrd.50548.

EUMETSAT (2019). Preliminary assimilation of IASI L2 profiles in NWP [Online]. Available at: https://www.eumetsat.int/IASI-assimilations.

EUMETSAT (2020a). Hyperspectral sounding for severe storm predicting [Online]. Available at: https://www.eumetsat.int/severe-storm-predicting-lab.

EUMETSAT (2020b). Instability monitoring with IASI [Online]. Available at: https://www.eumetsat.int/hyperspectral-instability-monitoring-using-iasi.

EUMETSAT (2021). MTG-IRS level 2 algorithm theoretical baseline description. EUM/RSP/TEN/17/935387 [Online]. Available at: https://www.eumetsat.int/media/45439.

EUMETSAT (2022a). IASI L2 monitoring pages [Online]. Available at: https://www.eumetsat.int/iasi-level-2-geophysical-products-monitoring-reports.

EUMETSAT (2022b). Assimilation of satellite sounder profiles in NWP [Online]. Available at: https://www.eumetsat.int/assimilation-satellite-sounder-profiles-nwp.

Feltz, M.L., Borg, L., Knuteson, R.O., Tobin, D., Revercomb, H., Gambacorta, A. (2017). Assessment of NOAA NUCAPS upper air temperature profiles using COSMIC GPS radio occultation and ARM radiosondes. *Journal of Geophysical Research*. doi:10.1002/2017JD026504.

Gambarcota, A. (2013). The NOAA unique CrIS/ATMS processing system (NUCAPS): Algorithm theoretical basis documentation. NOAA.

Holmlund, K., Grandell, J., Schmetz, J., Stuhlmann, R., Bojkov, B., Munro, R., Lekouara, M., Coppens, D., Viticchie, B., August, T. et al. (2021). Meteosat third generation (MTG): Continuation and innovation of observations from geostationary orbit. *Bulletin of the American Meteorological Society*. https://doi.org/10.1175/BAMS-D-19-0304.1.

Hungershofer, K. and Herold, C. (2018). Are the EARS-IASI L2 products useful for nowcasting? *Satellite User Conference*, Tallinn.

Immler, F.J., Dykema, J., Gardiner, T., Whiteman, D.N., Thorne, P.W., Vömel, H. (2010). Reference quality upper-air measurements: Guidance for developing GRUAN data products. *Atmospheric Measurement Techniques*. doi:10.5194/amt-3-1217-2010.

Klaes, K.D., Cohen, M., Buhler, Y., Schlüssel, P., Munro, R., Luntama, J.-P., von Engeln, A., Clérigh, E.Ó., Bonekamp, H., Ackermann, J. et al. (2007). An introduction to the EUMETSAT polar system. *Bulletin of the American Meteorological Society*. doi:10.1175/bams-88-7-1085.

Kocsis, Z. (2021). Combination of IASI L2 profiles with synop measurements. Convection Working Group.

Léonarski, L. (2021). Restitution du contenu en glace des nuages à partir de mesures à haute résolution spectrale dans l'infrarouge des instruments satellitaires IASI et IASI-NG. PhD Thesis, Université de Lille.

Mastro, P., Masiello, G., Serio, C., Cimini, D., Ricciardelli, E., Paola, F.D., Hultberg, T., August, T., Romano, F. (2022). Combined IASI-NG and MWS observations for the retrieval of cloud liquid and ice water path: A deep learning artificial intelligence approach. *IEEE Journal of Selected Topics in Applied Earth Observations and Remote Sensing*. doi: 10.1109/JSTARS.2022.3166992.

McNally and Watts (2003). A cloud detection algorithm for high-spectral-resolution infrared sounders. *Quarterly Journal of Royal Meteorological Society*. https://doi.org/10.1256/qj.02.208.

Rodgers, C.D. (2000). *Inverse Methods for Atmospheric Sounding: Theory and Practice*, Vol. 2. World Scientific, Singapore.

Silveira, B.B., Fourrié, N., Guidard, V., Chambon, P., Mahfouf, J.-F., Brousseau, P., Moll, P., August, T., Hultberg, T. (2022). Preliminary assessment of Metop-based temperature and humidity statistical retrievals within the 3D-Var AROME-France prediction system. *Monthly Weather Review*, 150(4). https://doi.org/10.1175/MWR-D-20-0189.1.

Strabala, K.I., Ackerman, S.A., Menzel, W.P. (1994). Cloud properties inferred from 8–12-μm data. *Journal of Applied Meteorology*, 33, 212–229. doi: https://doi.org/10.1175/1520-0450%281994%29033%3c0212:CPIFD%3e2.0.CO;2.

Susskind, J., Barnet, C., Blaisdell, J. (1998). Determination of atmospheric and surface parameters from simulated AIRS/AMSU/HSB sounding data: Retrieval and cloud clearing methodology, *Advances in Space Research*. https://doi.org/10.1016/S0273-1177(97)00916-2.

Toporova, M. (2021). Combining satellite observations with a virtual ground-based remote sensing network for monitoring atmospheric stability. PhD Thesis, University of Cologne.

Zhang, P., Hu, X., Lu, Q., Zhu, A., Lin, M., Sun, L., Chen, L., Xu, N. (2021). *FY-3E:* The first operational meteorological satellite mission in an early morning orbit. *Advanced Atmospheric Science*. https://doi.org/10.1007/s00376-021-1304-7.

2

Wind Observations

Régis BORDE[1] and Jean PAILLEUX[2]

[1] *European Organisation for the Exploitation of Meteorological Satellites (EUMETSAT), Darmstadt, Germany*

[2] *Météo et Climat, France*

2.1. Introduction

In meteorology, wind represents the displacement of a particle of air in the atmosphere. Its origins are often multifold and extend over a wide range of spatial scales; this movement can be described mathematically by Navier–Stokes equations in fluid mechanics. On a large scale, atmospheric circulation movements originate in the temperature difference between the equator and the poles, generating North–South displacements, as well as in the Coriolis force induced by the rotation of the Earth which deflects air masses to the right or to the left along the hemisphere. The force of the wind is generally directly related to the pressure gradient between the high and low pressure zones. Its direction naturally goes from high to low pressures near the equator where the Coriolis force is zero, while at higher latitudes, it remains close to the perpendicular to the pressure gradient. The force can be very variable, leading to winds that are sometimes calm and refreshing and sometimes violent and destructive. Winds are therefore a major component of atmospheric systems, and they are essential information for making weather predictions. As a result, winds are measured every day, wherever possible with radars, radio soundings, anemometers and also with airliners, so as to use all these data in real time in the numerical weather prediction (NWP) models. However, these

For a color version of all the figures in this chapter, see www.iste.co.uk/phulpin/satellites2.zip.

Satellites for Atmospheric Sciences 2,
coordinated by Thierry PHULPIN, Didier RENAUT, Hervé ROQUET and Claude CAMY-PEYRET.

in situ measurements are mainly located in the northern hemisphere and above land, and large areas of the globe are therefore not very well documented or not documented at all.

The idea of using satellite images to extract information on the movement of the air appeared as early as 1960, with Testuya Fujita and the first images of the polar satellite Television Infrared Observation Satellite (TIROS) (Fujita 1961, 1968; Menzel 2001). It was not until the launch of the geostationary Application Technology Satellite (ATS-1) in 1966 that Verner Suomi (University of Wisconsin) developed the first analytical tools to estimate cloud displacement from a series of ATS-1 images (Suomi 1969). Although the principle of cloud tracking seems quite simple, it took nearly 10 years to see the appearance, in the mid-1970s, of the first wind vectors (Cloud Motion Vectors, CMVs) extracted from the infrared channel of the American Geostationary Operational Environmental Satellite (GOES). CMV extraction methods were gradually automated in the 1980s, generalized to other geostationary satellites as well as extended to visible channels and water vapor. Observation of water vapor structures at high altitude with the channel at 6.7 microns thus provided additional information to that of the CMVs, which mainly represented the movement of clouds at low altitude. No longer limited to cloud tracking alone, wind vectors were then renamed atmospheric motion vectors (AMVs).

Currently, several space agencies are operationally extracting AMV products using images from their own satellites: EUropean Organisation for the Exploitation of METeorological SATellites (EUMETSAT), National Oceanic and Atmospheric Administration Satellite and Information Service (NOAA/NESDIS), Cooperative Institute for Meteorological Satellite Studies (CIMSS), Japan Meteorological Agency (JMA), Korean Meteorological Agency (KMA), Chinese Meteorological Agency (CMA) and Indian Meteorological Department (IMD). These wind vectors have been derived in real time from geostationary satellites for three decades and from polar satellites since the early 2000s. They represent a significant fraction of the observations assimilated daily in numerical weather prediction models, and are in fact the only observations of the movement of air particles that are available at high altitudes, and completely covering polar regions, tropical and mid-latitude areas, and in particular the vast expanses of ocean (Figure 2.1). They are among the five types of observations assimilated into numerical models that have the greatest impact on the quality of the prediction (English et al. 2013).

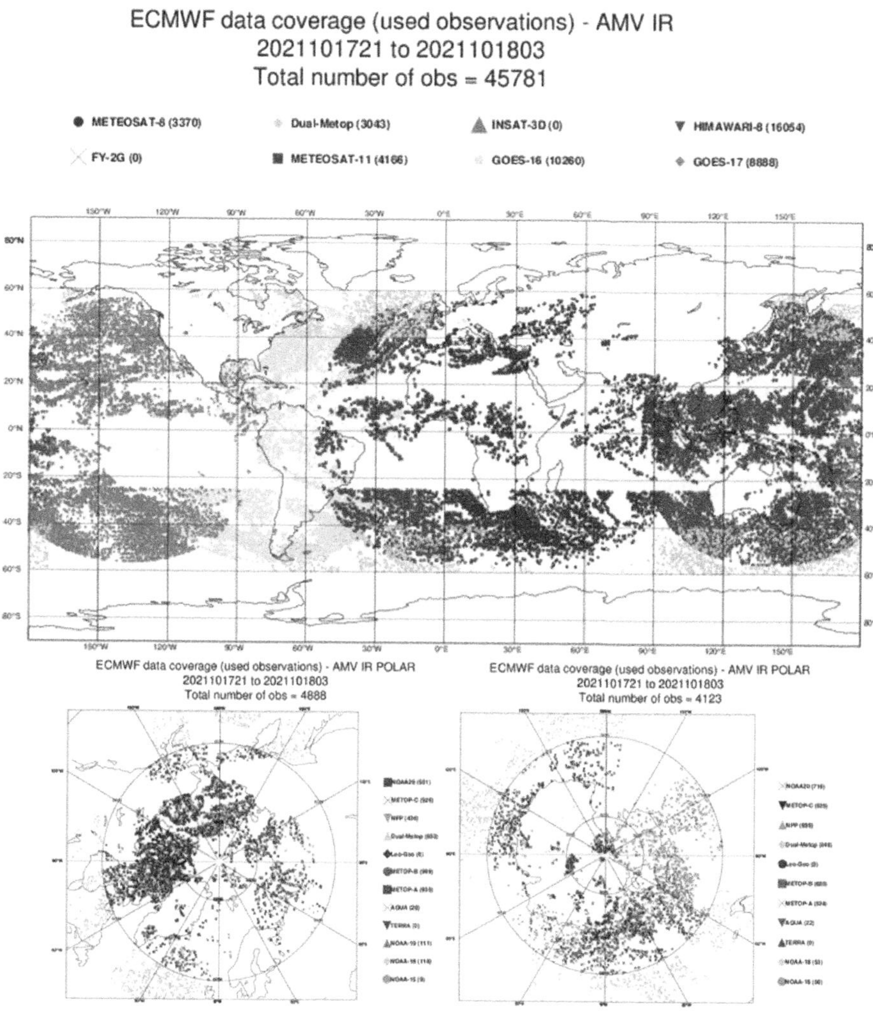

Figure 2.1. *All infrared winds extracted from geostationary and polar satellites, which were used in the ECMWF weather prediction model on October 17, 2021, at 00:00 UTC (source: ECMWF[1])*

In addition to observing winds in the atmosphere, surface winds are estimated over the oceans by scatterometers carried on moving satellites. Surface winds from ASCAT instruments on board MetOp satellites have a very significant positive impact on numerical predictions (see Chapter 4). Scatterometers are active instruments, of the radar type, which make it possible to retrieve the force and direction of wind on ocean

1. www.ecmwf.int.

surfaces from variations in amplitude of the backscattered signal caused by small waves (of the order of 1 cm) created by the wind. Winds restituted by scatterometers are particularly useful for the visualization of the surface atmospheric circulation and for the determination of the force and direction of wind in areas with low observations, such as those where cyclones originate. Surface winds are also estimated with altimeter radars such as those of Jason satellites.

The importance of these observations is such that a specific working group on winds has been created by the Coordination Group for Meteorological Satellites (CGMS) within the World Meteorological Organization (WMO). Established in 1991 and formalized in 1994, the International Winds Working Group (IWWG) is one of the CGMS's oldest scientific groups[2]. Through its bi-annual workshops, it provides a unique forum to discuss and coordinate research and development activities needed to optimize wind remote sensing from satellites and to improve the use of this information in NWP and for climate monitoring.

However, although they are indispensable observations for numerical prediction, AMVs also suffer from intrinsic conceptual weaknesses that limit their potential to fully document the dynamics of the atmosphere. The whole extraction process assumes that clouds are passive tracers, that is, they move exactly at the speed of the local wind. This assumption is undoubtedly incorrect or at least unsuitable for certain weather situations, as illustrated in particular by the case of orographic clouds. Above all, the information given by AMVs is located at a single level in the atmosphere, and therefore does not describe the vertical wind profile at a given geographical location. Such vertical wind profiles can be obtained with onboard lidars or with infrared sounders. These two types of more recent observations are dealt with in the last two sections of this chapter, after those devoted to AMVs.

2.2. AMVs

2.2.1. *Extraction of AMVs*

Although considerable differences exist, the general scientific principle of extracting AMVs from satellite images is fairly homogeneous for all the algorithms used by space agencies. The technique essentially comprises four successive steps that begin by selecting a cloud or a water vapor structure in the first image, monitoring this target in the following images, estimating the altitude of the detected movement and finally assigning a quality index associated with each wind vector (Schmetz et al. 1993; Borde et al. 2014). A cloud or a water vapor structure is identified by calculating

---

2. wisc.edu.

quantities such as entropy, contrast and the number of cloudy pixels in boxes of 16×16 or 24×24 pixels. The selected box is then searched for in the second image with a cross-correlation process. To limit the computing time and avoid the detection of unrealistic winds, the search for the target in the second image is limited within a box of 80×80 pixels, centered on the position of the target box in the first image.

All possible correlations between the target box of the first image and the boxes present in the search area of the second image are calculated, and the maximum correlation identifies the final position of the target box in the second image. The speed and direction of movement can then be easily calculated from the respective positions of the target box in the first and second images and the time between the acquisition of the two images.

Estimating the altitude of the detected motion is the most difficult step and also the greatest source of error. An initial difficulty consists of precisely identifying which particular structure was followed during the detection of the movement in the scene studied. Indeed, it is often the case that several types of cloud evolving at different altitudes coexist inside the target box of 24×24 pixels. The method operationally implemented at EUMETSAT uses individual contributions to the correlation of each pixel in the box to accurately identify clouds that were tracked (Borde et al. 2014), as well as the cloud product associated with each cloud pixel. The altitude of the wind is finally assigned to the altitude of the top of the clouds of the pixels considered. A wide variety of clouds exist, such as opaque, semi-transparent, fractional, multilayer, etc., and different techniques are used to calculate the altitude of their top according to their characteristics. While recent methods certainly allow for better estimation of the height of the top of semi-transparent clouds, they remain very sensitive to local atmospheric and microphysical characteristics. The reliability of this estimate often remains difficult to assess, particularly in the case of very thin clouds, fragmented clouds and in the case of multilayer clouds.

Most wind vector extraction algorithms use three consecutive satellite images to extract AMVs, thus generating two intermediate vectors that are then used to derive the final product. A quality index (QI) with a value between 0 and 1 is calculated for each wind vector by applying a series of spatial and temporal coherence tests to intermediate vectors (Holmlund 1998). This index facilitates the automatic selection of AMVs that are assimilated into prediction models.

2.2.2. *Current production and outlook*

AMVs are derived from all geostationary satellites to a maximum latitude of about 60° north and south, beyond which the viewing angle becomes too large to allow reliable use of observations. The production of AMVs at the highest latitudes is provided by satellites in a Sun-synchronous orbit passing over the polar regions approximately every 100 minutes. The first wind vectors above the poles were extracted in the 2000s from the images of the MODIS instrument on board the Terra and Aqua satellites (Key et al. 2003). Today, the AVHRR imagers installed onboard the American NOAA and European MetOp satellites (Klaes et al. 2007) provide the bulk of this production, as does the VIIRS instrument on board the Suomi NPP and NOAA-20 satellites. The latest version of the wind vector extraction algorithm with AVHRR developed in 2015 at EUMETSAT allows for the extraction of three different products (Hautecoeur and Borde 2017; Horváth et al. 2017), using either a pair or a triplet of AVHRR images taken from one or two MetOp satellites, with different covers and time intervals between successive images. The assimilation of global AVHRR wind products (Borde et al. 2016) in NWP models showed a significant positive impact on prediction scores of several weather agencies, as it makes it possible to document in particular the latitude band 40–70 degrees, where few wind observations are generally available for assimilation.

With the arrival of new generation satellites, the overall production of AMVs available for numerical prediction services has exploded in recent years. Improving the spatial, spectral and temporal resolutions of the new instruments allows for Earth observation with satellite products of unprecedented quality and frequency. The "generalization" of imagers with 16 spectral bands on board the new generation geostationary satellites (HIMAWARI, GeoKOMPSAT, GOES-R, MSG and the future MTG-I) also allows for greater homogenization of inter-agency AMV products, using the same spectral bands to extract wind vectors. Thus, for MSG/SEVIRI, wind vectors are derived from the visible spectral channels at 0.8 μm and HRVIS, from two water vapor channels at 6.2 and 7.3 μm and from one infrared channel at 10.8 μm. Visible channels are only used during the day, when the sun illuminates the Earth's surface. Since radiation at 10.8 and 0.8 μm interacts very little with gases and other constituents of the atmosphere, the targets observed are located at a lower altitude near the Earth's surface. Conversely, since the two channels at 6.2 and 7.3 μm are situated in the water vapor absorption bands, they are used to observe the movement of water vapor structures situated in the upper and middle tropospheres, respectively, at 250 and 550 hPa in a standard atmosphere.

The overall operational production of AMVs should logically continue to increase in the coming years, enriched by the availability of new satellites, new products and new extraction techniques. Examples include the launch of the future European geostationary satellite MTG-I in 2022, which will allow for semi-hourly extraction of AMVs over Europe and Africa, the operational production of AMVs from SLSTR instruments of Sentinel-3 satellites planned in 2022 (Barbieux et al.

2021), the launch of the METImage instrument on the MetOp-SG-A satellite in 2024, which will allow for the extraction of AMVs in water vapor channels over the North and South poles (Borde et al. 2019) and the production of AMV products over the Arctic from the Russian Arktika-M satellites placed in a highly elliptical orbit. In addition, new AMV products are being developed, based on stereoscopic techniques from several satellites in order to improve estimation of the altitude of the detected movements (Carr et al. 2020). Finally, new mathematical and statistical approaches such as optical flows, or using artificial intelligence and machine learning are being investigated to extract denser wind fields from satellite images and thus support nowcasting.

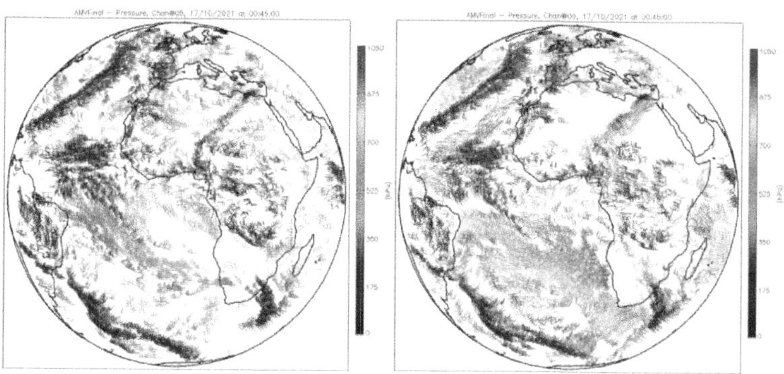

Figure 2.2. *Examples of wind vectors extracted from water vapor channels at 6.2 µm (left) and 10.8 µm Meteosat 11 infrared on October 17, 2021, at 00:45 UTC. The color scale corresponds to the altitude of wind vectors*

2.3. 3D winds derived from hyperspectral sounders

3D wind profiles can in theory be obtained from the vertically resolved observations of infrared sounders. A demonstration product is currently being proposed by CIMSS using data from the AIRS atmospheric infrared sounder on board the AQUA satellite (Santek et al. 2019). The area of overlap between consecutive AIRS images makes it possible to extract wind only at high latitudes, beyond 70 degrees of latitude. The overall impact of this product was considered neutral in the US Goddard Earth Observing System Model version 5 (GEOS-5). However, as the number of observations assimilated into the model was small, the impact by observation proved very good for AMVs extracted from the AIRS humidity fields.

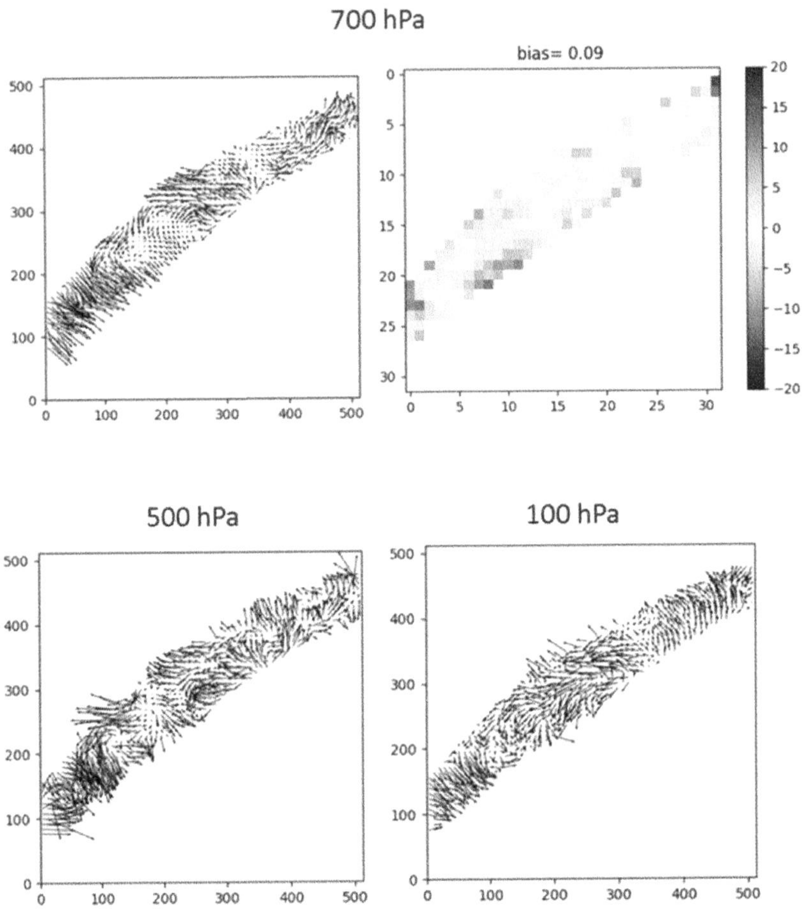

Figure 2.3. *Example of IASI 3D winds observed at 700, 500 and 100 hPa from the MetOp-A and MetOp-B satellites over the North Pole on July 4, 2018 at 00:46:09 UTC. Winds are illustrated here on a grid of 512 x 512 pixels (1 pixel ~80 x 80 km). Velocity bias in relation to the European Centre prediction model is illustrated for 700 hPa (source: Borde et al. (2019))*

A scientific processor based on 3D optical flow methods has recently been developed at EUMETSAT to extract wind profiles from infrared sounders. It is based on an algorithm that initially worked with three layers of information on cloud top height (Héas and Mémin 2008), and has been modified to ingest vertical moisture, temperature and ozone profiles from the IASI infrared atmospheric sounding interferometer (August et al. 2012). Wind profiles are extracted from pairs of IASI data sets taken by the MetOp-B and MetOp-C satellites on high-latitude zones (Hautecoeur et al. 2021; Borde et al. 2019). Although still at a preliminary stage, the first results seem very promising, providing dense multilevel wind fields in the atmosphere over the areas observed by the IASI instrument, beyond 45 degrees of latitude (Figure 2.3). Operational production is planned for the second part of 2023. The algorithm will then be adapted to process MetOp-SG/IASI-NG products operationally around 2024. However, the main objective is the MTG/IRS instrument on board the MTG-S satellites, the first launch of which is planned for 2024. The production of 3D winds over Europe could then reach an observation frequency of 30 min with a pixel sampling of 4 km. This could be a major step forward in observing atmospheric dynamics from satellite data.

2.4. Measuring wind from space using Doppler lidar

2.4.1. *Introduction*

The launch of the Aeolus satellite, carried out on August 22, 2018, is the culmination of a European Space Agency (ESA) mission to test the use of an onboard lidar to measure vertical wind profiles from space. This Aeolus mission (also called ADM-Aeolus; ADM: Atmospheric Dynamics Mission) belongs to the ESA series called "Earth Explorer" (see Chapter 4 of Volume 1). Aeolus is a 1,360 kg satellite placed in a Sun-synchronous orbit at an altitude of 320 km. It is equipped with a lidar designed to observe the wind by the Doppler effect, an instrument called ALADIN (Atmospheric LAser Doppler INstrument). While other onboard systems (imagers, diffusiometers) have routinely provided many wind measurements for decades, either at a single level (sea level) or at a few irregular levels (characterized by clouds or water vapor structures), ALADIN is the first instrument that can measure vertical wind profiles from space.

In order to validate this new type of data and thus this spatial wind-lidar concept effectively, data is processed and then disseminated in NWP centers, which use them in a global assimilation system. After a monitoring and evaluation period of a few months, Aeolus data has been used from January 2020, in the operational analysis and forecast of the ECMWF. At the beginning of 2022, the same validation operations were carried out in other meteorological centers using prediction models. Aeolus data will continue to be used in prediction models until the end of its instrument's life. This one could arrive quite quickly since the life of the ALADIN instrument has been estimated at three years from its orbit. However, the various evaluations already carried out, including on the improvement of forecasts provided by new wind measurements, seem sufficient to envisage other satellite missions built on the same concept to succeed the Aeolus experimental mission in a fully operational context (Stoffelen et al. 2021).

2.4.2. Measurements from ALADIN lidar on board Aeolus

The principle of lidar, the different types of instruments and the different meteorological applications are described in Chapter 9 of Volume 1. To measure winds from space, Aeolus uses a high-powered laser that probes atmospheric layers from about 30 kilometers above sea level to the ground. It sends pulses from the satellite at a single wavelength set at 355 nm (in ultraviolet), according to a viewing angle deviating by 35° from nadir, in a vertical plane perpendicular to its trajectory, on the right with respect to the satellite's movement. The main measurement of the ALADIN lidar is therefore a set of wind values along this line of sight (LOS) inclined at 35°, obtained from echoes received by the instrument from different layers of the atmosphere and Doppler shifts measured on these echoes. The main targets of the laser are essentially the molecules of the main air components (nitrogen, oxygen) in clear skies. However, its high spectral resolution also allows it to identify several types of aerosols as well as the top of clouds.

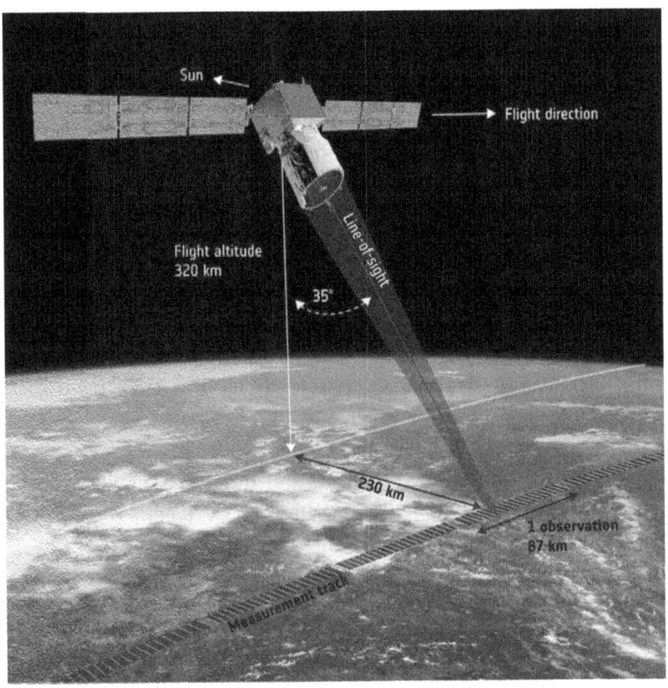

Figure 2.4. *Principle of wind measurement by the Aeolus space mission (source: ESA[3])*

3. www.esa.int.

By this Doppler measurement technique, only the component of wind along the LOS can be obtained. It is called "LOS Wind". In order to obtain good accuracy, Doppler signals are accumulated and averaged over a distance of the order of 87 km (see Figure 2.4) along the satellite's trajectory (or rather its track on ground according to the LOS) in order to produce an LOS wind profile. On this spatial scale, the vertical wind is generally negligible compared to the horizontal wind, which makes it possible to extract, every 87 km or so, horizontal winds along the line of sight called "HLOS winds" (Horizontal Line Of Sight winds). This horizontal wind is the product of greatest interest to users, particularly those involved in assimilation for NWP models and climate reanalyses.

2.4.3. *Culmination of a long process*

Since the 1970s, lidars have been used in meteorology to study the chemical composition of the atmosphere, aerosols, clouds and wind from the ground. Since the 1980s, most atmospheric measurement campaigns have used different types of lidars, first installed on the ground, and then often also on board research aircraft. By the beginning of the 21st century, ground-based lidars began to function regularly in the context of operational meteorological profiler networks. In 2020, several dozen profiler-lidar stations were operated in Europe to observe aerosols (volcanic ash in particular) or wind.

Lidar technology took much longer to develop so as to be used on board a satellite, especially to observe wind, which explains why the first wind-lidar in space was not seen until 2018. And it was only at the beginning of the 21st century that the first space missions such as CALIPSO, a joint NASA-CNES satellite launched in 2006 with the CALIOP aerosol cloud lidar on board, were implemented, together with two other instruments (see Chapters 4, 6 and 9 of Volume 1). Technical difficulties associated with space-borne lidar missions are much more severe than for the same ground-based instruments. There are at least two reasons for this:

– a lidar is an active instrument that consumes a lot of energy. This explains in particular why the estimated lifetime of the Aeolus ALADIN lidar (due to its laser source) is only three years, while, for example, the first European MetOp satellite, launched in 2006, still has instruments working 14 years later;

– ALADIN is a type of instrument very sensitive to the environment in which it is immersed and to its variations. It is therefore very difficult to ensure that it will function well in space by means of ground tests simulating the space environment.

While previous projects exploring wind-lidars for space failed to lead to a space mission, around 1990 ESA launched a project called the ADM (Atmospheric

Dynamics Mission). The project group brought together specialists in lidar technology as well as potential scientific users of wind data for meteorology and climate, especially those who assimilate observations into numerical prediction models (either for daily weather prediction or for climate reanalyses). The objective of the ADM project was to converge users' needs with designing a global wind-observing space mission. The 1990s were marked by a very rapid evolution of research on lidar technology and data assimilation methods, and it is in this context that the preliminary ADM studies took place. In 1999, they led to the design of the ALADIN lidar operating by direct detection in ultraviolet (355 nm), which was selected by ESA to constitute its second Earth Explorer mission (see Chapter 4 of Volume 1).

This space mission was entrusted to Airbus Defence and Space, which developed the instrument and the satellite in its Toulouse center. As early as 2003, a model of the instrument was designed that produced correct wind observations from the ground. But as soon as the model was put in space conditions, it no longer worked. It took several years of environmental reconfiguration, testing and readjustment to arrive at data that met expected specifications. Aeolus was finally launched 11 years behind the original plan. It is now the fifth satellite in the ESA/Earth Explorer series when it was initially in the second position. However, these years of delay have been used by scientists both to refine mission specifications and to better prepare the processing of actual data and their use in models (see, for example, Dabas et al. (2008) or Tan et al. (2007)).

2.4.4. *Situation in 2022 and outlook*

During the 2020s, HLOS wind data from Aeolus were successfully assimilated into several NWP centers. As of the beginning of 2022, the ECMWF had been using them for two years (operational on January 9, 2020) after demonstrating significant improvement in predicting, especially in the tropics and near the poles. Météo-France has been using them operationally since June 30, 2020, for its Arpège global prediction model. Figure 2.5 shows the geographical distribution and the number of data actually used over a six-hour period (on January 12, 2021 from 15:00 to 21:00 UTC).

It is noted that data coverage and the number of observation points are relatively small compared to, for example, equivalent data coverage in AMV winds (Figure 2.1). This is due to the unique LOS (at 35° from nadir) of the ALADIN instrument onboard Aeolus. The advantage of this instrument compared to other systems observing the wind from space is that at each observation point appearing on the map, there is multilevel wind data on the vertical: a profile extending from the stratosphere (25–30 km altitude) to clear sky ground, if not to the top of clouds targeted by the lidar. Compared with the Aeolus experimental mission, there is therefore considerable room for improvement that could be obtained by improving the coverage of the planet in observations, either by placing several satellites of the same type in orbit, or by replacing the fixed sight instrument with a scanning instrument.

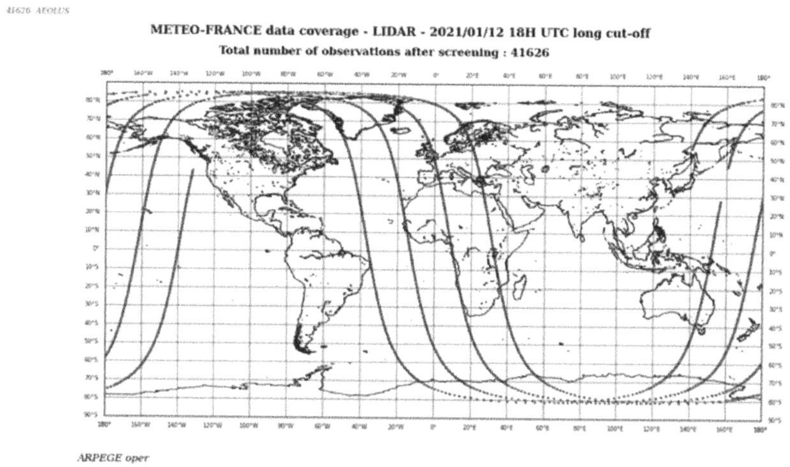

Figure 2.5. *Geographical distribution of Aeolus wind data used at Météo-France in the Arpège prediction model over a period of six hours at the beginning of 2021 (source: Météo-France)*

At the date of publication (2023), the Aeolus mission can be considered to have demonstrated the feasibility and usefulness of a space-based wind-lidar for meteorology and climate. Launched in 2018, with an expected lifetime of three years, it was operational until May 2023, i.e about 5 years. Planning for future satellites with wind-lidars has already begun.

2.5. References

August, T., Klaes, D., Schlussel, P., Hulberg, T., Crapeau, M., Arriaga, A., O'Carroll, A., Coppens, D., Munro, R., Calbet, X. (2012). IASI onMetop-A: Operational level 2 retrievals after five years in orbit. *J. Quant. Spectrosc. Radiat. Transfer*, 113, 1340–1371.

Barbieux, K., Hautecoeur, O., De Bartolomei, M., Carranza, M., Borde, R. (2021). The Sentinel-3 SLSTR atmospheric motion vectors product at EUMETSAT. *Remote Sens.*, 13, 1702. https://doi.org/10.3390/ rs13091702.

Borde, R., Doutriaux-Boucher, M., Dew, G., Carranza, M. (2014). A direct link between feature tracking and height assignment of operational EUMETSAT atmospheric motion vectors. *J. Atmos. Oceanic Technol.*, 31, 33–46. doi: http://dx.doi.org/10.1175/JTECH-D-13-00126.1.

Borde, R., Hautecoeur, O., Carranza, M. (2016). EUMETSAT global AVHRR winds product. *J. Atmos. Oceanic Technol.*, 33, 429–438. doi: http://dx.doi.org/10.1175/JTECH-D-15-0155.1.

Borde, R., Carranza, M., Hautecoeur, O., Barbieux, K. (2019). Winds of change for future operational AMVs at EUMETSAT. *Remote Sens.*, 11(18), 2111. https://doi.org/10.3390/rs11182111.

Carr, J.L., Wu, D.L., Daniels, J., Friberg, M.D., Bresky, W., Madani, H. (2020). GEO–GEO stereo-tracking of atmospheric motion vectors (AMVs) from the geostationary ring. *Remote Sens.*, 12, 3779. https://doi.org/10.3390/rs12223779.

Dabas, A., Denneulin, M.L., Flamant, P., Loth, C., Garnier, A., Dolfi-Bouteyre, A. (2008). Correcting winds measured with a Rayleigh Doppler lidar from pressure and temperature effects. *Tellus*, 60A, 206–215.

English, S., McNally, T., Bormann, N., Salonen, K., Matricardi, M., Horanyi, A., Rennie, M., Janisková, M., Di Michele, S., Geer, A. et al. (2013). Impact of satellite data, ECMWF research department technical memorandum [Online]. Available at: www.ecmwf.int/sites/default/files/elibrary/2013/9301-impact-satellite-data.pdf.

Fujita, T. (1961). Outline of a technique for precise rectification of satellite cloud photographs. Mesoscale research project paper 3, University of Chicago.

Fujita, T. (1968). Present status of cloud velocity computations from ATS-1 and ATS-3. *COSPAR Space Res.*, 9, 557–570.

Hautecoeur, O. and Borde, R. (2017). Derivation of wind vectors from AVHRR Metop at EUMETSAT, *J. Atmos. Oceanic Technol.*, 34, 1645–1659. doi:10.1175/JTECH-D-16-0087.1.

Héas, P. and Mémin, E. (2008). 3D motion estimation of atmospheric layers from image sequences. *IEEE Trans. Geosci. Remote Sens.*, 46(8), 2385–2396.

Holmlund, K. (1998). The utilization of statistical properties of satellite-derived atmospheric motion vectors to derive quality indicators. *Weather Forecast.*, 13, 1093–1104.

Horváth, Á., Hautecoeur, O., Borde, R., Deneke, H., Buehle, S.A. (2017). Evaluation of the EUMETSAT global AVHRR wind product. *J. Appl. Meteor. Climatol.*, 56, 2353–2376. https://doi.org/10.1175/JAMC-D-17-0059.1.

Key, J., Santek, D., Velden, C.S., Bormann, N., Thepaut, J.-N., Riishojgaard, L.P., Zhu, Y., Menzel, W.P. (2003). Cloud-drift and water vapor winds in the polar regions from MODIS. *IEEE Trans. Geosci. Remote Sens.*, 41(2), 482–492.

Klaes, K.D., Cohen, M., Buhler, Y., Schlüssel, P., Munro, R., Luntama, J.-P., von Engeln, A., Clérigh, E.Ó, Bonekamp, H., Ackermann, J. (2007). An introduction to the EUMETSAT Polar System. *Bull. Am. Meteorol. Soc.*, 88, 1085–1096.

Menzel, W.P. (2001). Cloud tracking with satellite imagery: From the pioneering work of Ted Fujita to the present. *Bull. Am. Meteorol. Soc.*, 82, 33–47.

Santek, D., Nebuda, S., Stettner, D. (2019). Demonstration and evaluation of 3D winds generated by tracking features in moisture and ozone fields derived from AIRS sounding retrievals. *Remote Sens.*, 11, 2597. https://doi.org/10.3390/rs11222597.

Schmetz, J., Holmlund, K., Hoffman, J., Strauss, B., Mason, B., Gärtner, V., Koch, A., van de Berg, L. (1993). Operational cloud motion winds from METEOSAT infrared images. *J. Appl. Meteorol.*, 32, 1206–1225.

Stoffelen, A., Benedetti, A., Borde, R., Dabas, A. Flamant, P., Forsythe, M., Hardesty, M., Isaksen, L., Källen, E., Könich, H., et al. (2020). Wind profile satellite observation requirements and capabilities. *Bul. Am. Meteorol. Soc.*. doi:10.1175/BAMS-D-18-0202.1.

Suomi, V.E. (1969). Recent developments in satellite techniques for observing and sensing the atmosphere. In *The Global Circulation of the Atmosphere*, Corby, G.A. (ed.). Royal Meteorological Society, Reading.

Tan, D.G.H., Andersson, E., Fisher, M., Isaksen, L. (2007). Observing system impact assessment using a data assimilation ensemble technique: Application to the ADM-Aeolus wind profiling mission. *Q. J. R. Meteorol. Soc.*, 133, 381–390.

3

Surface Variables

Jean-François MAHFOUF
CNRM, Université de Toulouse, Météo-France, CNRS, France

3.1. Observation of the Earth's surface from space

In this chapter, we provide an overview of some properties of the Earth's surface deduced from space-borne remote sensing. We will stick to those that are of particular importance for the atmosphere, either because their knowledge allows us to better estimate some properties of the atmosphere, or because they participate in energy, matter or momentum exchanges between the two media, and thus modulate the coupling between these compartments of the Earth system. As a result, they are involved in the evolution of the atmosphere and must be taken into account in weather prediction. We will also restrict ourselves in terms of spatio-temporal scales. Indeed, depending on the frequency range explored within the electromagnetic spectrum, it is possible to access spatial scales from a few tens of meters (visible domain) to several tens of kilometers (microwave domain). Revisiting the same point of the planet can take less than an hour for geostationary satellites, while some instruments on board low orbit satellites can only see the same scene several weeks apart. We choose here to mainly describe measurements made from instruments resolving kilometric spatial scales and time scales allowing for a certain description of the diurnal cycle (several measurements per day). We will thus propose some illustrations from the field of numerical weather prediction where the operational models from national meteorological services describe the spatio-temporal scales chosen. It is important to remember that the quality of these models, which make predictions a few days ahead, depends mainly on their initial conditions

---

For a color version of all the figures in this chapter, see www.iste.co.uk/phulpin/satellites2.zip.

Satellites for Atmospheric Sciences 2,
coordinated by Thierry PHULPIN, Didier RENAUT, Hervé ROQUET and Claude CAMY-PEYRET.

and that certain surface parameters, having slow changes on these time scales, are often prescribed (i.e. values imposed by an external specification); this is the case, for example, with the sea surface temperature (SST) or the sea ice extension. This assumption is no longer true when we are interested in longer time scales (seasonal or climate predictions) for which a coupling is necessary between the different compartments of the Earth system, in particular the atmosphere and oceans. Space-borne remote sensing is therefore of particular importance in numerical weather prediction for prescribing slowly changing surface parameters.

The contribution of terrestrial surfaces to a spectral radiance L_λ for a wavelength λ measured by a radiometer from space (at nadir) in infrared or microwave range can be expressed schematically[1] as follows (for a non-scattering and cloud-free atmosphere):

$$L_\lambda = \varepsilon_s B_\lambda(T_s)\tau_o + \int \varepsilon_\lambda(T) B_\lambda(T) d\tau_\lambda \qquad [3.1]$$

where $B_\lambda(T)$ is the Planck function and τ is the atmospheric transmittance (depending on temperature, pressure and amount of gases). There are thus two components: a first term representing the contribution of the surface (with its temperature T_s and its emissivity ε_s) attenuated by atmospheric transmission τ_o and a second term expressing the contribution of all layers of the atmosphere with their own emission ε at a certain temperature T. This equation shows that both the composition of the atmosphere and the properties of the surface (temperature and emissivity) will contribute to the measurement made from space. We will see how to remove this ambiguity when it exists (depending on the spectral domain).

The characteristics of terrestrial surfaces from space are generally acquired either in the solar domain (visible and near-infrared) through their reflective properties of the energy from the sun, or in the infrared and microwave domains by the own thermal emission of the surface in application of Planck's law. In the microwave field, the reflective properties of surfaces are also exploited by means of active instruments (spaceborne radars) emitting radiation to Earth. Active measurements are more frequently used in this domain of the electromagnetic spectrum (frequencies below 15 GHz) because of the relative transparency of the atmosphere, making it easier to isolate the effects of the surface in the measured signal.

1. The part of the descending radiation reflected from the surface is ignored here (see Chapter 9 of Volume 1, on the physics of measurement for a more complete description).

3.2. Energy balances at the surface and at the top of the atmosphere

Surface temperature is at the heart of transfers between the Earth's surface and the atmosphere, and can be deduced from an energy balance that expresses the equilibrium between inputs by incident radiative fluxes in the solar and thermal domains of the electromagnetic spectrum and the losses associated with these incident radiative fluxes (reflection and transmission), turbulent flows, the fluxes transmitted in the environment (soil, ocean, sea ice, etc.) as well as the own thermal emission of the surface, as schematized in Figure 3.1.

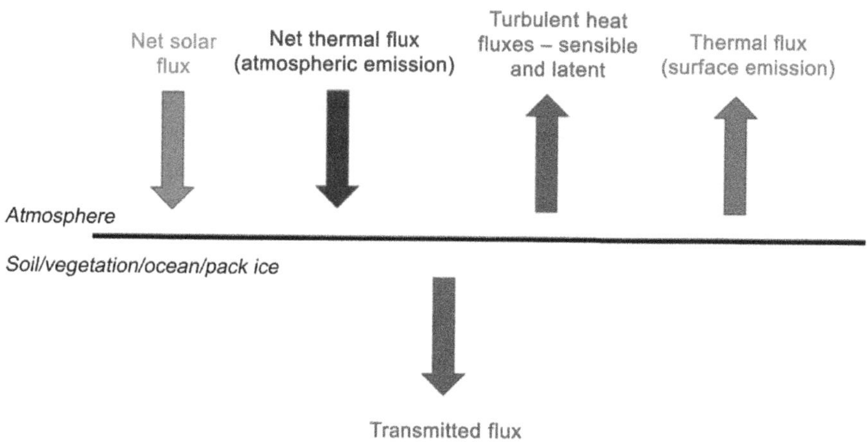

Figure 3.1. *Schematic representation of the surface energy balance for estimating the surface temperature. Energy inputs are represented by the orange (solar radiation) and black (thermal radiation) arrows, and energy losses are represented by blue (turbulent fluxes for sensible and latent heat), gray (surface emission in the thermal domain) and green (heat flux transmitted in the medium; conduction flow and/or attenuated solar flux) arrows*

In the solar domain, the reflection efficiency of the surface is defined by its **albedo**, which represents the ratio between the incident energy and reflected energy (in a certain frequency range of the solar spectrum). Some dependencies with the direction of incident and reflected radiation must sometimes be taken into account (this property is called **bidirectional reflectance**). They can be used as a signature of particular surfaces. In the thermal field, surface emission is generally compared to that of a *black body*. Thus, **emissivity** represents the ratio between the energy emitted by a surface and that of a black body having the same temperature. As in the solar domain, some surfaces have directional dependencies that can also help to characterize them from space.

The turbulent fluxes F_x for a quantity X (wind, temperature, humidity, carbon dioxide concentration) between a surface and the atmosphere often take the following form:

$$F_x = \rho_a C_D U (X_s - X_a) \quad [3.2]$$

where ρ_a is the density of the air in the vicinity of the surface, C_D is a transfer coefficient describing the intensity of exchanges and depending mainly on atmospheric stability (i.e. the vertical temperature gradient), U is the modulus of wind near the surface, X_s is the value of the quantity at the surface and X_a is the value in the air near the surface.

These fluxes therefore depend directly on surface values. Thus, in addition to temperature, it is necessary to consider a variable informative on humidity, such as the mixing ratio or the **specific humidity at the surface**[2], as well as the **wind near the surface.** The C_D coefficient also depends on the heterogeneities of the surface characterized by a parameter called the **roughness length** upon which certain radiative properties may also depend.

3.3. Ocean surfaces

3.3.1. *Surface temperature*

SST has been available thanks to space-borne remote sensing on a global scale for about 50 years (Minnett et al. 2019) with an uncertainty less than 0.5 K for kilometer spatial scales nowadays. The first multi-channel radiometers in the infrared domain (in particular the AVHRR imager available since 1981 on operational meteorological satellites) made it possible to detect the presence of clouds in an observed scene and to correct the signal from atmospheric effects (by gases and aerosols), by comparing spectral bands in *atmospheric windows*. This measurement is quite easy because it results, after atmospheric corrections, from an inversion of the Planck function, ocean surfaces behaving like a black body with a very good approximation. The high *thermal inertia* of oceans means that their temperature varies slowly over time, with a small amplitude of the diurnal cycle (< 2 K). It is thus possible to use persistence as a very good approximation (previous observations) when an observed scene is affected by clouds. The combination of satellites in geostationary and Sun-synchronous orbits (offering resolutions between 1 and 5 km) now makes it possible to have daily SST products on a global scale at a

2. This estimate is quite simple over oceans because the surface is saturated and thus it depends only on its temperature and atmospheric pressure.

resolution of 0.05° x 0.05°, meeting the needs of numerical weather prediction at various scales (Donlon et al. 2012). An example of a daily multi-satellite SST product (OSTIA analysis) developed by the UK Meteorological Service is shown in Figure 3.2.

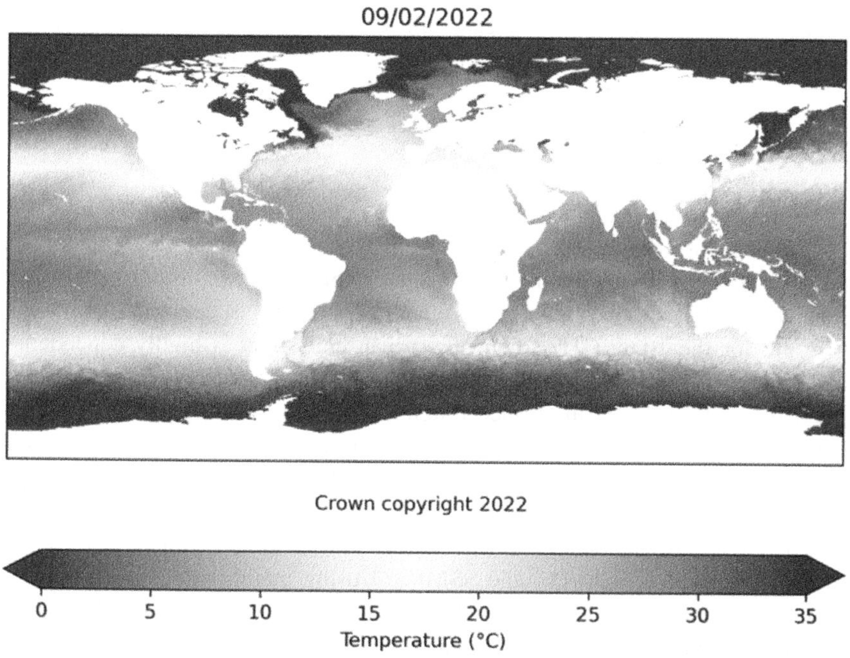

Figure 3.2. *Example of a daily product (OSTIA analysis) of sea surface temperature at a spatial resolution of 1/20th degrees combining estimates from several satellites and in situ data (February 9, 2022) (© Crown 2022)*

The microwave domain is also interesting for estimating SST because low frequencies (< 10 GHz) show a high sensitivity of radiances measured from space to this quantity, partly due to a weak interaction of radiation with gases and particles in the atmosphere. However, the conversion of radiance into a physical temperature is more delicate than in the thermal domain because it involves surface emissivity, which is much lower than one. In addition, it depends, as explained below, on the shape of the ocean surface and its nature (such as the presence of foam), and is also highly polarized (directional dependencies). The restitution of SST is based on algorithms using physical radiative transfer models (for atmosphere and surface) or statistical regressions. SST estimates are then available in regions where the surface is not accessible to visible and infrared instruments due to the frequent occurrence of clouds (such as the intertropical convergence zone). Microwave radiometers with channels below 10 GHz are few (currently AMSR-2**, GMI**, HY-2B**, HY-2C**) and are only available on satellites in low Earth orbit, thus penalizing temporal revisiting. Finally, spatial resolutions associated with these frequencies are much lower (pixel size between 30 and 50 km) than in the optical domain. Therefore, microwave measurements allow for more robust estimation of multi-sensor products where infrared measurements are less reliable.

3.3.2. *Surface wind*

The microwave domain leads to particular interactions of radiation with ocean surfaces. They are deformed by the wind that generates structures (capillary and gravity waves) modifying their electromagnetic properties (emission, backscatter) compared to a flat surface (where the reflection is called "specular"). As shown in Figure 3.2 for the 10.69 GHz frequency, ocean emissivity is highly dependent on signal polarization, *angle of incidence* and wind speed. This dependence on speed is explained by the increase in surface roughness leading to a signal emitted by the slopes of the waves (instead of a flat surface) as well as to the presence of foam, generated by wave breaking, which behaves in this frequency range like a black body. These properties have been exploited with conical scanning radiometers (with a fixed angle of incidence to distinguish between horizontal and vertical polarizations) such as SSM/I** and SSMI/S** using mainly atmospheric windows of low frequencies between 10 and 37 GHz. As previously explained for SST, in this frequency domain, inversion algorithms rely on radiative transfer codes including surface emissivity modeling or statistical regressions. Access to wind direction requires more information. This was achieved by the WindSat instrument on board the Coriolis satellite between 2003 and 2020, by measuring four polarized signals (but unique in their kind) or active remote sensing for which the same scene is seen by the instrument from different angles (both in incidence and azimuth).

Active space-borne remote sensing by radar for measuring wind on ocean surface was initiated by the NASA SEASAT** mission launched in June 1978, and continued with numerous instruments launched by the European (AMI**/ERS** and then ASCAT**/MetOp**), American (NSCAT**), Indian (OSCAT**) and Chinese (SCAT on HY-2 and CFOSAT**) space agencies. The ASCAT instrument (called a **scatterometer**) operating in C-band (frequency 5.225 GHz) is particularly interesting because it has been operational since 2006, thanks to two other launches in 2012 and 2018 on board MetOp satellites. It has two sets of three fixed antennas (Figure 3.4) to examine the same ground scene in three azimuths for several angles

**See Appendix 3.

of incidence. Thus, for each triplet of backscatter coefficients, it is possible to estimate wind direction and intensity near the surface. Geophysical models (GMF) have been developed (Hersbach et al. 2007) to achieve this inversion, however with an ambiguity of 180° concerning wind direction (the instrument cannot discriminate whether the structure examined on sea moves forward or backward). Spatial resolutions imposed by the frequency of the instrument are quite low (between 25 and 50 km) but remain very useful thanks to a certain homogeneity of ocean surfaces, given the small number of in situ data (some drifting and anchored buoys) and the central role played by this quantity on the intensity of fluxes at the interface.

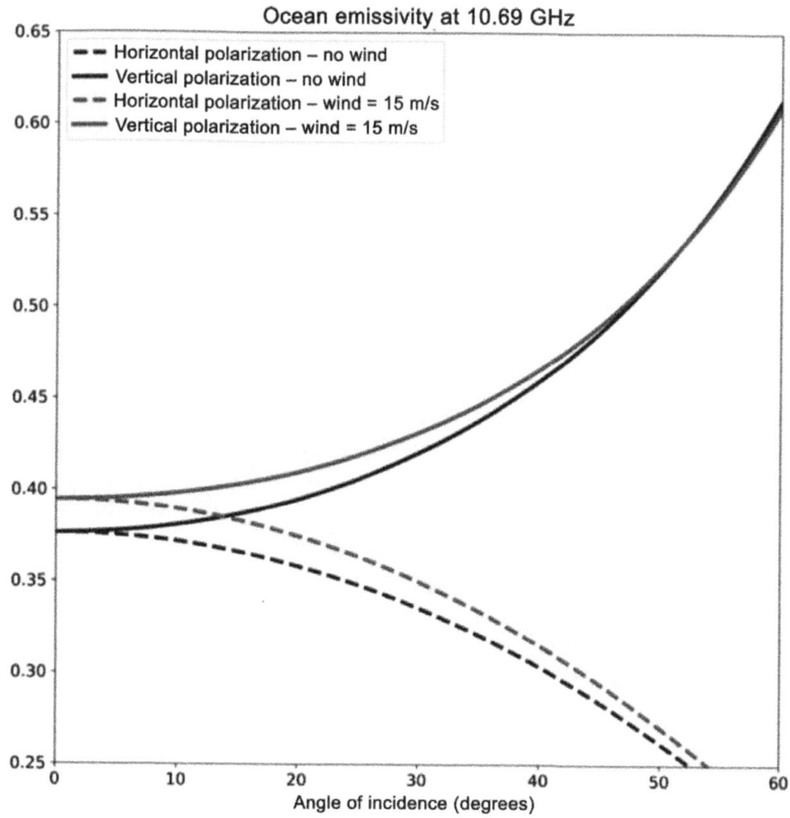

Figure 3.3. *Microwave emissivity of an ocean surface at 10.69 GHz without wind (black curves) and with wind of 15 m/s (red curves) as a function of the angle of incidence of the signal for horizontal (dashed curves) and vertical (solid line curves) polarizations according to the semi-empirical FASTEM model*

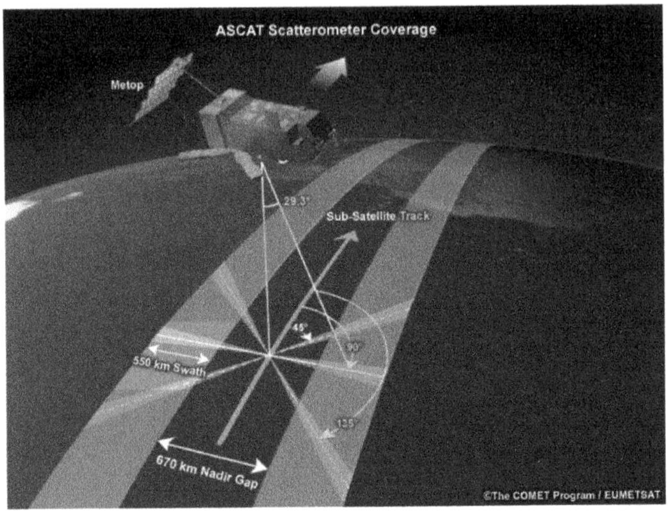

Figure 3.4. *Representation of surface swaths (pink bands) of the ASCAT scatterometer on board the MetOp satellite. A set of three antennas allows access to the same scene from three different azimuths (45°, 90°, 135°) to provide information on wind direction at the ocean surface. (© EUMETSAT)*

A first NASA experimental mission was launched in 2016 to provide wind measurements at the surface of tropical oceans and especially within cyclones. Indeed, wind measurements by scatterometers can be highly contaminated in the presence of intense rainfall. They are therefore not very useful within cyclones. By using a lower frequency such as the L-band (1.4 GHz), the atmosphere becomes almost transparent. A set of eight small satellites was thus set at an altitude of 520 km with an orbit inclined at 35°, each of them measuring the signal reflected by the ocean surface from satellites of GNSS constellations** located at approximately 20,000 km altitude. Reflection properties of the surface contain the same information as emissivity, including the surface wind dependency discussed earlier. This innovative mission showed the ability of this new technology called "GNSS-R" (R for reflectometry) to provide useful information on wind at the ocean surface (Ruf et al. 2019).

3.3.3. *Sea ice*

Sea ice or pack ice must be characterized by its surface temperature, its vertical extension and also its horizontal extension, which has strong seasonal variations in

**See Appendix 3.

polar regions. Other properties such as its albedo or emissivity are also relevant for the quantification of the surface energy budget and its contribution to measurements from space. Microwave measurements have very different signatures between open water and sea ice for both passive (emissivity) and active (backscatter coefficients) instruments. Sea ice has a more isotropic backscatter behavior (so-called Lambertian surface) than open sea and ice caps, as well as a lower dependency with incidence angle. Thus, algorithms based on parameters measuring anisotropy and changes with the incidence angle allow for the detection of the sea ice edge, and therefore its spatial extension with scatterometers such as ASCAT. These isotropy properties can also be exploited using radiometers such as SSMI/S or AMSR-2 by examining differences between horizontal and vertical polarizations of brightness temperatures at several frequencies (19, 37, 85 GHz). Multi-sensor approaches generally provide better quality estimates. Passive instruments can be used to estimate the sea ice thickness because the emission comes from a volume and no longer simply from the surface as with radar backscatter. Nevertheless, radar altimeters (ERS-2, Envisat**, Cryosat-2**) can provide an estimate of this thickness by measuring the height of emerged ice and using the hydrostatic balance that exists between snow-covered ice and the ocean ("freeboard" technique). Estimating properties such as surface temperature or albedo from space is quite difficult in the optical domain because sea ice, being a cold and reflective surface, can be easily confused with clouds, and therefore lead to significant errors when they are misidentified.

When a numerical prediction model (providing fairly accurate information on the atmosphere) and a radiative transfer model (simulating brightness temperatures[3]) are available, it is possible to inverse the radiative transfer equation (equation [3.1]) in order to estimate a surface emissivity on sea ice assuming that the surface temperature is known, as proposed by Karbou et al. (2014) using window channels of the AMSU-A** and AMSU-B** microwave sounders. Such an approach makes it possible to assimilate a number of surface-sensitive sounding channels into numerical models by ensuring that the contribution of the surface in equation [3.2] is close to observations. However, an error in modeling the surface temperature will be compensated by an erroneous restitution of surface emissivity to lead to a radiance in accordance with the observation. This methodology is also used above continental surfaces to produce climatological atlases by instruments (Prigent et al. 2006) because surface emissivity is very variable in space and time, making it difficult to model.

**See Appendix 3.

3. Product of the physical temperature by the emissivity. In the microwave domain, this quantity is proportional to the radiance (Rayleigh–Jeans approximation).

3.4. Continental surfaces

Continental surfaces show much greater spatio-temporal variability, at all scales, than ocean surfaces (open water and sea ice) just described. Space-borne remote sensing makes it possible to characterize them quite precisely for properties that evolve slowly on the time scales of meteorological phenomena such as topography and dominant types of surfaces (bare soils, natural vegetation, crops, urban areas, lakes, etc.). Here, we describe surfaces not covered with snow. Snow-covered surfaces will be discussed in the next section.

3.4.1. *Surface temperature*

Surface temperature on continents is characterized by a high amplitude of the diurnal cycle that can reach several tens of degrees due to their low thermal inertia. However, this amplitude is very variable depending on the energy inputs by radiation and the properties of the surface (dry or wet soil, presence or absence of vegetation, smooth or rough surface). Techniques used to estimate the surface temperature over continents by space-borne remote sensing are very similar to those considered for oceans. The combination of two thermal infrared measurements between 10 and 13 μm corrected for the atmospheric effect (so-called "split-window") technique is often used. It is also necessary to specify or estimate surface emissivity, which has strong geographical variations (between 0.85 for a desert to 0.99 for a tropical forest). Focusing on instruments on board geostationary satellites (such as the SEVIRI** radiometer on MSG**) results in a good sampling of the diurnal cycle (every 15 minutes in the absence of clouds). For high latitudes, however, polar-orbiting satellites must be used (more frequent revisiting than at low latitudes in areas inaccessible to satellites in geostationary orbit). Instruments on polar-orbiting satellites generally have better spectral resolution than those on geostationary satellites for improved atmospheric corrections (such as MODIS**/Terra, AATSR**/Envisat, SLSTR**/Sentinel-3). Precisions obtained are generally between 2 and 3 K (comparisons against in situ measurements).

Despite the central role played by this quantity in the surface energy balance, its estimation from space on continental surfaces is hardly used in atmospheric modeling because of its low predictability. Indeed, its value at a given time depends mainly on radiative fluxes at that time and less on values a few hours before. However, comparisons between observations and modeling, particularly for the amplitude of the diurnal cycle, have highlighted systematic errors in surface modeling related to the specification of some properties (such as roughness length) (Trigo et al. 2015). We show in the following that the amplitude of the diurnal cycle can be used as an indicator for the hydric state of soil.

**See Appendix 3.

3.4.2. *Water content of soil*

Unlike ocean surfaces, which evaporate at a maximum rate (called "potential") since water is always available, evaporation over continental surfaces is driven both by atmospheric demand (forcing conditions: radiation, wind, humidity and air temperature) and by the surface water availability (dry, wet or even frozen soil; presence or absence of vegetation, etc.). The water content of soil is a dominant parameter that will modulate the evaporation of continental surfaces (outside urban areas and open water surfaces). When the surface is sufficiently wet (e.g. after rain), it can evaporate at the potential rate. On the contrary, when soil is very dry (as in a desert or vegetation undergoing hydric stress), the quantity of water that can evaporate is insignificant. The regimes between these two extremes being harder to quantify.

Since the 1980s, space-borne remote sensing has used the dependency of the diurnal cycle amplitude of surface temperature in the infrared domain on soil water as shown in Figure 3.5. The root system of plants modulates the link between surface evaporation and soil water. Indeed, on bare soil, evaporation comes from water located in layers near the surface; on vegetated surfaces, evaporation (called evapotranspiration) comes from water from deeper layers where plant roots are located. The measurement of the amplitude of the diurnal cycle thus gives information on surface water in the absence of vegetation and on water in the root zone in the presence of vegetation. It has widespread applications for agricultural practices (thermal infrared allowing access to hectometric resolutions of fields for irrigation purposes using specific sensors). However, the high dependency of this amplitude on factors other than soil water (roughness, wind, air humidity) has limited its use in the field of atmospheric modeling.

Evapotranspiration is the sum of the evaporation of bare soil, plant transpiration and evaporation of water intercepted by vegetation. **Potential evapotranspiration** is the maximum amount of water that can be evaporated by a surface in a given climate. For example, evaporation is always potential over a water-saturated surface such as oceans and lakes.

Stomata are small orifices at the surface of plant leaves through which gas exchanges with the atmosphere take place: carbon dioxide absorption necessary for photosynthesis and plant transpiration (water loss) by evaporation. The opening of stomata is regulated by biological and environmental factors. **Stomatal resistance** is a quantity expressing the limitation imposed by a plant cover of leaf surface unit to the water vapor transfer from the stomata to the surrounding air.

Box 3.1. *Evaporation and the role of vegetation*

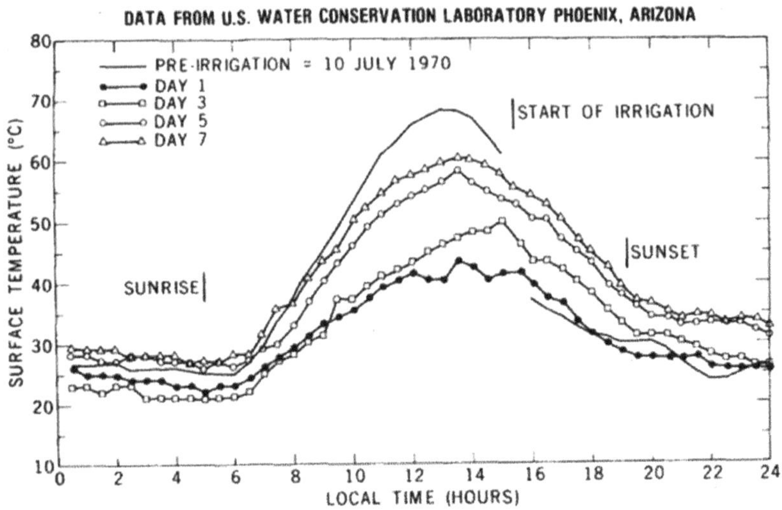

Figure 3.5. *Diurnal cycle of surface temperature before (curve without symbols) and after an irrigation period (curves with symbols) highlighting the dependency of its amplitude on soil water content (source: Schumgge (1978), © 1978 American Meteorological Society)*

Remote sensing in the microwave domain has allowed more robust quantitative estimates of soil water over the past 20 years. It is based on the fact that the dielectric constant of liquid water is much higher than that of other materials that constitute continental surfaces (organic and mineral matter). In the presence of liquid water in the soil, the reflectivity r of the surface increases, thus decreasing its emissivity ε (since $\varepsilon = 1 - r$). This can be seen in Figure 3.6 at a frequency of 1.4 GHz for the brightness temperature T_b and backscatter coefficient σ_0.

This dependency of σ_0 on soil water has been used with C-band scatterometers (already mentioned for estimating wind at the ocean surface), in particular the ERS-1 instrument in the 1990s and then the ASCAT instrument in the 2000s. One of the difficulties associated with measuring σ_0 is that it is also sensitive to other parameters such as roughness or water content of vegetation (Figure 15.6). In addition, information comes only from the first millimeters of the soil layer at this frequency. Wagner et al. (1999) proposed an algorithm based on minimum and maximum values of normalized backscatter coefficients at an incidence of 40° to provide soil water indices (ranging from 0 for dry soil to 1 for saturated soil). The seasonal effects of vegetation (slow variations) can be separated from the contribution of soil water that has higher temporal variability. The robustness and simplicity of this algorithm led EUMETSAT to propose a surface soil water product operational since 2008 from the ASCAT scatterometer. This product is operationally assimilated into several global numerical prediction models to improve the initial soil states, as at the ECMWF since 2015 (de Rosnay et al. 2013), after adjusting its climatology to that of the numerical model.

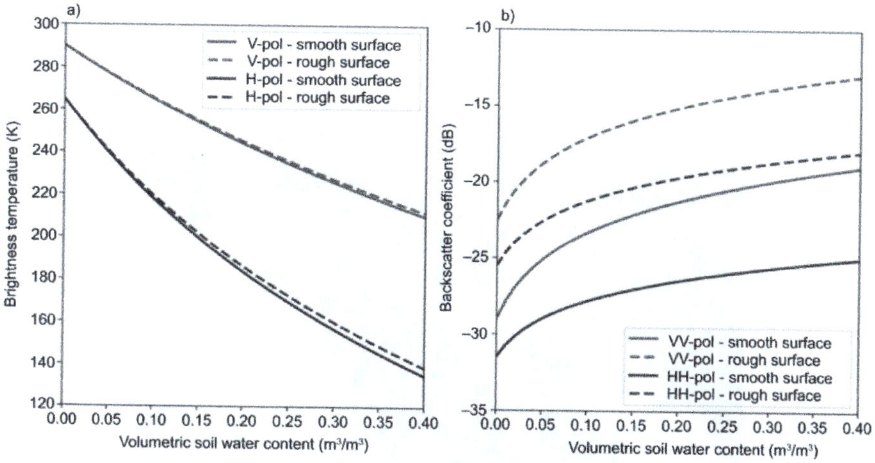

Figure 3.6. *Schematic dependency of the brightness temperature (a) and the backscatter coefficient (b) at 1.4 GHz with soil water content. The effects of signal polarization are shown by the color of the curves. The effects of surface roughness are shown by the type of curve (smooth surface: solid line; rough surface: dashed line)*

The C-band (frequencies between 6.6 and 6.9 GHz), where there is a significant dependency of brightness temperature on soil water, has also been exploited since the late 1970s thanks to the SMMR**, AMSR-E**, WindSat and AMSR-2 radiometers. With an antenna of 2 m in diameter, the most recent instrument (AMSR-2) produces pixels on the ground of about 50 kilometers, while the first SMMR instrument with an 80 cm antenna only allowed resolutions greater than 100 km. Passive remote sensing is less affected by roughness effects but, like radar, it lacks sensitivity in the presence of dense vegetation. It can also be affected by radio frequency interferences caused by radiocommunication antennas whose power is several orders of magnitude greater than the natural emission of surfaces.

By considering lower frequencies, namely the L-band (1.4 GHz), it is possible to obtain a signal that is much more sensitive to soil water (down to a depth of several centimeters) and also less sensitive to vegetation and atmosphere. The technological

**See Appendix 3.

difficulty lies in the size of the antenna, as it must be very large (6 m) to lead to a pixel size on the ground of the order of 50 km. It was bypassed by ESA with the SMOS** mission launched in 2009 using the interferometry technique (a set of radiometers arranged in a star shape examining the same scene from different angles) (Figure 3.7). NASA launched a similar SMAP** mission in 2015 with a large physical antenna deployed in space. The radio frequency interference problems noted with C-band instruments also appeared in the L-band but in other parts of the world.

Figure 3.7. *Artist's impression of the SMOS satellite launched in 2009 by ESA to measure brightness temperature at 1.4 GHz (soil water and ocean salinity). Each arm of the Y-shaped antenna with a length of 4 m includes a set of 23 radiometers to obtain from interferometry ground measurements with a resolution of 40 km (© ESA)*

While this estimate of surface soil water content is very interesting in regions with little vegetation, it is less directly useful in those with high vegetation density, since it is the deep soil water content that controls surface evapotranspiration. A number of studies have shown that it is possible to extract information about deep soil moisture from near-surface measurements, including through data assimilation techniques that combine the temporal dimension of observations with the dynamics of soil schemes describing water movements in the first few meters (Kumar et al. 2009). L-band brightness temperature assimilations using relatively simple surface

**See Appendix 3.

emissivity models requiring many input parameters have been attempted in numerical prediction models with mixed results (Munoz-Sabater et al. 2019). Statistical approaches or approaches based on neural network techniques seem better able to take into account the effects of nonlinearities between measurement and soil water as well as biases between observations and modeling (Rodriguez-Fernandez et al. 2019).

3.4.3. *Surface albedo*

Surface albedo is a property that can be deduced quite easily from space-borne remote sensing in the visible and near-infrared domains, from raw measurements of bidirectional reflectance. It is first necessary to correct the measurement from atmospheric effects (gases, aerosols, clouds), based on information from a numerical model (atmosphere and/or radiative transfer) or other spectral bands of the instrument. It is then necessary to perform an angular integration of the reflectance based on the inversion of a semi-empirical model of BRDF describing and simply separating the different physical components involved in the reflection process (isotropic, surface and volume diffusions) in the presence of bare soil and vegetation (Roujean et al. 1992). Finally, spectral integration over broad frequency bands (linear regressions using radiative transfer simulations) leads to albedo values. With geostationary satellites, the variation of the zenith angle during the day makes it possible to sample many directions to lead to a daily albedo. Based on specific thresholds, it is possible to identify snow-covered surfaces characterized by much higher values (but can sometimes be confused with clouds). The most recent algorithms lead to estimates of surface albedo with an uncertainty of approximately 20%. Figure 3.8 shows an example of daily albedo obtained from the SEVIRI radiometer on MSG. The highest values (> 0.40) appear over the Sahara Desert, while the lowest values are associated with the equatorial forest areas of Congo and Brazil (< 0.10).

It is also possible to separate the contributions of bare soil and vegetation to the total albedo measured by the satellite using a Kalman filter and assuming the fraction of vegetation in the pixel to be known (Carrer et al. 2014). The bare soil albedo can in turn provide information about the surface soil water content, as it is usually much lower for wet soils.

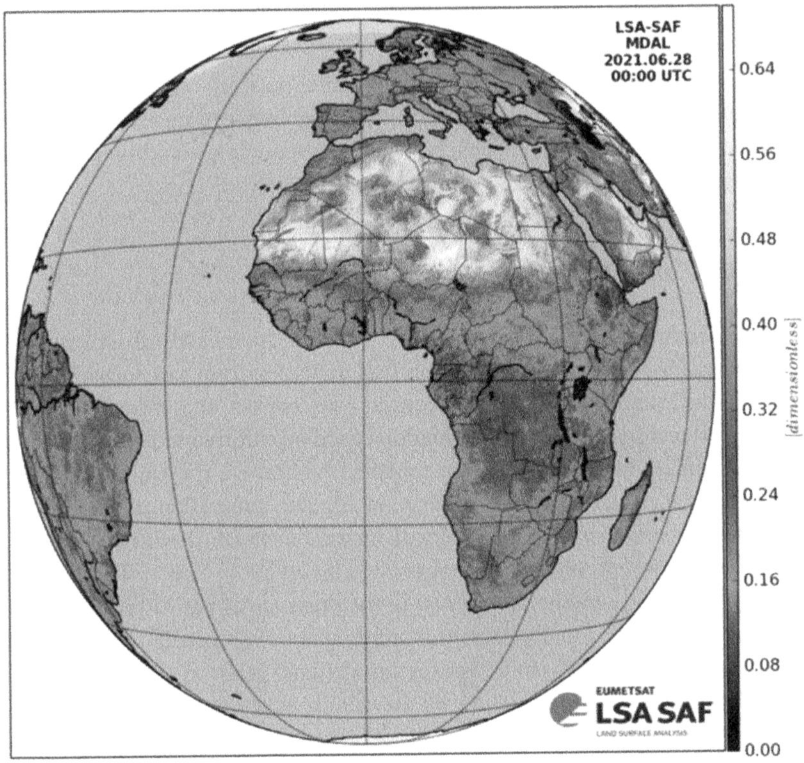

Figure 3.8. *Daily mean albedo produced by EUMETSAT's LSA SAF for June 28, 2021*[4] *(© EUMETSAT 2021)*

3.4.4. *Vegetation properties*

Vegetation has a particular spectral signature in the visible and near-infrared domains due to the fact that leaf pigments absorb a significant fraction of the radiation in the blue and red parts of the visible spectrum to allow photosynthesis of atmospheric carbon dioxide. Thus, the spectral band between 0.4 and 0.7 μm is referred to as the photosynthetically active radiation (PAR). In this range, a peak reflection can be noted at 0.55 μm (explaining the green color perceived by the human eye for most plants) (Figure 3.9). At 0.7 μm, the reflection increases sharply and then gradually decreases to 2.0 μm. Beyond this value, absorption dominates, with plant water being the main absorber. The difference in behavior of the reflected spectrum of vegetation on both sides of 0.7 μm led to the definition of a normalized

3. Data available at: https://landsaf.ipma.pt.

difference vegetation index (NDVI) from reflectance in the visible ρ_{VIS} (< 0.7 μm) and in the near-infrared ρ_{PIR} (> 0.7 μm):

$$NDVI = (\rho_{PIR} - \rho_{VIS})/(\rho_{PIR} + \rho_{VIS})$$

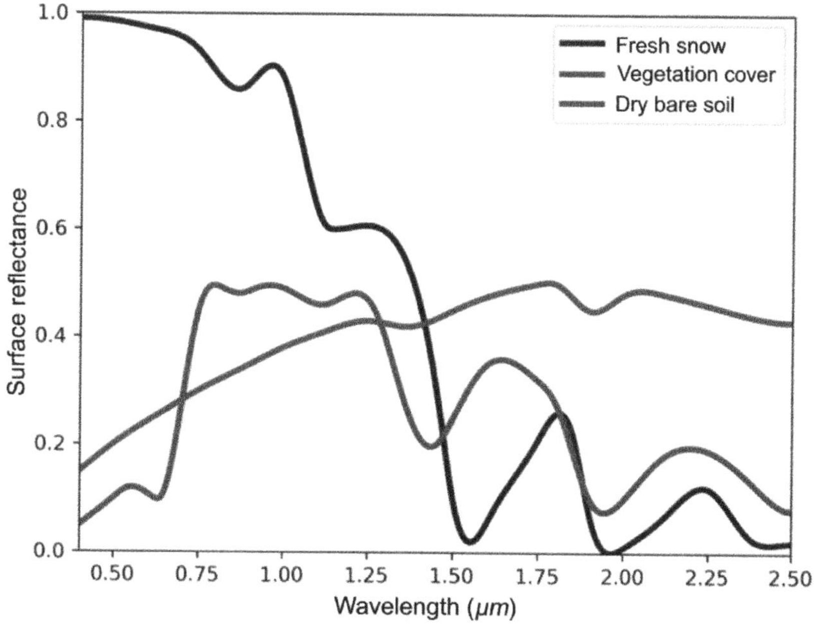

Figure 3.9. *Schematic representation of the spectral dependency of the surface reflectance in the visible and near-infrared domains for three types of surfaces: bare and dry soil (brown curve), vegetation cover (green curve) and snow cover (blue curve)*

This quantity is much higher for vegetation than its value over bare ground, whose reflectance varies continuously on either side of the frequency 0.7 μm (Figure 3.9). It has often been used to provide information on horizontal (FCover) and vertical (LAI) densities of vegetation through empirical relationships, but whose validity remains limited to a particular geographical location and ecosystem (Wang et al. 2005). A more physical and therefore more general approach, based on a BRDF model, has been proposed by Roujean and Lacaze (2002). These two parameters (FCover and LAI) strongly control evapotranspiration of vegetation, and estimates from space-borne remote sensing allow for their specification in meteorological models, either in the form of climatologies or by considering more frequent updates to better take into account inter-annual variabilities, such as those induced by summer droughts (Boussetta et al. 2015). They are also very important for estimating natural carbon fluxes emitted or captured by biomass. Indeed, it is through the same mechanism of opening of stomata (pores on leaf surface) that vegetation absorbs the carbon dioxide necessary for its growth by photosynthesis and loses part of its water in the transpiration process. Spatial observations of LAI and soil water content can simultaneously control water and carbon fluxes over continental surfaces by assimilating data into land surface models (Albergel et al. 2017). While advances in space-borne remote sensing over more than 30 years have resulted in a better specification of a number of parameters characterizing continental surfaces, it is important to recall that currently this technique does not provide access to quantities that also regulate water flows at the interface, such as soil texture, root zone depth and minimum stomatal resistance to transpiration of plants, which are only available from in situ measurements.

3.5. Snow-covered surfaces

3.5.1. *Spatial coverage and albedo*

Since snow reflectivity in the visible range is much higher than that of bare ground or vegetation (Figure 3.9), this property is used to estimate the presence of snow on ground from space, from radiometers previously mentioned like MODIS, AATSR and SEVIRI. Other spectral characteristics of snow are used to distinguish it from clouds, which from a radiative point of view are also cold and reflective surfaces, in particular its significant absorption in the near-infrared beyond 1.5 μm (Figure 3.9) or the shape of its BRDF. The spatial coverage of snow on ground is more complex over forest areas because it is masked by the canopy (whose albedo is generally very low), which only temporarily retains snow by interception after episodes of precipitation. Several global spatial snow cover products exist on a daily scale with resolutions of a few kilometers to validate or constrain numerical atmospheric models (GlobSnow products from ESA**, IMS** from NOAA** or H-SAF** from EUMETSAT). They are used in numerical weather forecasting to improve the initial states of snow schemes, particularly to remove predicted snow where it is not observed (Drusch et al. 2004). Spatial coverage of snow cover is a central parameter for a realistic estimation of surface energy balance. However, no information on snow mass is provided by measurements in the solar domain. It is during periods of melting snow cover, when energy and water balances are strongly coupled, with consequences on runoff and river flows, that errors in modeling the thickness of the snow cover may appear. It is therefore necessary, in addition to its horizontal extension, to obtain information on its vertical extension.

**See Appendix 3.

As with continental surfaces, snow albedo is deduced from angular and spectral integrations of bidirectional reflectance. Thanks to its variations (decrease) in the visible part, it is also possible to access information on the presence of impurities in the snow cover as well as in the near-infrared on the size of grains.

3.5.2. *Equivalent water content*

Microwave measurements provide quantitative information on the snowpack mass, in the form of the snow water equivalent (SWE, which is the useful quantity for the surface water balance). The spectral signature of snow in this domain allows such an estimation, since the emission comes from a volume that is more important for low frequencies (generally, channels below 37 GHz are used) and when the snow is dry. Indeed, as discussed previously for soil water, the presence of liquid water in the snowpack leads to a strong reduction in surface emissivity, which can lead to an erroneous interpretation of the signal received from space. Estimation of the SWE requires knowledge on snow density to lead to a physical height in order to compare it with in situ estimates. Snow emission models in the microwave range (Pulliainen et al. 1999) for calculating brightness temperature also require snow density, grain size and temperature. The lack of knowledge of some parameters leads to significant uncertainties about SWE products. Operational products rely on conical scanning microwave radiometers (to distinguish polarizations) such as SSMI/S and AMSR-2. Finally, it is interesting to note that the representation of the snow cover in atmospheric models is evolving towards multi-layer approaches. However, information on the vertical structure of snowpack is currently beyond the reach of spaceborne remote sensing.

Other properties of the snowpack can be obtained thanks to active and passive remote sensing in the microwave domain, such as the "sastrugi" (mesoscale snow dunes formed by wind) on ice caps, due to the strong azimuthal dependence of the backscattered or emitted signal.

3.6. Expected changes

The wealth of the electromagnetic spectrum from the visible range to that of microwaves has, for about 50 years, made it possible to gain access to an increasingly precise knowledge of the properties of Earth's surfaces (oceans, sea ice, continents), particularly to better constrain energy, water and carbon budgets in atmospheric modeling for which they are the lower limit. Increasing the spatial resolution of atmospheric models will require the use of observations from space at increasingly finer scales (less than 100 meters). This is already proposed by the Sentinel satellites of the European Copernicus program. Future missions by space agencies, notably ESA and EUMETSAT for Europe, will make it possible to explore new spectral bands (BIOMASS**, ICI**, FLEX**, FORUM**) that will provide new information on the vegetation and polar region surfaces. Atmospheric models are evolving towards integrated systems including all the compartments of the Earth system (atmosphere, ocean, sea ice, continental surfaces). Since space-borne remote sensing measurements are most often sensitive to several compartments of the Earth system (as described above), it will be possible to use these observations in such models in a more direct and combined manner without having to carry out atmospheric corrections in order to derive surface properties or to know the surface properties in order to retrieve information on the atmosphere located above.

3.7. References

Albergel, C., Munier, S., Leroux, D.J., Dewaele, H., Fairbairn, D., Laviania Burbu, A., Gelati E., Dorigo, W., Faroux, S., Meurey, C. et al. (2017). Sequential assimilation of satellite-derived vegetation and soil moisture products using SURFEX_v8.0: LDAS-Monde assessment over the Euro-Mediterranean area. *Geosci. Model Dev.*, 10, 3889–3912. https://doi.org/10.5194/gmd-10-3889-2017.

Boussetta, S., Balsamo, G., Dutra, E., Beljaars, A., Albergel, C. (2015). Assimilation of surface albedo and vegetation states from satellite observations and their impact on numerical weather prediction. *Rem. Sens. Env.*, 163, 111–126. https://doi.org/10.1016/j.rse.2015.03.009.

Carrer, D., Meurey, C., Ceamanos, X., Roujean, J.-L., Calvet, J.-C., Liu, S. (2014). Dynamical mapping of snow-free vegetation and bare soil albedos at global 1 km scale from 10-year analysis of MODIS satellite products. *Rem. Sens. Env.*, 140, 420–432.

De Rosnay, P., Drusch, M., Vasiljevic, D., Balsamo, G., Albergel, C., Isaksen, L. (2013). A simplified extended Kalman filter for the global operational soil analysis at ECMWF. *Quart. J. Roy. Meteor. Soc.*, 139, 1199–1213.

Dolon, C.J., Martin, M., Stark, J., Roberts-Jones, J., Fiedler, E., Wimmer, W. (2012). The operational sea surface temperature and sea ice analysis (OSTIA) system. *Rem. Sens. Env.*, 116, 140–158.

Drusch, M., Vasiljevic, D., Viterbo, P. (2004). ECMWF global's snow analysis. Assessment and revision based on satellite observations. *J. Appl. Meteor.*, 43, 1282–1294.

Hersbach, H., Stoffelen, A., de Haan, S. (2007). An improved C-band scatterometer geophysical model function: CMOD5. *J. Geophys. Res.*, 112, C03006, doi:10.1029/2006/2006JC003743.

**See Appendix 3.

Karbou, F., Rabier, F., Prigent, C. (2014). The assimilation of observations for the advanced microwave sounding unit over sea ice in the French global numerical weather prediction system. *Mon. Wea. Rev.*, 142, 125–140. https://doi.org/10.1175/MWR-D-13-00025.1.

Kumar, S.V., Reichle, R.H., Koster, R.D., Crow, W.T., Peters-Lidard, C.D. (2009). Role of subsurface physics in the assimilation of surface soil moisture observations. *J. Hydromet.*, 10, 1534–1547. https://doi.org/10.1175/2009JHM1134.1.

Minnett, P.J., Alvera-Azcarate, A., Chin, T.M., Corlett, G.K., Gentemann, C.I., Karagali, I., Li, X., Marsouin, A., Marullo, S., Maturi, E. et al. (2019). Half a century of satellite remote sensing of sea-surface temperature. *Rem. Sens. Env.*, 233, 111336. https://doi.org/10.1016/j.rse.2019.111336.

Munoz-Sabater, J., Lawrence, H., Albergel, C., Rosnay, P., Isaksen, L., Mecklenburg, S., Kerr, Y., Drusch, M. (2019). Assimilation of SMOS brightness temperatures in the ECMWF integrated forecasting system. *Quart. J. Roy. Meteor. Soc.*, 145, 2524–2546, https://doi.org/10.1002/qj.3577.

Prigent, C., Aires, R., Rossow, W.B. (2006). Land surface microwave emissivities over the global for a decade. *Bull. Amer. Meteor. Soc.*, 87, 1573–1584. doi:10.1175/BAMS-87-11-1573.

Pulliainen, J., Grandell, J., Hallikainen, M.T. (1999). HUT snow emission model and its applicability to snow water equivalent retrieval. *IEEE Trans. Geosci. Remote Sens.*, 37, 1378–1390.

Rodriguez-Fernandez, N., de Rosnay, P., Albergel, C., Richaume, P., Aires, F., Prigent, C., Kerr, Y. (2019). SMOS neural network soil moisture data assimilation in a land surface model and atmospheric impact. *Remote Sens.*, 11, 1334. https://doi.org/10.3390/rs11111334.

Roujean, J.-L. and Lacaze, R. (2002). Global mapping of vegetation parameters from POLDER multiangular measurements for studies of surface-atmosphere interactions: A pragmatic method and its validation. *J. Geophys. Res.*, 107D, 10129–10145.

Roujean, J.-L., Leroy, M., Deschamps, P.-Y. (1992). A bidirectional reflectance model of the Earth's surface for the correction of remote sensing data. *J. Geophys. Res.*, 97, 20455–20468.

Ruf, C., Asharaf, S., Balasubramaniam, R., Gleason, S., Lang, T., McKague, D., Twigg, D., Waliser, D. (2019). In-orbit performance of the constellation of CYGNSS hurricane satellites. *Bull. Amer. Meteor. Soc.*, 100, 2009–2023. https://doi.org/10.1175/BAMS-D-18-0337.1.

Schumgge, T. (1978). Remote sensing of surface soil moisture. *J. Appl. Meteor.*, 17, 1549–1557.

Trigo, I., Boussetta, S., Viterbo, P., Balsamo, G., Beljaars, A., Sandu, I. (2015). Comparison of model land skin temperature with remotely sensed estimates and assessment of surface-atmosphere coupling. *J. Geophys. Res. Atmos.*, 120, 12096–12111. doi:10.1002/2015JD023812.

Wagner, W., Lemoine, G., Rott, H. (1999). A method for estimating soil moisture from ERS scatterometer and soil data. *Rem. Sens. Env.*, 70, 191–207.

Wang, Q., Adiku, S., Tenhunen, J., Granier, A. (2005). On the relationship of NDVI with leaf area index in a deciduous forest site. *Rem. Sens. Env.*, 95, 244–255.

4

The Assimilation of Satellite Data in Numerical Weather Prediction Systems

Bill BELL[1], **Jean-Noël THÉPAUT**[1] **and John EYRE**[2]

[1] *Copernicus Climate Change Service, European Centre for Medium Range Weather Forecasts, Shinfield Park, Reading, UK*

[2] *Met Office, Fitzroy Road, Exeter, UK*

4.1. Introduction

Over the last 40 years, observations from satellites have become increasingly important in numerical weather prediction (NWP). They are now the dominant source of observational information determining the skill of forecasts from global NWP systems, and they are an important source of information for regional and convective-scale NWP. NWP systems are also exploited for climate reanalyses (Hersbach et al. 2020) in which modern data assimilation systems are used to analyze the atmospheric state over multi-decadal timescales. Such reanalyses have found a broad range of application, spanning the study of specific historical synoptic events to the analysis of long-term climate trends. Satellite data play a key role in determining the quality of such reanalyses. Figure 4.1 shows that with a modern NWP system, the assimilation of increasing volumes of observations, largely from satellites, results in steadily improving forecast quality over the continental northern hemisphere. Over the southern hemisphere, the improvements are more dramatic still, particularly following the introduction of satellite data from the TOVS instruments in 1979.

For a color version of all the figures in this chapter, see www.iste.co.uk/phulpin/satellites2.zip.

Satellites for Atmospheric Sciences 2,
coordinated by Thierry PHULPIN, Didier RENAUT, Hervé ROQUET and Claude CAMY-PEYRET.

From the earliest missions in the mid-1960s (see Chapters 1–4), the satellite observing system has developed rapidly. Current NWP systems typically assimilate more than 20 million observations per day from around 90 satellite instruments operated by US, Asian and European agencies.

In this chapter, we review the history of developments in the use of satellite observations in NWP from the earliest experiments around 1970 to the present day, with a focus on the assimilation of satellite data in atmospheric analysis schemes, and briefly review the theoretical principles underpinning these analysis schemes.

Following closely the analysis by Eyre et al. (2020, 2021), we first describe the early history of satellite data assimilation, spanning the period up to 2000 (sections 4.2–4.4, covering the key missions, early work on assimilation and a brief review of relevant theory) followed by a description of progress in recent years, from 2000 onwards, and current status in section 4.5. Some concluding remarks are given in section 4.6.

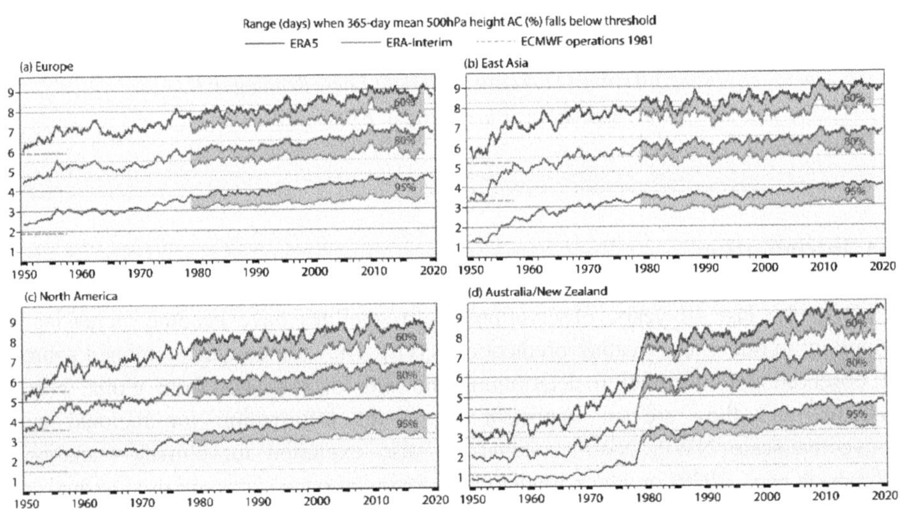

Figure 4.1. *Increase in reforecast skill in the ERA-Interim and ERA5 reanalyses from 1979 to 2019 and 1950 to 2020, respectively. Range (in days) at which running 365-day mean anomaly correlations of 500 hPa height forecasts at 00:00 and 12:00 UTC from 1950 to 2020 reach 95% (green), 80% (orange) and 60% (blue) for (a) Europe, (b) East Asia, (c) North America, and (d) Australia/New Zealand. Also shown (dashed) is the average skill of ECMWF operational forecasts for 1981. The heavy lines denote ERA5, the thin lines denote ERA-Interim. Shading denotes the difference between ERA5 and ERA-Interim during the period for which both are available (1979–2019)*

4.2. Early meteorological satellites

The important milestones in the development of meteorological satellites are described in detail in Chapters 1–4 of Volume 1. Sounding instruments, providing information on vertical profiles of atmospheric temperature, were first deployed on Nimbus-3 in 1969, and by 1972 were operational on NOAA-2. A new generation of operational polar-orbiting satellites started in 1978 with TIROS-N and continued with subsequent NOAA satellites and then EUMETSAT satellites.

By the time of the First GARP Global Experiment (FGGE), which ran from December 1978 to November 1979, there was a complete ring of geostationary satellites providing atmospheric motion vectors (AMVs), together with a new generation of polar-orbiting satellites providing temperature and humidity soundings. This was an important milestone as it effectively marked the beginning of the space-based component of the global observing system of WMO (see Chapter 7 of Volume 1). It also provided operational NWP with a new wealth of observational data. Although some years of progress were needed to exploit their potential, this allowed the state of the global atmosphere to be analyzed with an accuracy and completeness not previously possible. Furthermore, it provided a template for subsequent reanalyses of the global atmosphere (see Hersbach et al. 2020 and references therein).

4.3. Assimilation of satellite soundings 1970–2000

4.3.1. *Early sounding instruments*

Table 4.1 summarizes the early history of the atmospheric sounding instruments of the US's polar-orbiting satellites – both the Nimbus research series and the TIROS/NOAA operational series. These instruments were the most relevant to NWP because, in the main, their data were available internationally in near real-time for operational use.

The combination of HIRS, MSU and SSU is known as TOVS, and flew on many satellites starting with TIROS-N and ending with NOAA-14, until superseded by the Advanced TOVS (ATOVS) system from NOAA-15 onwards. TOVS therefore provided operational satellite soundings for over 20 years and was the focus of much research and development on the use of satellite data in NWP in the 1980s and 1990s. It still provides invaluable information for atmospheric reanalyses.

Nimbus series		
Nimbus-3	1969–1970	SIRS, IRIS
Nimbus-4	1970–1971	SIRS, IRIS, SCR
Nimbus-5	1972	ITPR, SCR
Nimbus-6	1975	HIRS, SCAMS, PMR, LRIR
Nimbus-7	1978–1994	LIMS, SAMS
TIROS/NOAA series		
NOAA 2-5	1972–1979	VTPR
TIROS-N	1978–1980	TOVS = HIRS+MSU+SSU
NOAA-6/14	1979–	TOVS = HIRS+MSU+SSU
NOAA-15+	1998–	ATOVS = AMSU-A+AMSU-B+HIRS

Table 4.1. *Early satellite sounding instruments*

TOVS provides complete global coverage at mid- and high latitudes every 12 hours from one satellite. With a nominal two-satellite system, near-global coverage is therefore available every six hours.

The main limitation of the TOVS system, even in cloud-free conditions, is its poor vertical resolution. Individual weighting functions have a width of 5–10 km. Combining channels with overlapping weighting functions increases the vertical resolution but only to ~3 km. It is important to realize that this also limits the effective horizontal resolution of the TOVS system to many times the spatial sampling distance of HIRS (at least for mid-latitude baroclinic structures); atmospheric perturbations with small scale in the vertical also tend to have small scale in the horizontal.

The USA's Defense Meteorological Satellite Program (DMSP) also carried sounding instruments during this period: nine satellites in this series carried SSM/T, an MSU-like instrument although with seven channels; three satellites carried SSM/T-2, an AMSU-B-like instrument.

Sounding capability from geostationary satellites began in 1980 with the VISSR Atmospheric Sounder (VAS) on GOES. VAS continued in operation until 1996. They were used, mainly in the USA, for nowcasting with particular application in hurricane monitoring. Efforts to use VAS data in NWP had limited success, mainly because of their limited vertical resolution and because of the presence of other observations of high quality.

A more detailed summary of early sounding instruments and their capabilities is given by Menzel et al. (2018).

4.3.2. *Assimilation experience: 1970s*

Nimbus 3, carrying the first satellite temperature sounders, was launched in April 1969, and data from its SIRS instruments were first used very shortly afterward. In initial experiments, retrievals from SIRS data were "bogussed" into the objective analysis of the USA's National Meteorological Center. These experiments demonstrated cases of significant changes to analyses over the Pacific and subsequent forecasts over the USA.

Important early work on the assimilation of satellite data was performed in Australia, motivated by the relative scarcity of other observations in the southern hemisphere. Before satellite sounding data became available, the Australian Bureau of Meteorology demonstrated the benefit of using satellite cloud imagery interpretation to estimate and then assimilate surface pressures (PAOBs) and 1,000–500 hPa thicknesses in the southern hemisphere. As early as 1972, investigators assimilated VTPR data from NOAA-2, -3 and -4 in the form of retrievals from cloud-cleared radiances. By 1976, the benefits of assimilating VTPR retrievals and PAOBs were demonstrated within a continuous data assimilation (DA) system.

The results of several DA experiments in the late 1970s are summarized by Ohring (1979). Ohring's paper summarized the state of the art at the time and concluded that the assimilation of satellite data delivered, "on average, a small improvement in numerical forecasts". Ohring identified several problems that needed to be addressed before more impact could be expected, including: differences between retrievals and collocated radiosondes of 2–3 K; analyses using satellite retrievals had lower eddy potential energy; satellite soundings are not point observations – they have their own error characteristic, and improved analysis schemes may enhance impact; improvements in retrieval methods were likely but the basic problem was poor vertical resolution, and the statistical/climatological nature of the retrieval techniques of the time may suppress horizontal structure.

4.3.3. *Assimilation experience: early 1980s*

The observational dataset from FGGE was the focus of several observing system experiments (OSEs) in the early 1980s. Typical data coverage during most of the

FGGE period is shown in Figure 4.2. In the later part of FGGE, following the launch of NOAA-6 to complement TIROS-N, the coverage of satellite sounding data was considerably enhanced. Investigators concluded that the assimilation of satellite data gave measurable benefits in both hemispheres, with larger benefits realized in the data sparse southern hemisphere. Experiments at ECMWF led to the conclusion that AMVs were particularly important for the analysis of the tropics, and that satellite soundings were of paramount importance for extra-tropical analysis over ocean areas.

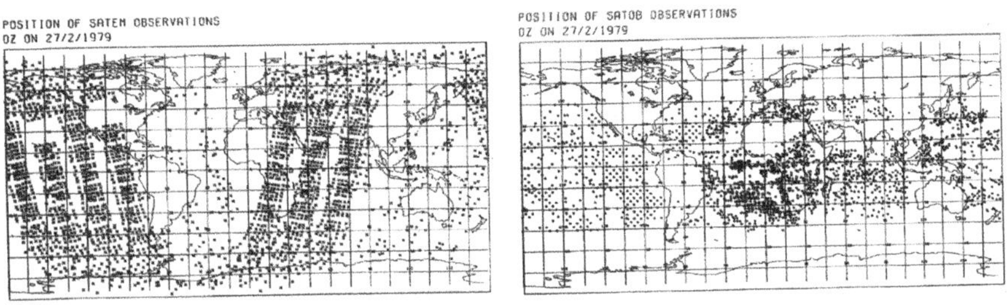

Figure 4.2. *Satellite data coverage during FGGE: temperature retrievals from TOVS and VTPR (left) and AMVs (right) (source: Bromley (1984))*

4.3.4. *Problems arising in the late 1980s*

By the late 1980s, the general performance of NWP systems had improved considerably. As a result, the short-range forecasts were more accurate and the potential for damage through erroneous observations, or observations assimilated into an inappropriate way, was much greater. Some of the problems identified several years before (Ohring (1979), listed in section 4.3.2) now became dominant and started to lead to satellite sounding data having either an average negative impact or at least an impact that was very variable from case to case.

The problems encountered during this period were summarized in the report of an ECMWF/EUMETSAT Workshop in 1989 on the use of satellite data in operational NWP (ECMWF 1989), highlighting the difficulties caused by the limited vertical resolution of the available satellite sounding data. It transpired that a significant problem with the assimilation of satellite sounding data lay in the initial attempts to treat the sounding-based retrievals as low resolution, poor-quality radiosonde observations, an approach that failed to make optimal use of the generally high-quality radiance observations.

4.4. Relevant aspects of data assimilation theory

To appreciate the challenges of assimilating satellite sounding data, and how they have been tackled over the years, it is necessary to consider some basic ideas in data assimilation and the history of how they have been implemented. The basic concepts of data assimilation and satellite sounding retrieval are, of course, closely related (see, for example, Rodgers (1976) and Lorenc (1986)).

Ideally, in order to define the initial conditions in NWP, we might wish to solve the Bayesian problem: What is the probability of atmospheric state, $\mathbf{x}$, given observations, $\mathbf{y^o}$? Or, equivalently, evaluate:

$$P(\mathbf{x}|\mathbf{y^o}) = P(\mathbf{y^o}|\mathbf{x})\,P(\mathbf{x})/\,P(\mathbf{y^o}) \qquad [4.1]$$

It is far beyond the scope of present-day computer power to estimate the full probability density function (PDF) and it probably always will be. Ensemble methods may be able to explore sufficient aspects of it to be useful.

At present, most operational NWP systems use a variational (or equivalent) approach, in which we ask the question: What is the most probable atmospheric state, $\mathbf{x}$, given observations, $\mathbf{y^o}$? Or, equivalently, maximize $P(\mathbf{x}|\mathbf{y^o})$, which is equivalent to maximizing

$$\ln\{P(\mathbf{x}|\mathbf{y^o})\} = \ln\{P(\mathbf{y^o}|\mathbf{x})\} + \ln\{P(\mathbf{x})\} + \text{constant} \qquad [4.2]$$

If all the PDFs are Gaussian, then this is the same as minimizing a penalty or cost function:

$$J(\mathbf{x}) = \tfrac{1}{2}(\mathbf{x}-\mathbf{x^b})^T\,\mathbf{B}^{-1}(\mathbf{x}-\mathbf{x^b}) + \tfrac{1}{2}\{\mathbf{y^o}-H(\mathbf{x})\}^T(\mathbf{E}+\mathbf{F})^{-1}\{\mathbf{y^o}-H(\mathbf{x})\} \qquad [4.3]$$

where:

$\mathbf{x^b}$ is the background field;

$\mathbf{B}$ is the background error covariance;

$H(\mathbf{x})$ is the observation operator;

$\mathbf{E}$ and $\mathbf{F}$ are the error covariances of observations and their operator.

The optimal analysis can be found by minimizing (3) or by solving its gradient equation:

$$\nabla_{\mathbf{x}} J(\mathbf{x})^T = \mathbf{B}^{-1}(\mathbf{x} - \mathbf{x}^{\mathbf{b}}) + \tfrac{1}{2}\nabla_{\mathbf{x}} H(\mathbf{x})^T(\mathbf{E} + \mathbf{F})^{-1}\{\mathbf{y}^{\mathbf{o}} - H(\mathbf{x})\} = 0 \quad [4.4]$$

If we linearize the variational equation, equation [4.4], we can derive the optimal interpolation (OI) formula:

$$\mathbf{x}^{\mathbf{a}} = \mathbf{x}^{\mathbf{b}} + \mathbf{K}.\{\mathbf{y}^{\mathbf{o}} - H(\mathbf{x})\} \quad [4.5]$$

where:

$$\mathbf{K} = \mathbf{B}.\mathbf{H}^{\mathbf{T}}(\mathbf{H}.\mathbf{B}.\mathbf{H}^{\mathbf{T}} + \mathbf{E} + \mathbf{F})^{-1}; \quad [4.6]$$

$\mathbf{H}$ is the Jacobian of the observation operator $H(\mathbf{x})$; $\mathbf{H} = \nabla_{\mathbf{x}} H(\mathbf{x})$.

Equation [4.6] provides the "optimal" weights for the analysis, that is, those that minimize the analysis error. Sub-optimal analyses can be obtained using other empirical weights, which may be simpler to derive and to apply, that is, equation [4.5] but where **K** contains empirically derived weights.

Although the above derivation provides a framework for understanding the different approaches and the relationship between them, in a historical context they were actually developed in the reverse order. In general terms, empirical methods were superseded by optimal interpolation (OI, for example, operational at ECMWF from 1980) and this was superseded by three-dimensional variational schemes (3D-Var, for example, operational at NCEP in 1995 and at ECMWF in 1996).

The move to radiance assimilation, starting in the 1990s, tends to be associated with the introduction of variational (or equivalent) methods of DA in operational NWP: first within 3D-Var systems and later within 4D-Var systems (Rabier et al. 2000). An important intermediate step towards direct radiance assimilation was the assimilation of 1D-Var retrievals which used NWP short-range forecasts as background information. This removed large components of the climatological background from the retrieved profiles, and was much closer to direct radiance assimilation than the assimilation of retrievals based on climatological information.

4.5. The modern era (2000 to present)

4.5.1. *Assimilation strategies*

Having selected a data assimilation method, the practical details of the implementation need to be considered. For each satellite observation type to be assimilated into an NWP model, we are usually faced with a choice as to the form in which the data are best assimilated, and particularly as to the amount of pre-processing they undergo prior to assimilation. The common features of this choice can be illustrated using Figure 4.3, which schematically shows the main elements of the DA process. At the bottom of the figure, the continuous cycle of forecast–assimilate–forecast operations is shown. For each assimilation cycle, the NWP model background fields are interpolated to the position and time of each observation and then passed through an "observation operator" to create "forecast observations". The observations undergo some form of pre-processing, which can be more or less complicated. For example, satellite sounding radiances may be heavily pre-processed to generate a retrieved atmospheric profile, or they may be lightly pre-processed and presented to the assimilation system as radiances. The pre-processed observations are then compared with the "forecast observations"; their differences are called "observation increments" or "innovations". The DA procedure then converts these, through a process of mapping and weighting, into "analysis increments" which are added to the NWP background field to generate the final analysis.

It is important that the observation operator and the pre-processing of observations are carefully matched, so that the observation increments are the result of comparing like with like. This includes the very important treatment of biases in the observations and in the observation operators (see section 4.5.5). In general, if the observation operator is simple (e.g. not extending beyond simple interpolation of fields), then, for most satellite data types, the pre-processing must be complex. On the other hand, if raw or lightly pre-processed observations are assimilated, then the observation operator must be complex. This is because satellite instruments, whether passive or active, only measure radiation upwelling from the Earth–atmosphere system, and these measurements are related to the atmospheric geophysical variables and hence the variables of an NWP model (wind, temperature, humidity, etc.) in a complicated way.

There are two principal advantages in assimilating observations close to their raw form. Firstly, the "observation operator", $H(\mathbf{x})$, can be nonlinear and this is important for many remotely sensed observations. Variational methods no longer restrict assimilation to approximate linearization of this relationship, and this allows the true information content of the observation to be properly represented. Secondly, "raw" measurements, in the space of the observed variables, tend to have simpler error characteristics (see section 4.5.5).

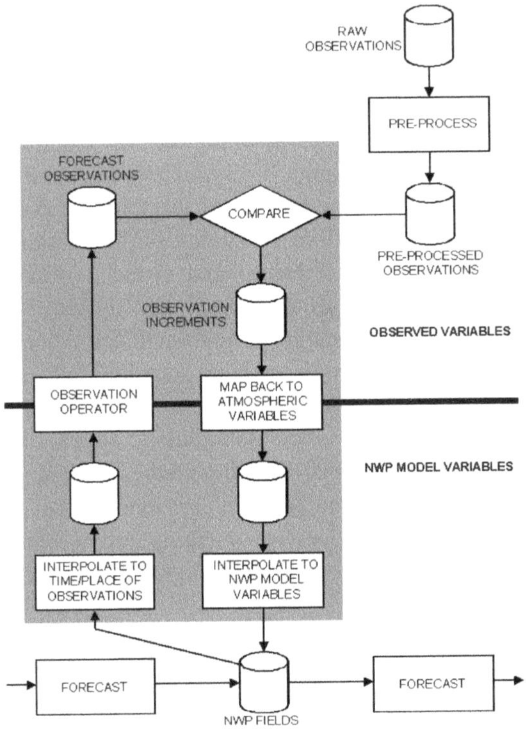

Figure 4.3. *Schematic of the assimilation of observations into an NWP model*

On the other hand, assimilation of raw observations presents some new problems. Firstly, raw observations tend to have more complex operators. Secondly, some observations are affected by physical variables not contained in the control variable (e.g. the stratosphere above the model top, or surface variables). Thirdly, there is a logistical problem: the need to develop and maintain expertise on all satellite observation operators and associated errors, requiring resources beyond the means of a single NWP center. It was the recognition of this problem in the mid-1990s that stimulated the search for new strategies to improve links between "assimilation centers" and "satellite centers" and to share the necessary work on observation operators and satellite data pre-processing software. New collaborative activities, such as the EUMETSAT Satellite Applications Facility for NWP

(NWP SAF) in Europe and the Joint Center for Satellite Data Assimilation (JCSDA) in the USA, were the outcome.

Preferred assimilation options for various data types were reviewed by Eyre (1997). A summary is given below.

– Passive temperature/humidity soundings – as radiances.

– Wind information – as AMVs where the observation relies on the tracking of small-scale features that cannot be represented by the NWP model fields, as well as by 4D assimilation of radiances for larger-scale features.

– Scatterometry – as retrieved "ambiguous" wind vectors, or as winds chosen to be the "best solution" during pre-processing.

– MW imagery (for information on water vapor, cloud water and ice, precipitation and wind speed) – direct assimilation of imager radiances, alongside sounder radiances; see section 4.5.3.

– Visible/IR imagery. In global NWP, IR "clear-sky" radiances are assimilated in a similar way to IR sounder radiances, primarily for information on water vapor as well as information on wind. In some high-resolution NWP systems, this strategy is used and, additionally for information on cloud itself, retrieved cloud variables (e.g. fractional coverage and cloud-top pressure) are assimilated.

– Radio occultation (RO) – retrieved refractivity or bending angle, for the reasons given in section 4.5.9.

These main types of satellite data assimilated into modern NWP systems are described in sections 4.5.2–4.5.9.

4.5.2. *Advanced infrared sounders*

In recent years, developments in the use of IR sounding data at NWP centers have been dominated by *advanced* or *hyperspectral* IR sounders – IR radiometers that measure radiation in thousands of spectrally narrow bands, providing significantly improved vertical resolution (by a factor of ~3) compared to filter radiometers such as HIRS.

The development of advanced IR sounder technology has been reviewed by Menzel et al. (2018). The Atmospheric Infrared Sounder (AIRS) was designed during the 1990s to address the requirement for high-quality temperature and water vapor retrievals in the context of increasingly accurate NWP forecasts. Launched on the NASA Earth Observing System (EOS) Aqua satellite in 2002, AIRS was the first advanced IR sounder to be used in operational NWP.

Several NWP centers demonstrated impact from assimilating AIRS radiances and experiences gained from using AIRS data played an important part in preparing the NWP community for rapidly exploiting the next generation of operational hyperspectral sounders.

The Infrared Atmospheric Sounding Interferometer (IASI) was developed jointly by EUMETSAT and France's *Centre National d'Etudes Spatiales* (CNES) (see Chapters 6 and 9 of Volume 1). It was launched onboard Europe's operational polar orbiting meteorological satellites, MetOp-A, -B and -C, in 2006, 2012 and 2018, respectively.

Following on from the success of the AIRS mission, the first Cross-track Infrared Sounder (CrIS) was launched on board NOAA's Suomi NPP satellite in 2011 and subsequently on NOAA-20 in 2017. CrIS, like IASI, is a Michelson interferometer. It has a somewhat lower spectral resolution than IASI, but the noise performance in the long-wavelength range used for temperature sounding significantly surpasses that of IASI and AIRS.

The observation errors associated with advanced IR sounder radiances are characterized by strong inter-channel correlations, with contributions from apodization, errors of representation and forward model errors. Initial NWP applications assumed diagonal observation error covariance matrices, ignoring inter-channel correlations and requiring inflation of the observation errors to account for unspecified error correlations. More recently, correlated observation errors have been accounted for in operational data assimilation systems leading to useful improvements to analysis and forecast skill.

At present, only a subset of channels is assimilated into NWP systems, typically in the range 100–200 channels for each instrument, due to computational cost considerations and the scientific challenges of optimally representing error correlations. Furthermore, certain spectral regions affected by variable trace gases that are not well-represented in NWP systems (e.g. ozone) are avoided. Consequently, subsets of "NWP-friendly" channels have been established by the NWP community, based on information content studies. It is anticipated that using the full information content of the spectrum should provide further benefits beyond those achieved using selected channels.

IR radiance observations are strongly affected by scattering and absorption by clouds. Initial implementations of advanced IR sounder data in NWP were limited to cloud-free scenes only, using cloud detection or cloud clearing techniques.

However, selecting cloud-free scenes results in the rejection of large numbers of channels with no sensitivity to cloud, thereby discarding useful information about the stratosphere and upper troposphere. Subsequently, techniques have been developed to assimilate IR observations in cloudy scenes, using only channels with little or no sensitivity to cloud.

In the longer term, it is expected that the use of more advanced cloudy radiative transfer (RT) models, including the effects of scattering from cloud particles, will allow the assimilation of more cloud-affected radiances.

The use of advanced IR sounder radiances in NWP initially avoided observations affected by land surfaces, due to uncertainties in the specification of land surface emissivity and skin temperature. In general, IR land surface emissivity was assumed to have a fixed value when, in reality, it exhibits substantial spectral and spatial variability. Initial work on estimates of land surface emissivity has led to improvements in the representation of emissivity in some operational systems, either using retrieval schemes or climatological atlases. The use of retrieved spectral emissivity and skin temperature estimates appropriate to each field of view, such as from the 1D-Var scheme implemented at the Met Office (Pavelin and Candy 2014), allows the assimilation of surface-sensitive channels over land.

Hyperspectral IR sounding from geostationary satellites is in its infancy. The Chinese Meteorological Agency's Geostationary Interferometric Infrared Sounder (GIIRS), flown on board the FY-4A satellite from 2016, was the first such instrument to be launched into space and will be followed by EUMETSAT's next-generation geostationary platform, MTG (Meteosat Third Generation) which will carry an advanced IR sounder (IRS) from 2023.

4.5.3. *Microwave sounders and imagers*

The four-channel MSU instrument, primarily a temperature sounder, on the TOVS satellites was superseded by the 15-channel AMSU-A instrument on the NOAA series of satellites from 1998 (NOAA-15–NOAA-19) as well as the three MetOp satellites from 2006 onwards. Each of these AMSU-A instruments was supplemented by a humidity sounding instrument, initially AMSU-B on NOAA-15 through -17, followed by the MHS instrument on NOAA-18 and -19 and the three MetOp satellites. These instruments were equipped with three channels spanning the water vapor absorption line at 183 GHz as well as two channels at lower frequencies to provide information on cloud liquid water and ice (89/150 GHz for AMSU-B and 89/157 GHz for MHS). From 2011, The U.S. Joint Polar Satellite System (JPSS) satellites carried the ATMS instrument, which effectively combined and enhanced the temperature and humidity sounding capabilities of AMSU-A and AMSU-B/MHS into a single instrument.

Regarding the strategy for assimilating these radiance observations, initial efforts focused entirely on a pre-selection of observations where the impact of cloud and precipitation could be safely neglected. This was possible for sounders such as AMSU-A and ATMS which have a range of window channels that can provide accurate estimates of the impact of hydrometeors on the sounding channels. As temperature sounding channels (51–64 GHz) are primarily only sensitive to liquid water clouds, with ice water only important with very large hydrometeors and deep columns, it was possible to use a large percentage of the MW temperature sounding channels in this "clear-sky" framework. However, low-peaking channels, such as AMSU channel 5 (peaking at 750 hPa) and even AMSU channel 6 (peaking at 400 hPa) had to be used more carefully, leading to rejection of a large number of observations. The situation was worse for the humidity sounding channels at 183 GHz because the impact of ice cloud is much larger. Therefore, significant amounts of 183 GHz observations were not used. For MW imagers, sensing at low levels and therefore sensitive to liquid water cloud and even precipitation, the situation was worse again. This led at ECMWF to the development of an "all-sky" assimilation capability, initially only for MW imagers. The all-sky approach has now been adopted or is under development at most NWP centers (Geer et al. 2018) and has been extended to the temperature sounding channels of AMSU-A. Attention is now turning to applying the all-sky approach to IR data and visible observations.

The observations used have increased significantly in number. By 2020, observations from a constellation of microwave instruments were available for assimilation: ATMS from the JPSS satellites (Suomi NPP and NOAA-20); AMSU-A on the three MetOp satellites as well as some older POES satellites (NOAA-15, -18 and -19); MHS from all three MetOp satellites and NOAA-19; SAPHIR on Megha-Tropiques; GMI on GPM; MWHS/-2 on FY-3B, -C and -D; MWTS-2 on FY-3D; MTVZA-GY on Meteor M N2 and N2-2; AMSR-2 on GCOM-W; MWRI on FY-3B, C and D; WindSat on Coriolis and SSMIS on DMSP F17 and F18.

In addition to providing operational resilience in the event of loss of a single sensor, it has been found that adding a single new humidity sensor in general reduces the standard deviation of the difference between the existing water vapor channel radiance observations and their NWP background equivalents by between 0.5 and 1.0%, showing that the system is not yet saturated. This lends support to the development of small-satellite constellations or geostationary MW sensors, so that we can continue to enjoy frequent observations in the future observing system.

The development of fast ocean surface emissivity models (e.g. FASTEM, Kazumori and English (2014)) has supported the exploitation of MW observations over oceans, although the development of a single physically based model working from low MW frequencies (L-band) through to sub-mm (over 300 GHz) through to the IR, complete with uncertainty characterization, remains a challenge (English et al. 2020).

By contrast, over land areas, there are significant challenges, including: poorly known skin temperature, ambiguity between changes in the surface and the radiative impact of clouds, and emissivity changing rapidly with snow cover, snow conditions and surface wetness, none of which is well known a priori. Therefore, studies aiming to improve modeling of these more difficult surface types are under way. For some surfaces, for example, desert and coastal areas, the emissivity can be inferred from an atlas, but uncertainty in defining the instantaneous skin temperature remains.

4.5.4. *Radiative transfer modeling*

Effective assimilation of radiances relies on fast, accurate RT models to simulate the radiances equivalent to the atmospheric state described by the NWP model. Two fast RT models have been used predominantly in NWP DA: the RTTOV model developed by the NWP SAF of EUMETSAT (Saunders et al. 2018) and the Community Radiative Transfer Model (CRTM) developed by JCSDA (Ding et al. 2011). These models have been developed to enhance their capabilities considerably and to improve their accuracy. They have also been enhanced to support a greatly increased number of satellite instruments, covering a wider range of the electromagnetic spectrum.

4.5.5. *Observation uncertainties*

Within DA theory, the uncertainties in the observations (and their operators) are represented by the error covariance matrix **R**. In principle, this is a full matrix representing not only the expected error variance in each observation but also the correlation between the errors in each pair of observations. In practice, it is not practical to implement a full matrix, both for computational reasons and for lack of knowledge of the relevant error correlations. Until recently, observation errors have all been treated as uncorrelated, with a diagonal **R** matrix. Where this has been known to be a poor approximation, the problem has been addressed empirically, either by inflating the error variances or thinning the observations, or both.

It is currently still the general practice to treat horizontally correlated errors by thinning. However, in recent years, progress has been made to handle the vertical problem, that is, the correlation of errors between different radiances in the same vertical sounding, through a full matrix (Weston et al. 2014). An important tool in the quantification of **R** has been the Desroziers technique (Desroziers et al. 2005). However, for practical implementation within an NWP DA system, it has proved necessary to improve the conditioning of matrices given by the Desroziers technique.

Another area of development has been in the introduction of scene-dependent observation errors, initially to distinguish between land and sea in surface-sensitive channels and to allow for scan-angle dependence of surface sensitivity. Recently, the development of all-sky assimilation techniques, where radiances are assimilated into both clear and cloudy scenes, has necessitated the use of observation error models that depend on the extent to which a given scene is affected by cloud (taking into account the cloud affecting the observations and the cloud as represented by the NWP model).

4.5.6. *Atmospheric motion vectors (AMVs)*

AMVs are produced by tracking cloud or areas of water vapor in consecutive satellite images. Geostationary imagery is ideal for this due to the frequent viewing of the same area of the Earth's atmosphere. In the early days, AMVs were produced by tracking primarily in the infrared window channel at 11 μm (IR). The technique is described in Chapter 12.

A major source of error was (and remains today) the assignment of cloud top heights in AMV processing. Initial approaches exploited the IR channel measurements in conjunction with forecast temperature profiles. The addition of water vapor channels in subsequent generations of geostationary imagers enabled new and complementary approaches to height assignment. The range of channels used in the tracking step was also extended to exploit the benefits of visible, IR and water vapor channel measurements in specific scenes.

The effect of image intervals was also investigated in the early days to optimize the impact of AMVs in the analysis of tropical cyclones, as well as more generally. Velden (2000) describes optimal image intervals of 5–10 minutes for cloudy IR, visible and water vapor imagery but 30 minutes for clear sky water vapor imagery. Although in the early years (1970–2000) standard image intervals were 30 minutes, AMVs were produced using 15 minute, or shorter, intervals over restricted domains.

Early experiments using data from the 1979 GARP experiment showed benefits from assimilating AMV data, as well as highlighting some problems, notably slow speed biases in the extra-tropical jet regions. Nevertheless, throughout the early 1990s, the quality of AMVs improved steadily. Using ECMWF's ERA-15 global reanalysis, it was shown that background fits (RMS) for high-level AMVs, for example, improved from 9–10 m/s to 6 m/s during the period 1979–1993. Alongside the general benefits of AMVs on forecast skill, several studies demonstrated the impact of AMVs on tropical cyclone track forecasts.

A key development soon after 2000 was the routine production of AMVs over the polar regions by tracking features in MODIS and AVHRR imagery. This is possible in the polar regions where images from successive polar orbiter overpasses overlap. The polar winds have been assimilated operationally in NWP centers since 2003. In recent years, new datasets have been developed to help close the gap between the coverage provided by polar winds and that provided by the geostationary imagers. The approaches developed involved using image pairs instead of image triplets (EUMETSAT MetOp winds), and deriving AMVs from multiple satellites (e.g. LEO–GEO winds or dual-MetOp winds); the resulting coverage from all datatypes is illustrated in Figure 4.4. Improvements were seen in assimilation experiments using these new datasets.

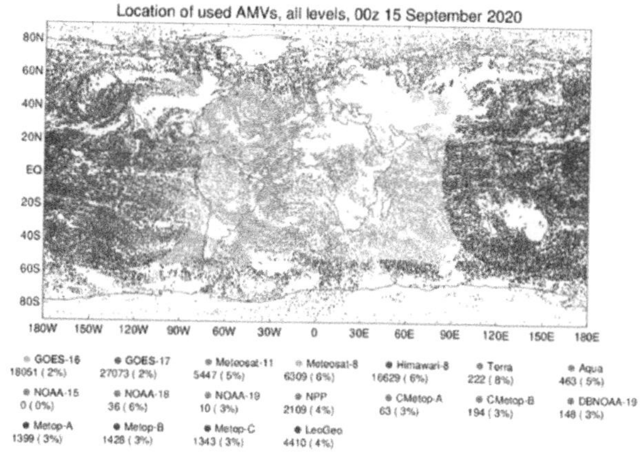

Figure 4.4. *Data coverage plot showing the AMVs assimilated into the Met Office global model for a single model cycle*

Another key change has been an increase in the frequency of AMV products. GOES and MTSAT-2 (Himawari-7) AMVs changed from a three-hourly product to an hourly product in 2014 and 2011, respectively. Most geostationary AMV datasets are now provided hourly. This was an important step, evidenced by impact trials at the Met Office, which showed the biggest improvement in forecast impact from an AMV change since 2000 resulted from the assimilation of data more frequently in the time window.

With advances in the imagers and rapid scan operations, it has become possible to generate AMVs using imagery at 5–15-minute intervals. The pixel resolution has improved, enabling the tracking of smaller spatial targets providing higher resolution AMV products. In addition, the greater range of channels can be used for height assignment and tracking to enhance quality and coverage, for example, the introduction of the 3.9 μm channel for night-time, low-level winds, complementing the daytime visible capability.

It is also worth noting the development of AMVs with stereo-based heights. Stereo approaches can be applied when cloud features are viewed from at least two different viewing angles, with the heights derived using geometry. The best example is the MISR AMVs produced from multi-angle radiometer imagery. The MISR winds provide only modest impact due to the narrow MISR swath, but future missions have been proposed that have the potential to provide larger AMV datasets with more accurate stereo-based heights (Mueller et al. 2017).

Most AMV products capture large-scale to synoptic-scale flow but do contain information at much smaller scales. Several AMV producers have adapted their AMV schemes to derive winds for nowcasting and assimilation in high-resolution models, particularly for forecasting of high impact weather events.

4.5.7. *Scatterometers*

Wind scatterometers are radar instruments that measure the backscatter of radiation from the rough ocean surface, using two or more viewing directions. The backscatter provides information on the ocean surface wind speed and direction.

Different scatterometers have used various viewing geometries but their principle of operation is the same. The satellite views the same patch of ocean from a number of directions within a period of a few minutes. One challenge in assimilating scatterometer data is that there are usually two wind directions, separated by ~180 degrees and associated with similar wind speeds, consistent with each set of backscatter measurements. There have been two basic approaches to this problem. The first approach is to retrieve the two (or more) wind vectors consistent with the measurements, and then to select the "best" for presentation to the assimilation system. The procedure for selecting the "best" can use horizontal consistency or closeness to an NWP background field or both. The second approach is to present all possible "ambiguous" wind vectors to the assimilation system and to allow it, implicitly, to make the selection.

Scatterometers have continued to evolve in one of two families of instruments, depending on the radar frequency used: C-band (4–8 GHz) or Ku-band (12–18 GHz). Also, new concepts of scatterometers are starting to emerge.

In common with other surface observations, it is found that scatterometer observations only have a significant impact on the analysis up to 600 hPa. However, experiments assimilating a single ASCAT wind close to a tropical cyclone demonstrated that 4D-Var is able to propagate the scatterometer wind information to the upper troposphere. More recently, the impact of scatterometer winds on coupled ocean–atmosphere models has been investigated. Laloyaux et al. (2016) found an added benefit from scatterometer winds in the coupled ECMWF system compared to those in the uncoupled system, on both winds and ocean temperatures, plus a better estimation of the cold wake from a tropical cyclone. They noted that this benefit arises from implicit coupling in the observation operator and from improved atmospheric forcing on upper-level ocean mixing.

4.5.8. *Radio occultation observations*

The RO measurement technique was originally developed by the planetary science community in the 1960s, and the concept of RO using the constellation of Global Positioning Satellites (GPS) emerged in the late 1980s. GPS was the first network of Global Navigation Satellite Systems (GNSS) – from here on, we will refer generically to "GNSS-RO" for this type of observation. GNSS-RO measurements provide information on the vertical gradient of the atmospheric refractive index, which is related to the gradients of density (and hence of temperature and pressure) and of water vapor density (Kursinski et al. 1997). When a radio signal from a transmitter on a GNSS satellite passes through the limb of the atmosphere, the signal is delayed, and its path is bent by gradients in the refractivity field. If this signal is intercepted by a GNSS receiver on another satellite, behind the Earth's limb, then the time of receipt and the direction from which the signal is received are different from those that would have been obtained from a straight-line path through a vacuum. This is illustrated in Figure 4.5 in the geometrical optics approximation, where only one ray arrives at the receiver at any given time. The angle through which the ray is deflected – the "bending angle", α – can be related in a simple way to the measured Doppler shift and phase delay of the received signal by assuming a spherical symmetry.

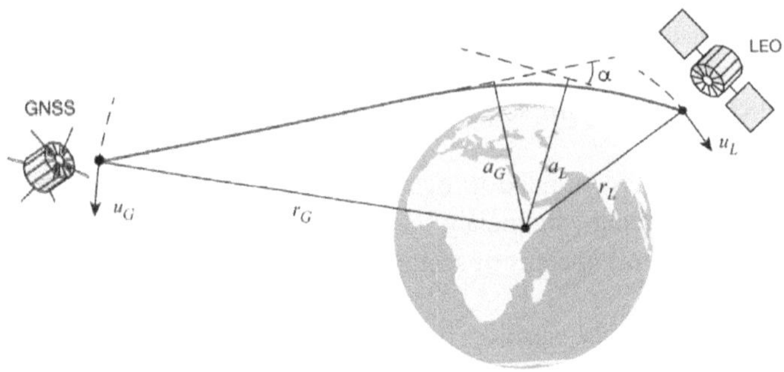

Figure 4.5. *The geometry of a radio occultation measurement showing the refracted ray from a transmitting satellite (on the left) to a receiving satellite (on the right)*

During an occultation event, a low Earth orbiting (LEO) satellite sets or rises relative to the GNSS satellite, and the ray between them sweeps out a path of changing tangent height through the atmospheric limb; a typical occultation takes ~1 minute to profile the atmosphere (0–60 km) and thus provides a profile of bending angle against height (or, more strictly, impact parameter – a_G in Figure 4.5). The profile of refractive index can be derived from the profile of bending angle, analytically in the limit of a spherically symmetric atmosphere, using the Abel transform (Kursinski et al. 1997).

GNSS-RO measurements from a LEO satellite in near-polar orbit are globally distributed in a quasi-random manner. They provide temperature information for the stratosphere and upper troposphere, and humidity information for the lower troposphere. They have a high vertical resolution of 0.5–1 km but a low horizontal resolution of ~200 km. They provide information of high accuracy, equivalent to random errors in a temperature of ~1 K (for height ranges for which humidity effects are negligible). Moreover, their systematic errors are very low: < 0.2 K, as required for climate monitoring. GNSS-RO measurements are often described as "all-weather" because they are negligibly affected by cloud and precipitation when compared to other satellite measurements. The space/time sampling of the measurements is determined by the number of GNSS receivers and the number of available GNSS constellations; a constellation of six LEO satellites can potentially provide ~3,000 occultations per day for each GNSS constellation. Furthermore, by the standards of space-borne remote sensing, they are relatively inexpensive.

The vertical information content from GNSS-RO data is highly complementary to that from advanced IR sounders such as IASI (Collard and Healy 2003).

GNSS-RO was first demonstrated for the Earth's atmosphere with the GPS/MET "proof of concept" mission from 1995 to 1997 and continued by the CHAMP mission from 2000. The COSMIC constellation of six LEO satellites is of particular interest because it resulted in an order of magnitude increase in the available data from 2006. More recently, there have been rapid developments in the use of constellations of small satellites to carry GNSS-RO receivers.

The assimilation options for GNSS-RO data were first discussed by Eyre (1994). Most global NWP centers have moved to assimilating bending angles with a one-dimensional operator (Eyre et al. 2021).

Early assimilation studies with GNSS-RO data showed that, as expected, the largest impacts are found in the upper troposphere and lower stratosphere, and particularly in the southern hemisphere. Surprisingly, large impacts have been found when comparatively small numbers of observations have been assimilated, confirming that GNSS-RO data are highly complementary to other global observing systems. The introduction of COSMIC GNSS-RO measurements had a significant impact correcting NWP temperature biases.

An important characteristic of GNSS-RO observations is that their biases are small, and smaller than the random errors assumed in the DA. Therefore, they can be assimilated without bias correction into an NWP model, meaning they become "anchor measurements" in the radiance bias correction schemes. This is important in both operational NWP and climate reanalyses. Further, the improved consistency of climate reanalyses in the stratosphere since the assimilation of COSMIC in 2006 has been noted by several, including Hersbach et al. (2020).

4.5.9. *Impacts*

The impacts of observations within NWP systems are commonly assessed in three different ways.

1) Observing System Experiments (OSEs), also called Data Denial Experiments (DDEs), measure the effect of removing a subset of observations from, or adding it to, an NWP DA system. Impact is assessed using a range of metrics available, using either observations or NWP analyses as the "truth" against which the forecast is verified.

2) More recently, it has become common to use another, very sensitive metric: the "fit to background". This measures the fit of the short-range forecast (e.g. six hours) to the next batch of observations, particularly to observation types that are independent of the change being tested.

3) A more recent addition to the armory of methods used to assess impact is the Forecast Sensitivity-based Observation Impact (FSOI) method. This method measures the role of observations (from large subsets of observations down to a single observation) in reducing the error of a short-range forecast (typically a 24 h forecast for a global NWP system).

For more details on these and other impact assessment methods, see Eyre (2018).

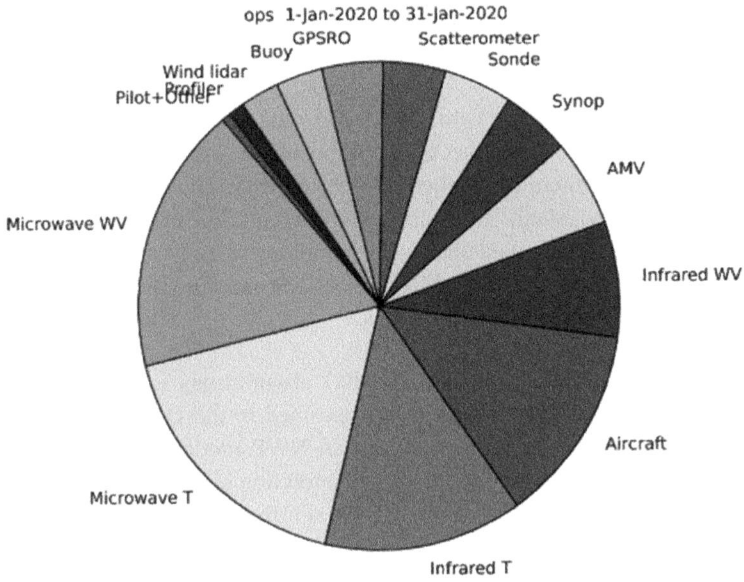

Figure 4.6. *Impact of the various components of the global observing system in the ECMWF operational NWP system in January 2020, as measured by the FSOI metric*

Figure 4.6 shows the FSOI metric for the 2020 ECMWF operational system, illustrating the dominant contribution from the microwave and infrared radiances.

All major NWP operational and research centers conduct DDEs, and several of them also generate FSOI assessments of impact. For the last 20 years, these experiments and their results have been reported and discussed through a series of WMO workshops on the impact of observations in NWP, and these results are available via the presentations at and reports of these workshops (e.g. see WMO (2016)).

4.5.10. *Reanalyses*

Atmospheric reanalyses, in which modern NWP systems are used to analyze retrospectively the atmospheric state over multi-decadal timescales, have found a wide application in atmospheric sciences. Reanalyses provide spatially coherent and continuous representations of the atmosphere, supporting applications requiring an accurate representation of synoptic scales at sub-daily resolution, as well as those examining long-term trends and variability in the Earth's climate. Several groups have produced a series of reanalyses over the last four decades, most notably ECMWF, NOAA, NASA and JMA. The state-of-the-art in (full-observing system) global reanalyses is currently represented by ECMWF's ERA5, NASA's MERRA-2 and JMA's JRA-55 (see Hersbach et al. (2020) and references therein). These reanalyses typically assimilate a diverse set of satellite observations, similar to that used in operational NWP, and benefit from sustained and ongoing efforts to reprocess satellite data in a consistent way throughout the lifetime of missions.

Although the incremental improvement in analysis quality in successive reanalyses is a product of developments in forecast models, data assimilation systems and observations for a given reanalysis improvements in quality over time result mainly from the evolving observing system, as illustrated in Figure 4.1. This shows that the evolving observing system, largely the increasing availability of satellite data during the period 1979–2019, accounts for a one to two day improvement in forecast skill over the period, based on re-forecasts from the ERA5 reanalysis.

4.6. Summary and conclusion

Following the advances of the early years, the assimilation of satellite data into NWP models has continued to progress in many directions. The number of satellite instruments from which data have been available has increased substantially, and the size of the R&D community active in this field is much greater than it was in 2000.

A new generation of hyperspectral IR sounders has become available, giving much enhanced vertical resolution and correspondingly increased impact on NWP skill. The impact of data from MW sounders and imagers has also increased, both because of a large increase in the number of sensors available for operational use and through advances in the ability to exploit cloud-affected radiances. Continued progress on RT modeling, on the correction of observation biases and on the characterization of observation uncertainty have all fed through into improved assimilation of IR and MW radiance data.

Progress on the assimilation of AMVs has also continued through an increase in the number of AMV datasets available, improvements in their quality and improved characterization of their uncertainties. Assimilation of scatterometer data has benefitted from increases in the number of instruments and their performance, and from improvements in quality control and assimilation methods. Observations from a new type of technology, RO, have become available and their usefulness for global NWP has been demonstrated.

Beneficial impacts of satellite data on NWP performance have been demonstrated by all major NWP centers using a range of metrics. OSEs/DDEs have continued to play a crucial role, and these have been complemented by assessments using the FSOI technique. In addition to their important role in operational NWP, satellite observations are playing a key role in atmospheric re-analyses, which provide valuable data sets for atmospheric and climate research and climate monitoring.

4.7. References

Aravéquia, J.A., Szunyogh, I., Fertig, E.J., Kalnay, E., Kuhl, D., Kostelich, E.J. (2011). Evaluation of a strategy for the assimilation of satellite radiance observations with the local ensemble transform Kalman filter. *Mon Wea Rev*, 139, 1932–1951. https://doi.org/10.1175/2010MWR3515.1.

Auligné, T., McNally, A.P., Dee, D.P. (2007). Adaptive bias correction for satellite data in a numerical weather prediction system. *Q J R Meteorol Soc*, 133, 631–642. https://doi.org/10.1002/qj.56.

Candy, B., Cotton, J., Eyre, J. (2021). Recent results of observation data denial experiments. Met Office forecasting research technical report 641 [Online]. Available at: https://library.metoffice.gov.uk/Portal/DownloadImageFile.ashx?objectId=438.

Collard, A.D. and Healy, S.B. (2003). The combined impact of future space-based sounding instruments on numerical weather prediction. *Q J R Meteorol Soc*, 129, 2741–2760. doi.org/10.1256/qj.02.124.

Desroziers, G., Berre, L., Chapnik, B., Poli, P. (2005). Diagnosis of observation, background and analysis-error statistics in observation space. *Q J R Meteorol Soc*, 131, 3385–3396. https://doi.org/10.1256/qj.05.108.

Ding, S., Yang, P., Weng, F., Liu, Q., Han, Y., Van Delst, P., Li, J., Baum, B. (2011). Validation of the community radiative transfer model. *J Quant Spectrosc Radiat Transfer*, 112, 1050–1064. https://doi.org/10.1016/j.jqsrt.2010.11.009.

ECMWF (1989). *Proceedings of ECMWF/EUMETSAT Workshop on "The Use of Satellite Data in Operational Numerical Weather Prediction"*, 9–12 May 1989, ECMWF, Reading.

English, S., Prigent, C., Johnson, B., Yueh, S., Dinnat, E., Boutin, J., Newman, S., Anguelova, M., Meissner, T., Kazumori, M. et al (2020). Reference-quality emission and backscatter modeling for the ocean. *Bull Amer Meteorol Soc*, 101(10), E1593–E1601 [Online]. Available at: https://journals.ametsoc.org/view/journals/bams/101/10/bamsD200085.xml.

Errico, R.M., Yang, R., Privé, N.C., Tai, K.-S., Todling, R., Sienkiewicz, M.E., Guo, J. (2013). Development and validation of observing-system simulation experiments at NASA's global modeling and assimilation office. *Q J R Meteorol Soc*, 139, 1162–1178. https://doi.org/10.1002/qj.2027.

Eyre, J.R. (1994). Assimilation of radio occultation measurements into a numerical weather prediction system. ECMWF Technical Memorandum 199 [Online]. Available at: https://www.ecmwf.int/sites/default/files/elibrary/1994/9331-assimilation-radio-occultation-measurements-numerical-weather-predictionsystem.pdf.

Eyre, J.R. (1997). Variational assimilation of remotely sensed observations of the atmosphere. *J Meteorol Soc Japan*, 75, 331–338. https://doi.org/10.2151/jmsj1965.75.1B_331.

Eyre, J.R. (2018). Observing system studies proposed for GNSS-R ocean surface wind data. Met Office forecasting research technical report 629. [Online]. Available at: https://digital.nmla.metoffice.gov.uk/IO_85344683-f3a2-4644-93a3-1472bf9b9534/.

Eyre, J.R., English, S.J., Forsythe, M. (2020). Assimilation of satellite data in numerical weather prediction. Part I: The early years. *Q J R Meteorol Soc*, 146, 49– 68. https://doi.org/10.1002/qj.3654.

Eyre, J.R., Bell, W., Cotton, J., English, S.J., Forsythe, M., Healy, S.B., Pavelin, E.G. (2021). Assimilation of satellite data in numerical weather prediction. Part II: Recent years. *Q J R Meteorol Soc*. http://doi.org/10.1002/qj.4228.

Geer, A.J. (2016). Significance of changes in medium-range forecast scores. *Tellus A: Dyn. Meteorol. Oceanogr.*, 68, 1. doi: 10.3402/tellusa.v68.30229.

Geer, A.J., Lonitz, K., Weston, P., Kazumori, M., Okamoto, K., Zhu, Y., Liu, E.H., Collard, A., Bell, W., Migliorini, S. et al. (2018). All-sky satellite data assimilation at operational weather forecasting centres. *Q J R Meteorol Soc*, 144, 1191–1217. https://doi.org/10.1002/qj.3202.

Hersbach, H., Bell, B., Berrisford, P., Hirahara, S., Horányi, A., Muñoz-Sabater, J., Nicolas, J., Peubey, C., Radu, R., Schepers, D. et al. (2020). The ERA5 global reanalysis. *Q J R Meteorol Soc*, 146, 1999–2049. https://doi.org/10.1002/qj.3803.

Hilton, F., Armante, R., August, T., Barnet, C., Bouchard, A., Camy-Peyret, C., Capelle, V., Clarisse, L., Clerbaux, C., Coheur, P.-F. et al. (2012). Hyperspectral Earth observation from IASI. *Bull Amer Meteorol Soc*, 93, 347–370. https://doi.org/10.1175/BAMS-D-11-00027.1.

Kazumori, M. and English, S.J. (2014). Use of the ocean surface wind direction signal in microwave radiance assimilation. *Q J R Meteorol Soc*, 141, 1354–1375. https://doi.org/10.1002/qj.2445.

Kursinski, E.R., Hajj, G.A., Schofield, J.T., Linfield, R.P., Hardy, K.R. (1997). Observing Earth's atmosphere with radio occultation measurements using the global positioning system. *J Geophys Res*, 102(D19), 23429–23465. https://doi.org/10.1029/97JD01569.

Laloyaux, P., Thépaut, J., Dee, D. (2016). Impact of scatterometer surface wind data in the ECMWF coupled assimilation system. *Mon Wea Rev*, 144, 1203–1217. https://doi.org/10.1175/MWR-D-15-0084.1.

Lazzara, M.A., Dworak, R., Santek, D.A., Hoover, B.T., Velden, C.S., Key, J.R. (2014). High-latitude atmospheric motion vectors from composite satellite data. *J Appl Meteorol Climat*, 53, 534–547. https://doi.org/10.1175/JAMC-D-13-0160.1.

Lorenc, A.C. (1986). Analysis methods for numerical weather prediction. *Q J R Meteorol Soc*, 112, 1177–1194. https://doi.org/10.1002/qj.49711247414.

Lorenc, A.C. and Marriott, R.T. (2014). Forecast sensitivity to observations in the Met Office Global numerical weather prediction system. *Q J R Meteorol Soc*, 140, 209–224. https://doi.org/10.1002/qj.2122.

Menzel, W.P., Schmidt, T.J., Zhang, P., Li, J. (2018). Satellite-based atmospheric infrared sounder development and applications. *Bull Am Meteorol Soc*. https://doi.org/10.1175/BAMS-D-16-0293.1.

Mueller, K.J., Liu, J., McCarty, W., Gelaro, R. (2017). An adjoint-based forecast impact from assimilating MISR winds into the GEOS-5 data assimilation and forecasting system. *Mon Wea Rev*, 145, 4937–4947. https://doi.org/10.1175/MWR-D-17-0047.1.

Ohring, G. (1979). Impact of satellite temperature soundings on weather forecasts. *Bull Am Meteorol Soc*, 60, 1142–1147. https://doi.org/10.1175/1520-0477(1979)060<1142:IOSTSD>2.0.CO;2.

Pavelin, E.G. and Candy, B. (2014). Assimilation of surface-sensitive infrared radiances over land: Estimation of land surface temperature and emissivity. *Q J R Meteorol Soc*, 140, 1198–1208. https://doi.org/10.1002/qj.2218.

Pavelin, E.G., English, S.J., Eyre, J.R. (2008). The assimilation of cloud-affected infrared satellite radiances for numerical weather prediction. *Q J R Meteorol Soc*, 134, 737–749. https://doi.org/10.1002/qj.243.

Rabier, F., Järvinen, H., Klinker, E., Mahfouf, J.-F., Simmons, A. (2000). The ECMWF operational implementation of four-dimensional variational assimilation. I: Experimental results with simplified physics. *Q J R. Meteorol Soc*, 126, 1143–1170. https://doi.org/10.1002/qj.49712656415.

Rodgers, C.D. (1976). Retrieval of atmospheric temperature and composition from remote measurements of thermal radiation. *Rev Geophys Space Phys*, 14, 609–624. https://doi.org/10.1029/RG014i004p00609.

Saunders, R., Hocking, J., Turner, E., Rayer, P., Rundle, D., Brunel, P., Vidot, J., Roquet, P., Matricardi, M., Geer, A.J. et al. (2018). An update on the RTTOV fast radiative transfer model (currently at version 12). *Geosci Model Dev*, 11, 2712–2737. https://doi.org/10.5194/gmd-11-2717-2018.

Smith, A., Atkinson, N., Bell, W., Doherty, A. (2015). An initial assessment of observations from the Suomi-NPP satellite: Data from the cross-track infrared sounder (CrIS). *Atmos Sci Lett*, 16, 260–266. https://doi.org/10.1002/asl2.551.

Weston, P.P., Bell, W., Eyre, J.R. (2014). Accounting for correlated error in the assimilation of high resolution sounders. *Q J R Meteorol Soc*, 140, 2420–2429. https://doi.org/10.1002/qj.2306.

WMO (2016). Sixth Workshop on the Impact of Various Observing Systems on NWP, Shanghai, China, 10–13 May 2016 [Online]. Available at: https://www.wmo.int/pages/prog/www/WIGOS-WIS/meetings/NWP-6_May2016_Shanghai/WMO6-Impact-workshop_Shanghai-May2016.html.

5

Nowcasting

Thibaut MONTMERLE
Météo-France, Operations Department for Prediction, Nowcasting Division, Toulouse, France

5.1. Introduction

Nowcasting generally refers to methods for predicting weather conditions at times ranging from 0–6 hours. It includes real-time warnings and predictions of specific parameters to warn or inform with a few hours of anticipation. Within these short timeframes, high impact weather phenomena are mainly studied, often with explosive development, which can have dramatic consequences with regard to the safety of people and property. This is the case for convective systems such as thunderstorms (and the associated risks in terms of hail, electrical activity, gusts), exceptional precipitation, fog and more generally the type of hydrometeors present on ground (for road conditions) or at an altitude (presence of conditions conducive to icing). Users of nowcasting are, of course, the general public; public authorities; and agriculture, road and aeronautical sectors. Another common application is assistance in planning and monitoring of various sporting or cultural events.

The main difference with numerical weather prediction (NWP) is the need to produce forecasts as quickly and as frequently as possible. This will have direct consequences on the algorithms used, which must be numerically efficient, and on the type of observations that will be considered. These observations must be spatialized, mainly in the form of images, be carried out at high spatial and temporal resolutions, and be available in very short timeframes (typically within 10 minutes).

For a color version of all the figures in this chapter, see www.iste.co.uk/phulpin/satellites2.zip.

Satellites for Atmospheric Sciences 2,
coordinated by Thierry PHULPIN, Didier RENAUT, Hervé ROQUET and Claude CAMY-PEYRET.

The time taken by an observation from its measurement to its processing by computers of meteorological centers is therefore particularly important. Two main classes of observations are typically used: meteorological radars and satellite measurements. Radars are the only means of describing the intensity and location of precipitation and, for those with Doppler capability, air circulation within precipitating systems. The revisiting time is very short (of the order of five minutes) and spatial resolution, which is often sub-kilometric, is excellent. However, radars are still expensive instruments, which implies a very heterogeneous global coverage, and no data is available in the open sea. As far as satellite data is concerned, and as we will see in the first part of this chapter, sensors on board geostationary platforms are particularly interesting in this context. The latter allow for constant monitoring of zones included in the observed disk. Conversely, the wavelengths considered remain strongly linked to the technical constraints imposed by this type of platform and are limited to the visible and to the infrared (IR) domains.

WEATHER RADARS.– Although not the main focus of this book, it is difficult to talk about nowcasting without mentioning weather radars. Because of their ability to provide three-dimensional measurements of precipitation intensity and air circulation in precipitating systems, these instruments remain the preferred instruments for most nowcasting applications over land.

Operational radars are based on the emission of electromagnetic waves, generally between 2 and 12 GHz, in a given direction by an antenna. The rotation of the antenna and the use of different angles of elevation then make it possible to explore a volume around the radar. Analysis of the amount of energy backscattered by targets encountered in the path of the beam, such as raindrops, makes it possible to gain access to the reflectivity, a quantity that depends on the intensity of precipitation. The phase shift of the backscattered signal makes it possible, by virtue of the Doppler effect, to recover the speed of these targets along the beam. Finally, a new generation of radar, called dual polarization radar because it emits polarized waves in two orthogonal directions, gives additional indications on the shape of the targets, and therefore makes it possible to distinguish the type of precipitation: rain, ice, snow, hail, etc.

Nowcasting tools are mainly based on image extrapolation, as well as structure detection and tracking, to make very short-range predictions of some of the phenomena identified in imagery. In recent years, the emergence of methods based on artificial intelligence (AI) has made it possible to improve some of these forecasts. Since these approaches do not take into account the physics of the atmosphere, however, they remain limited in terms of the time range, particularly in situations of explosive storm development.

The significant progress made over the past two decades on kilometer-scale NWP models, both in terms of the quality of initial states derived from data assimilation algorithms and the realism of the simulations obtained, means that more and more meteorological centers are also using this type of model for nowcasting applications. Only one model makes it possible to provide trends of a particular parameter by virtue of its physical parameterizations and its representation of the topography, for example. NWP models can thus be directly used by considering valid forecasts at the time of interest, or by running versions dedicated to nowcasting that have the advantage of making the shortest time ranges available quickly. Several pragmatic compromises must, however, be made in the chosen configuration, in particular that of considering, in calculating initial states, only observations within a restricted time window. In order to make the link with observation-only approaches (extrapolation and AI), which are highly numerically efficient and often relevant in the first tens of minutes, methods of merging with the images simulated by NWP models are increasingly used.

We will see in this chapter that satellite data are involved in many aspects of nowcasting, whether for the identification and characterization of particular weather phenomena or parameters or for their consideration in the initial states of models. They thus play a leading role in warning and anticipating certain meteorologically critical situations.

5.2. Satellite data for nowcasting

As detailed in Part 1 of this book, the constellation of meteorological satellites and the characteristics of the instruments they carry are continually evolving. For a clearer view of current and future space programs in the specific field of meteorology, readers may refer to the WMO OSCAR database[1]. Only major classes of platforms and sensors used in nowcasting are mentioned here.

5.2.1. *Polar-orbiting satellites*

The main advantage of polar-orbiting satellites is the high spatial resolution that their low orbit allows (they are also commonly called LEO for "Low Earth Orbit"). This proximity to the Earth's surface also allows active or passive remote sensing instruments to be carried in microwaves, sensitive to clouds and low precipitation. This type of instrument could be of interest for nowcasting, but unfortunately their

1. https://www.wmo-sat.info/oscar/spacecapabilities.

spatial coverage, which depends on the size of the measurement swath as well as the revisiting time on the area, is not compatible with the need for continuous monitoring of rapidly developing phenomena.

However, when these data are considered in the data assimilation systems of NWP models (see Chapter 4), they may directly or indirectly benefit certain products derived from satellite observations used in nowcasting. In addition to the prediction itself, some of the algorithms used in the development of these products, such as those described in section 5.3, require simulated information, for example, to determine a snowfall limit or the altitude of cloud tops.

5.2.2. *Geostationary satellites*

5.2.2.1. *General information*

With the continuous area-based monitoring they enable, instruments on board geostationary satellites are clearly better suited to nowcasting applications. They meet their needs in terms of temporal resolution, spatial coverage and the time needed to make the measurement available. However, their position farther away from the Earth's surface in the equatorial plane has the following consequences:

– production of images with lower spatial resolution than those of LEO satellites (typically 3 km against 500 m). These resolutions, however, remain sufficient to track weather systems such as synoptic and mesoscale cloud structures, as well as more localized circulation over most tropical and mid-latitude zones. At present, global coverage (with the exception of the polar zones) is practically ensured by European (Meteosat Second Generation (MSG)), Indian (INSAT), Russian (ELECTRO-L), Chinese (FY-4), South Korean (COMs), Japanese (HIMAWARI) and American (GOES-R) constellations;

– possibility of providing a complete image of the disk in near real-time (conventionally every 10–15 min, with the possibility of rapid scans of a few minutes on certain target zones);

– deterioration of spatial resolution and increase in parallax problems with latitude, which implies poor coverage in high latitude regions (coverage that can, however, be compensated for by the convergence of many polar orbiting satellites over these zones);

– limitation of spectral coverage: observations in microwaves would, for example, require too large an antenna to reach spatial kilometer-scale resolutions at this altitude.

Observations from geostationary satellites thus make it possible to detect the appearance of certain parameters and to monitor their evolution in real time over most inhabited regions of the globe.

5.2.2.2. *Embedded instruments: imaging radiometer*

The main instrument used in a nowcasting frame is an imaging radiometer of the ABI[2], SEVIRI[3] and AHI[4] types on board the GOES, MSG and HIMAWARI series satellites, respectively. All of these instruments share a large part of their characteristics and have between 12 (for SEVIRI) and 16 channels in the visible, near-IR and thermal IR domains. As described in Chapter 9 of Volume 1, the atmosphere absorbs very little IR thermal radiation, which makes it possible, among other things, to measure the temperature of continental and ocean surfaces. However, the presence of water in the atmosphere in the liquid state, in the form of drops or droplets, or in the solid state, in the form of ice or snow crystals, becomes an opaque or semi-opaque obstacle to these wavelengths. The radiation emitted by a cloud will thus depend strongly on its main characteristics: altitude, thickness, liquid or solid phase, texture and size with respect to the pixel. In the visible domain, it will also depend heavily on the relative position of the sun with respect to the satellite.

The various channels of these imagers are used either individually or in the form of combinations of the RGB[5] (Red, Green, Blue) type that make it possible to highlight certain characteristics of clouds, such as their vertical extension or their phase. As we will see in section 5.3, combinations have been identified specifically for the detection of certain phenomena such as volcanic ash, dust, forest fires, fog, circulation in the upper troposphere (jet stream, potential swirl), cloud type, convective phenomena and some of their attributes, sea surface temperature, snow cover, etc. EUMETSAT thus offers a number of these products within the framework of the SAFNWC.

SAFNWC.– The Satellite Application Facility for NoWCasting[6] is a project that has been conducted by EUMETSAT since 1997, which contributes to the research and development of several product categories: clouds, precipitation, wind, meteorological systems, convection and clear air physical properties for

2. Advanced baseline imager.

3. Spinning enhanced visible and infrared imager.

4. Advanced Himawari Imager.

5. EUMETSAT provides a guide to interpreting common RGB combinations at http://www.eumetrain.org/RGBguide/rgbs.html.

6. https://www.nwcsaf.org/.

geostationary satellites, clouds and precipitation for low-orbit satellites. SAFNWC delivers a new version of software every two to three years, allowing users to generate nowcasting products in their area of interest. In this context, Météo-France is in charge of cloud and convection products for geostationary satellites.

Most of these instruments have a rapid scan mode, the purpose of which is to provide more frequent images, typically five minutes, on a particular domain. The frequency of images may indeed be particularly significant for detecting the initiation or rapid development of convective phenomena soon enough. A 1-min frequency, successfully tested on the GOES-14 imager, has, for example, shown a certain added value in terms of pre-convective environment analysis, convection identification and monitoring of mature storms (SRSOR[7], Schmit (2015)).

On MTG[8], which was launched in December 2022, the SEVIRI radiometer has been replaced by the FCI[9]. In addition to the fact that this new instrument has four additional channels (to achieve a configuration comparable to that of ABI), it also has a spatial resolution of 1 km at nadir for channels in the visible domain and 2 km for those in the IR domain, compared to 3 km for SEVIRI. Sampling at a 500 m resolution can also be enabled for two channels in the visible domain, thus improving the current resolution of the HRV[10] mode of SEVIRI, and at a 1 km resolution for two channels in the IR domain. As shown in Figure 5.1, such an increase in resolution will allow for, among other things, better detection of localized cooling atop convective clouds. The RSS mode[11] will also make it possible to obtain images on Europe in 2.5 minutes compared to five minutes currently. All of these characteristics in terms of spatial and temporal resolutions are particularly interesting for nowcasting applications, in particular those related to convective systems and to the early detection of fog (see section 5.3).

In the same frequency ranges, the FCI imager will be assisted, from another platform (see Chapter 5 of Volume 1), by the IRS sounder[12] whose spectral resolution is much higher. This last characteristic is of particular interest for NWP, since it will make it possible to better describe thermodynamic conditions in clear air of the initial states of models by virtue of the assimilation of numerous additional channels. For nowcasting, it will first be necessary to make the best use of this input of information in existing algorithms to characterize cloud structures. As observations are always

7. Super rapid scan operations for GOES-R.

8. Meteosat third generation.

9. Flexible combined imager.

10. High-resolution visible.

11. Rapid scanning service.

12. InfraRed Sounder.

made in the visible and near-IR domains, no new information on the internal structure of clouds is expected. However, the resulting products, known as level 2 products, will be more precise and richer, especially in clear air (see Chapter 10 of Volume 1). This will allow for a much better description of unstable air masses from which storms can develop and from which established storms draw their energy.

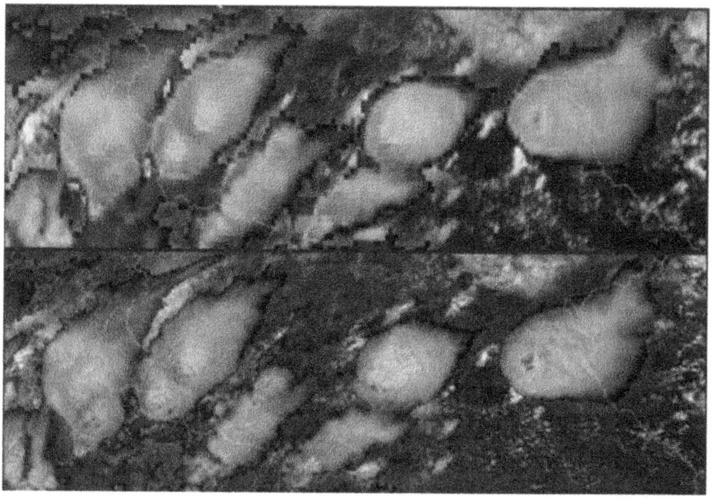

Figure 5.1. *Convective systems observed on the current MSG imagery (top), compared to an image simulation of the future MTG mission (bottom), which will allow for better visualization of cloud tops and thus better assess the intensity of thunderstorms (source: EUMETSAT (2019))*

5.2.2.3. *Onboard instruments: lightning detector*

Since 2016, the GOES-R series has been incorporating the GLM[13] (see Chapter 2 of Volume 1), which makes it possible to detect flashes due to lightning continuously, with a spatial resolution of 8 km over an area widely covering the American continent (Goodman 2013). As detailed in the next chapter, this type of instrument somehow compensates for the lack of intra-cloud information on IR imaging by helping to discriminate potentially stormy cells and warn of the likely presence of hail. The trend in electrical activity also provides information on the stage of development of the storm.

A comparable instrument, the LI, is on board the MTG. It will complete the network of European ground-based lightning sensors and will cover, with its four cameras, a large part of the disk observed by FCI.

13. Geostationary Lightning Mapper.

5.3. Observed phenomena

As we have seen in the introduction, a whole section of nowcasting is concerned with the detection of perceived weather phenomena. In its guide on nowcasting techniques (WMO 2017), the World Meteorological Organization places satellite data at the forefront of detecting and locating thunderstorms, particularly on oceans where there is no radar data. Their usefulness in characterizing the occurrence of rainfall and snowfall is also noted. The quasi-exclusive use of images in the visible and near-IR domains for alerts, however, limits the level of quantitative information on the constitution of observed clouds which, as we have seen in section 5.2.2.2, remain largely opaque at these wavelengths. Algorithms making it possible to find parameters related to the type of precipitation (size of hail, presence of icing clouds, etc.), to phenomena related to thunderstorms close to the ground (presence of tornadoes, gusts) or more generally any phenomenon of low layers (visibility, fog) exist but remain highly empirical. They also make extensive use of data from numerical forecasting.

5.3.1. *Air mass instability*

An unstable air mass tends to favor vertical or convective movements. Characterizing such conditions thus makes it possible to identify geographical zones conducive to the appearance and development of thunderstorms. Indices of instability, such as those of the SAFNWC product GEO-iSHAI[14], are generally based on the restoration of temperature and humidity profiles in clear air from comparisons between combinations of channels observed by an imager and those simulated by a radiative transfer model applied to variables predicted by NWP models. Such algorithms are very similar to what is done in satellite data assimilation systems (see Chapter 4). The relevance of the indices produced is therefore highly dependent on that of the forecasts used for their calculation.

5.3.2. *Convective systems*

Convective systems include all precipitating systems characterized by strong upward movements. Depending on the intensity of these movements, they can evolve into severe forms, typically in intense thunderstorms whose effects can be devastating: heavy precipitation, hail, gusts, lightning, tornado. They are thus the phenomenon primarily studied by nowcasting.

14. GEO imaging satellite humidity and instability.

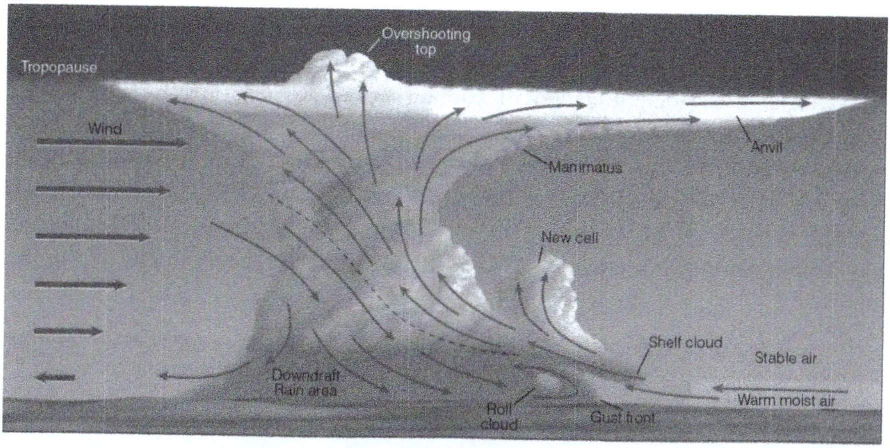

Figure 5.2. *Main characteristics of a storm cell (source: NOAA)*

As shown in Figure 5.2, a mature cumulonimbus has a very extensive cloud cover spatially at its apex, more commonly referred to as an "anvil". In the visible or near-IR domains, this anvil literally masks most underlying storm elements. Various algorithms using combinations of channels sensitive to different atmospheric layers will then try to characterize the altitude of the top of the storm from its birth to its dissipation.

5.3.2.1. *Convection initiation*

The detection of the birth of a thunderstorm from space is mainly based on the search, in several successive images, for sudden cooling of cloudy pixels in IR imaging within the same system. Such cooling is in fact the sign of a rapid vertical extension of the cloud due to localized heating, typically caused by a release of latent heat resulting from the condensation of raindrops. Where weather radars are available, exceedances of reflectivity thresholds may be used in addition (Roberts 2003; Mecikalski 2006). The CI product[15], proposed by SAFNWC and developed at Météo-France, is mainly based on IR imaging and uses an approach comparable to that used by SATCAST for GOES (Walker 2012): tracking cold pixels in a zone determined favorable in previous images, a multi-channel approach to quantify the vertical and horizontal extension of the cloud top. In Figure 5.3, it may be noted that pixels with a high probability of CI are often located upstream of mature convective cells, identified here by RDT objects (see the next section).

15. Convection initiation.

Although effective, the CI is nevertheless subject to several sources of error that sometimes generate false alarms or poor detections with respect to co-located radar data: errors in the determination of cooling linked to poor monitoring of cells, contamination by cirrus, insufficient spatial resolution to separate the signal given by the births of new cumulus at the infra-pixel level with the general cooling of the pixel. To filter out those false alarms, information from NWP, the use of new algorithms based on AI, as well as an increase in spatial and spectral resolutions of new generation imagers, such as MTG's FCI, will be particularly interesting in this context.

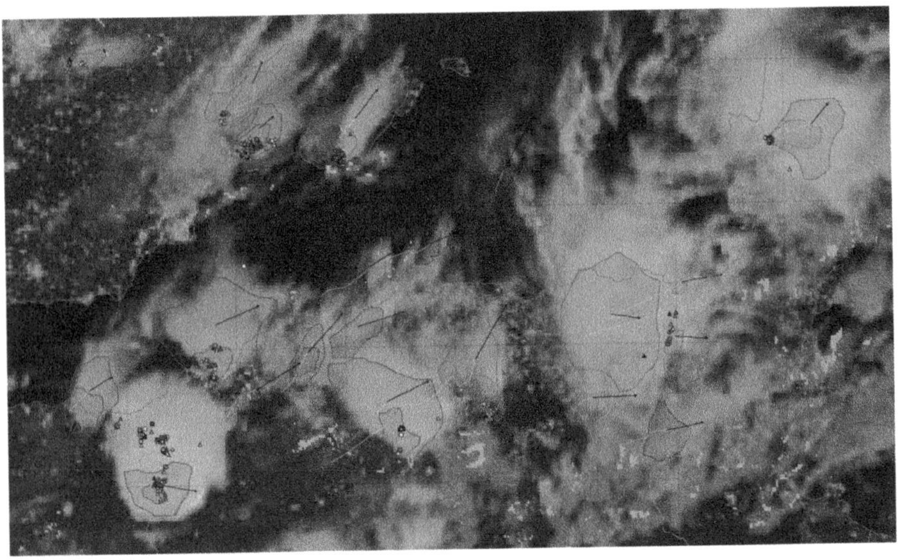

Figure 5.3. *RDT objects detected from SEVIRI channels in the western Mediterranean on September 21, 2021 at 22:00 UTC. Growth, mature and decaying cells are identified by red, purple and blue outlines, respectively. Lightning impacts accumulated over 10 min observed by the Météorage ground sensor network are represented by squares or triangles colored according to the intensities measured. The probability pixels of occurrence of CI within 30 min are marked in orange (+50%) and purple (+75%) (source: Météo-France)*

5.3.2.2. *Mature storm systems*

As illustrated in Figure 5.2, seen from space, a mature storm system will first be characterized by a cluster of high opaque clouds, clouds that are therefore cold in terms of IR brightness temperatures. An image segmentation will allow for isolating clouds of a certain spatial extension whose top is colder than a certain threshold. The tracking of the object thus detected in successive images will then provide the forecaster with the trends of some of its attributes: cooling (therefore intensification of upward movements), area, displacement vector, etc. The RDT product[16], developed at Météo-France as part of the SAFNWC and widely used worldwide, is based on this approach. An example of the visualization of this product is given in Figure 5.3. One of its peculiarities is that the thresholding is done adaptively, so as to encompass cloud cells belonging to the same system more effectively. RDT also makes it possible to follow the detected object in successive images thanks to a matching methodology, which also manages cell fusions and fissions. This monitoring then makes it possible to propose a forecast of the object's location by extrapolation (this point will be discussed in section 5.4.1).

The detection of localized cooling within a cumulonimbus, called "Overshooting Top" (OT below; see Figure 5.2), is an indicator of strong upward movements within the storm itself, movements that are sometimes intense enough to penetrate the stratosphere. It indicates the most active convective part that is often associated with large turbulence over a wide perimeter (Heymsfield 1991), high electrical activity, the presence of hail and accumulated heavy precipitation on the ground. The OT is generally co-located with maximum radar reflectivity if available (Bedka 2015).

Products such as RDT now incorporate, in addition to the detection of OT, lightning data when available (see section 5.2.2.3). This data is either provided by instruments on board the same geostationary platform as the imager (GLM, for example, for the new generation of GOES), or derived from measurements collected from the ground. Not only are they used to validate the detection of stormy objects, but they also make it possible to signal the most intense storms thanks to measured flash density. Studies such as that of Sieglaff (2013) have also shown that a sudden increase in the number of lightning jumps generally precedes the peak of thunderstorm activity by about 30 minutes, as indicated by the maximum cooling of the cloud top observed by satellite or by maximum radar reflectivity.

In this context, several studies have shown the value of a high temporal frequency of satellite observations. Decreasing to 1 min, for example, as was done in the United States in the years 2012–2016 during the SRSOR experiment, made it possible to demonstrate an acceleration in the vertical development of cumulonimbus when iced hydrometeors begin to be produced. This production is indeed accompanied by a release of latent heat which intensifies the upward movements at the heart of the system (Mecikalski 2016).

16. Rapidly developing thunderstorm.

5.3.3. *Characteristics of clouds*

5.3.3.1. *Cloud types*

Before a particular type of cloud can be determined, clear air zones must be determined with confidence. In these zones, brightness temperatures depend on soil type, atmospheric conditions and the position of the satellite and the sun, respectively. Observed values must also remain constant within a short period of time. The SAFNWC CMA[17] product (Derrien 2005) thus uses different multi-spectral approaches depending on the time of day, in order to use channels visible during the day, and analyzes the temporal consistency of results. In addition, it makes it possible to detect surfaces covered with snow or ice, as well as the presence of dust or volcanic ash in the atmosphere.

Figure 5.4. *Example of cloud types deduced from the SEVIRI imager onboard MSG, developed as part of the EUMETSAT SAFNWC*

For pixels determined as cloudy, additional multi-spectral algorithms will then provide the cloud type (fractional, semi-transparent, opaque) as well as the altitude slice in which they develop (low, medium, high). Low clouds, including fog, can thus be detected with confidence when no altitude cloud masks them. In some cases,

17. Cloud mask.

information on the base or top of clouds predicted by an NWP model is used to assist in this discrimination. An example of the restoration of cloud types is given in Figure 5.4.

5.3.3.2. *Altitude and temperature of cloud top*

For this type of information requiring a description of the thermodynamic conditions of the atmosphere, it is difficult to dispense with data predicted by NWP models. The degree of empiricism and the type of predicted data used can vary greatly depending on the nature of the cloud to be identified. We will confine ourselves here to opaque clouds. The interested reader should refer to scientific documentation available on the SAFNWC website[18] for other cloud types (semi-transparent or split).

For opaque clouds, therefore, the calculation is based on comparisons, at several wavelengths, between the radiances observed and the radiances calculated by a RTTOV-type[19] radiative transfer model using co-located temperature and humidity profiles simulated by an NWP model. The presence of dummy opaque clouds at increasing vertical levels is then imposed. The height and temperature of the cloud whose simulated radiance is closest to the observed one are then assigned to the observed pixel. It is this principle that is used in the CTTH[20] product of SAFNWC. This information, whose trend measured on successive images gives information on the stage of evolution of the cloud, can then be attributed to RDT-type objects. In the aeronautical sector, it is particularly used to determine the flight level.

5.3.4. *Hydrometeors*

Hydrometeors include all the liquid and frozen microphysical species present in the atmosphere. One part is composed of cloudy particles, another of precipitating particles. As we have seen in section 5.2.2, nowcasting algorithms that are based on satellite measurements have access only to imager measurements in the visible, near-IR and thermal IR domains. At these wavelengths, no quantitative measurement of hydrometeor content is possible from space.

Empirical algorithms, based on cloud top temperatures retrieved according to the method described in the previous section and on brightness temperatures of certain IR channels, make it possible to determine the phase of the cloud (liquid or icy).

18. https://www.nwcsaf.org/.

19. Rapid transmission for TOVs (Eyre 1991).

20. Cloud top temperature and height.

This information is all the more important as it provides information on the presence of icing clouds, which are of particular interest to the aviation sector. Other more complex methods, using among others simulations of radiances in the visible domain, allow access, by day, to more elaborate parameters such as vertically integrated water or ice contents.

In areas not covered by radars or rain gauges, the estimation of ground precipitation from space also remains very difficult from IR measurements (see Chapter 15). For convective systems, simple empirical relationships between ground rain and CTTH are sometimes used but are unreliable. More recently, AI algorithms, trained over several years in co-located radar reflectivity data and satellite measurements in the IR domain, have been implemented (COSPARIN[21]). However, because (i) the overall radar coverage is restricted and localized to continental areas, mainly mid-latitudes, and (ii) ground rain and IR brightness temperature are weakly correlated, the quantitative exploitation of rainfall thus restored remains difficult. In zones not covered by radars, however, rains thus restored remain very informative in terms of location.

5.3.5. *Wind*

The restoration of AMV[22] using IR imaging was developed as early as the 1990s (Schmetz 1993). It is based on the tracking of "tracers", generally consisting of a group of cloudy pixels, in several successive images (typically three) provided from a geostationary platform (see Chapter 2). At this stage, only the displacement of cloud structures between these images is calculated. It is then assumed that this displacement reflects the circulation of air at a certain altitude over the period of time between images. Because of the spectral width of channels used and the contamination of the signal by clouds, the determination of this altitude remains very uncertain. In the presence of wind shear (wind blowing in different directions at different altitude levels), lack of contrast in images, or underlying phenomena strongly affecting the evolution of the cloud system (relief, gust front near the ground), the restored wind can also be marred by high errors. In other cases, especially during established wind conditions, AMVs remain very informative. Moreover, they are routinely assimilated into NWP models (see Chapter 4) and used in the extrapolation of images (section 5.4).

Because of their better spectral and spatial resolutions, the new generation of imagers will allow for more accurate and more frequent estimates of this apparent

21. https://business.esa.int/projects/cosparin.

22. Atmospheric motion vectors.

wind. Certain phenomena, such as the divergence at the top of clouds which characterizes the horizontal expulsion of hydrometeors from convective towers, or the rotation phenomena linked to the presence of mesocyclones, will moreover probably be detectable thanks to the highest frequency of image acquisition.

5.4. Nowcasting of detected phenomena

Since this book focuses on satellite measurements, the main focus here will be on approaches that make it possible to predict the evolution of some of the structures detected in satellite imagery described in the previous section. Those based on imagery from ground-based weather radars will not be addressed.

5.4.1. *Method based on the tracking of structures*

This method is the one used by the RDT product. It is based on the matching of an object detected in the last available image with one (or more if merged) object detected in previous images. A combination of the displacement deduced during this step, that resulting from an overall AMV-type cloud movement analysis, and wind predicted in the mid-troposphere by an NWP model then provides a prediction of the future location of the object. This method, which is widely used in the aeronautical sector, proves to be particularly effective numerically and suitable for monitoring storm cells. A similar method is used to follow the convection initiation by CI objects which, being generally much less spatially extended, are more difficult to match.

On the other hand, this type of forecast gives no information on the future trend of the phenomenon. Only the past trend of attributes attached to the object (electrical activity, surface, top temperature, etc.) is available. For a mature system developing at sea in an established flow, however, the forecast may prove to be relevant for up to a few hours. Conversely, in the event of explosive storm development or in the presence of phenomena greatly influencing its evolution (relief, convergence of wind in the lower layers of the atmosphere, etc.), the forecast will only be exploitable for a few tens of minutes. This, however, is sufficient in many situations to prevent harsh conditions in a certain zone. The forecaster's examination of the evolution of the attributes of the object and of the environment in which they develop, for example, through the use of clear-air satellite products such as GEO-iSHAI or products derived from the NWP, will potentially enable them to confirm the criticality of the situation.

5.4.2. *Method based on image extrapolation*

This method can theoretically be applied to any image, such as cloud type, brightness temperature or ground rain rate. Its relevance depends strongly on the quality of the image initially available, on the shape and size of the structures represented, and obviously on the time range, in relation to observed weather conditions. The error of the forecast will also depend on the stage of development of the observed system. Indeed, for the same reasons as for the method based on the detection of objects, no physical trend is considered during the extrapolation.

The methodology is based first of all on a diagnosis of the movement between the last two images received. Such a displacement may be provided directly by AMV-type vectors (see section 5.3.5), or may come from image processing methods, for example, based on optical flow algorithms. The pixels of the initial image will then be displaced (or "advected") in time by applying the displacement field at regular intervals. For the same reasons as those mentioned above for object extrapolation, the relevance of such a forecast rarely exceeds one to two hours.

The SAFNWC EXIM[23] product thus offers image extrapolations for most of the cloud-related products described in the previous chapter.

5.4.3. *Method based on artificial intelligence*

Because of the amount of archived data available and their speed of execution, AI-based methods, which do not use any knowledge a priori of the physics of the atmosphere, are particularly suitable for certain nowcasting applications. Deep learning algorithms, or "machine learning", for example based on neural networks, appear to be particularly interesting for the statistical prediction of certain parameters.

The first step is to assemble a consistent database of successive observations. Co-located observations of different types can be considered at this stage, for example, observations coming from different channels of an imager or others coming from radars. Machine learning algorithms will then analyze this learning base by performing statistics and recognizing recurring patterns at different scales. Different parameters are generally adjusted during a second validation step applied to a new set of data that will serve as a reference. On the basis of a series of images, the algorithm thus calibrated will then be able to generate a probabilistic prediction. The longer the time range, the lower the probability that a pixel in the original image

23. EXtrapolated Imagery.

will be at a certain point in the image, which will actually result in blurring of the image. Various post-processings are then generally applied so as to translate this probability into binary information that is more easily interpreted.

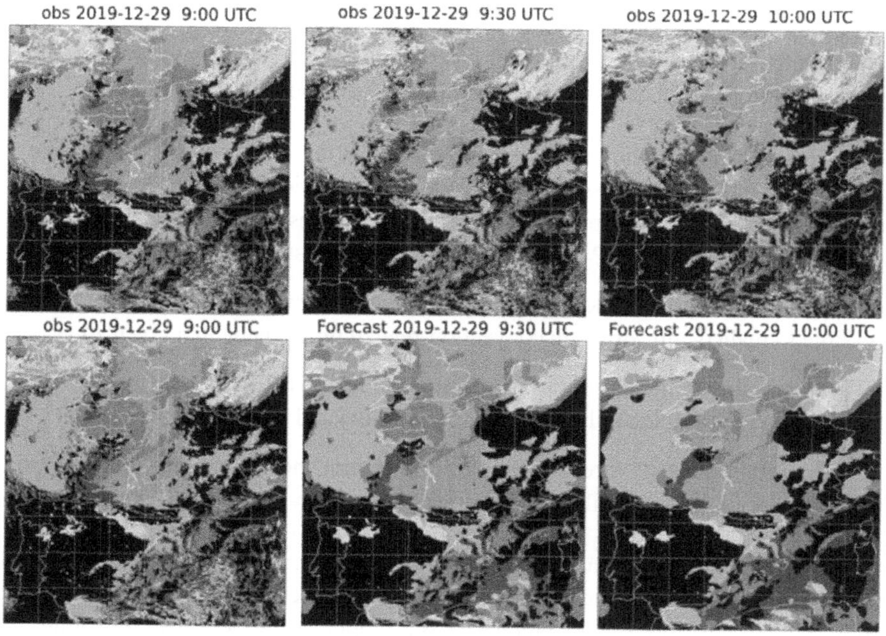

Figure 5.5. *Observations (top line) and forecasts by machine learning (bottom line) of cloud types for December 29, 2019. The statistical forecast is initialized from the observation of 09:00 UTC. Colors correspond to the types of clouds described in the legend of Figure 5.4 (source: L. Berthomier, Météo-France)*

The more the structures to be expected are present statistically in the learning base, the more effective this type of method is. For a simple cloud mask, the results are, for example, much better than those obtained from the image extrapolation methods described in the preceding section. In Figure 5.5, where the method is applied to cloud types, it can be seen that predicted low clouds (yellow-orange) remain well stationary, while the highest clouds (blue-green) follow the observed flux on a larger scale. Conversely, very localized structures, characterized, for example, by very intense values, will tend to fade rapidly during the forecasting due to the under-representation of such structures in the learning base. The increase in the contrast of the predicted image, for example, by histogram preservation methods, makes it possible in some cases to partially compensate for this defect.

5.4.4. *Use of numerical forecasting*

The first three methods described above only produce image forecasts; they do not make it possible to predict the evolution of the systems observed because of internal physical phenomena and thermodynamic characteristics of the environment in which these systems develop. After a certain deadline, typically after one to three hours, the information provided by NWP models becomes more relevant.

In this context, satellite data are mainly used for the search for the initial state of the forecast through data assimilation. As the assimilation of satellite data in NWP models is covered in Chapter 4, only the specificities related to nowcasting are described here. For an NWP model to be used in nowcasting, it is necessary that:

– it has a kilometer-scale resolution and physical parameterizations that make it possible to realistically simulate convection-related phenomena in a given zone;

– despite the numerical cost of such models, forecasts up to a time range of six hours are available frequently (typically every hour) and as quickly as possible;

– the initial state of forecasts takes into account the information derived from observations via data assimilation algorithms.

These prerequisites make it necessary to retain, in the assimilation, only observations collected in a very short period of time, typically of the order of 10 minutes. For more conventional NWP applications, the waiting time for observations can be much longer because target deadlines are farther away and the need for forecasters is less urgent. For nowcasting applications, on the other hand, extremely short data reception, processing and provision times are required. A vast majority of satellite measurements will therefore unfortunately not be available in time.

For some of them, however, such as the radiances of geostationary imagers in Rapid Scan mode, this will be possible. IR measurements will then be able to be assimilated, which will have the main impact of improving the initial thermodynamic description in clear air, in particular the average tropospheric humidity (Montmerle 2007). Some measurements of radiometers on polar-orbiting platforms could also be taken into account. However, for the same reasons as explained in section 5.2.1, their low spatial coverage and the fact that they are only occasionally available make them less attractive for nowcasting applications.

5.4.5. *OBS-NWP fusion*

In order to provide forecasters with the best possible prediction as frequently as possible, methods of merging the predictions of extrapolated images or images derived from AI algorithms with images simulated by NWP models have recently appeared. In this type of approach, it is sufficient for the model to be capable of simulating an image of interest (such as radar reflectivities or brightness temperatures) for such a fusion to be possible. The resulting forecast is a combination of these different solutions to which time-dependent weights are attached. More weight will thus be given in the first hours to the extrapolated images and then progressively favor the solution given by the NWP. In most algorithms, assigned weights are derived from statistics on the recent relevance of each of these predictors to what has actually been observed.

As the predictability of a meteorological phenomenon depends on its spatial scale, the transition between image extrapolation and NWP will be all the more relevant and continuous as the observed phenomenon is large scale. This is the case for rain fronts, squall lines and other medium-scale convective systems. For isolated short-lived storms, the fusion solution will be less relevant.

5.4.6. *Probabilistic forecast*

The marked spatial location of a thunderstorm as well as the complexity of the physical phenomena involved at different stages of its evolution, some of which are highly nonlinear, explain that errors present in the initial conditions or in the physical parameterizations of the model will amplify with the time range. The purpose of probabilistic forecasting is to take into account these various sources of error by generating a set of forecasts, each of whose members is obtained by artificially disturbing certain parameters. Conventionally, this relates to errors attributed to observations in data assimilation algorithms and those related to different variables used in the physics of an NWP model. Disturbances may also be applied to the extrapolation or fusion methods mentioned above.

The set of predictions produced will then provide a panel of possibilities that can be translated into the form of probabilities, for example, the probability that a thunderstorm will strengthen or that another remarkable weather phenomenon will develop at a certain place in the coming hours. In a manner comparable to that produced for cyclone trajectories, this type of method can also propose a range of preferred trajectories that thunderstorms will take at nowcasting time ranges.

5.5. Perspectives

In the future, nowcasting will benefit from advances in the large observation systems it uses. Switching the SEVIRI radiometer from MSG to MTG FCI will improve spectral, horizontal and temporal resolutions. The advent of the LI lightning detector will provide new information which, for convection products in particular, will be used in adjustment, real-time processing and verification phases. AI will also take an increasingly prominent place, whether in the algorithms for restoring unobserved parameters, detecting structures or statistical forecasting. Finally, the increase in the means of calculation makes it possible to consider forecasts in a probabilistic form, by taking into account various sources of error in ensembles of forecasts.

5.6. References

Bedka, K.M., Wang, C., Rogers, R., Carey, L.D., Feltz, W., Kanak, J. (2015). Examining deep convective cloud evolution using total lightning, WSR-88D, and GOES-14 super rapid scan observations within deep convective clouds. *Weather Forecasting*, 30, 571–590. doi: 10.1175/WAFD14-00062.1.

Derrien, M. and Le Gleau, H. (2005). MSG/SEVIRI cloud mask and type from SAFNWC. *International Journal of Remote Sensing*, 26, 4707–4732.

EUMETSAT (2019). Meteosat Third Generation, the case for preparing national users [Online]. Available at: https://www-cdn.eumetsat.int/files/2020-04/pdf_mtg_info-pack_v12.pdf.

Eyre, J. (1991). A fast radiative transfer model for satellite sounding systems. ECMWF Res. Dep. Tech. Mem 176. ECMWF, Reading.

Goodman, S.J., Blakeslee, R.J., Koshak, W.J., Mach, D., Bailey, J., Buechler, D., Carey, L., Schultz, C., Bateman, M., McCaul Jr., E. et al. (2013). The GOES-R geostationary lightning mapper (GLM). *Atmospheric Research*, 125–126, 34–49. https://doi.org/10.1016/j.atmosres.2013.01.00.

Heymsfield, G., Fulton, R., Spinhirne, J. (1990). Aircraft overflight measurements of midwest severe storms: Implications on geosynchronous satellite interpretations. *Monthly Weather Review*, 119, 436–456.

Mecikalski, J.R. and Bedka, K.M. (2006). Forecasting convective initiation by monitoring the evolution of moving cumulus in daytime GOES imagery. *Monthly Weather Review*, 134, 49–78. doi: 10.1175/MWR3062.1.

Mecikalski, J.R., Jewett, C.P., Apke, J.M., Carey, L.D. (2016). Analysis of cumulus cloud updrafts as observed with 1 min resolution super rapid scan GOES imagery. *Monthly Weather Review*, 144, 811–830. doi: 10.1175/MWRD14-00399.1.

Montmerle, T., Rabier, F., Fischer, C. (2007). Respective impact of polar-orbiting and geostationary satellites observations in the Aladin/France numerical weather prediction system. *Quarterly Journal of the Royal Meteorological Society*, 133, 655–671.

Roberts, R.D. and Rutledge, S. (2003). Nowcasting storm initiation and growth using *GOES-8* and WSR-88D data. *Weather Forecasting*, 18, 562–584.

Schmetz, J., Holmlund, K., Hoffman, J., Strauss, B., Mason, B., Gärtner, V., Koch, A., van de Berg, L. (1993). Operational cloud motion winds from meteosat infrared images. *Journal of Applied Meteorology*, 32, 1206–1225.

Schmit, T.J., Goodman, S.J., Gunshor, M.M., Sieglaff, J., Heidinger, A.K., Bachmeier, A.S., Lindstrom, S.S., Terborg, A., Feltz, J., Bah, K., et al. (2015). Rapid refresh information of significant events: Preparing users for the next generation of geostationary operational satellites. *Bulletin of the American Meteorological Society*, 96, 561–576. doi:10.1175/BAMS-D-13-00210.1.

Sieglaff, J.M., Hartung, D.C., Feltz, W.F., Cronce, L.M., Valliappa, L. (2013). A satellite-based convective cloud object tracking and multipurpose data fusion tool with application to developing convection. *Journal of Atmospheric and Oceanic Technology*, 30, 510–525. doi: http://dx.doi.org/10.1175/JTECHD12-00114.1.

Walker, J.R., MacKenzie, W.M., Mecikalski, J.R., Jewett, C.P. (2012). An enhanced geostationary satellite-based convective initiation algorithm for 0–2-h nowcasting with object tracking. *Journal of Applied Meteorology and Climatology*, 51, 1931–1949. doi:10.1175/JAMC-D-11-0246.1.

World Meteorological Organization (2017). Guidelines for nowcasting techniques Report, WMO, no. 1198, Geneva.

6

Observation and Monitoring of Tropical Cyclones from Space

Frank ROUX
Laboratoire d'Aerologie, Université de Toulouse, CNRS, France

6.1. Introduction

Tropical cyclones spend most of their lives over oceans where conventional weather measurements are rare. Also, satellite observation is the essential tool allowing forecasters of the regions concerned to identify emerging disturbances, determine their position, follow their trajectory, examine their structure, characterize their environment, estimate the intensity of rainfall and winds, and understand their possible evolution.

While numerical weather prediction models, on a global or regional scale, are now able to represent cyclones and the trajectories they follow fairly accurately, meteorological satellites provide essential information to validate analyses, refine intensity and control projections. Similarly, satellite observation, particularly from geostationary orbit, is the only one that can identify and monitor cyclones throughout the globe, and thus produce reliable statistics on possible changes in their frequency or intensity in relation to natural climate variability or anthropogenic global warming.

For a color version of all the figures in this chapter, see www.iste.co.uk/phulpin/satellites2.zip.

Satellites for Atmospheric Sciences 2,
coordinated by Thierry PHULPIN, Didier RENAUT, Hervé ROQUET and Claude CAMY-PEYRET.

6.2. Visible and infrared imagery

The era of space observation of tropical cyclones began modestly with the American experimental satellite TIROS-1. Its rudimentary scientific equipment (two infrared television cameras, wide angle and zoom) and its short lifespan (78 days) nevertheless made it possible to obtain, on April 10, 1960, the first image from space of a tropical cyclone of unknown existence over the South Pacific. During the 1960s, polar-orbiting weather satellites provided more accurate observations, both visible and infrared, with horizontal resolutions of 5–10 km. Such images identify cyclones over ocean regions that are not very accessible to meteorological measurements, except by aircraft dedicated to such missions over the North Atlantic and the Northeast Pacific.

Beginning in the 1970s, geostationary orbiting meteorological satellites provided good space coverage and much better time tracking (Figure 6.1). Forecasters were therefore able to better define cyclone trajectories and estimate their characteristics, first on the Atlantic, then on the Pacific. For the Indian Ocean, it was not until the late 1990s that effective satellite surveillance was conducted.

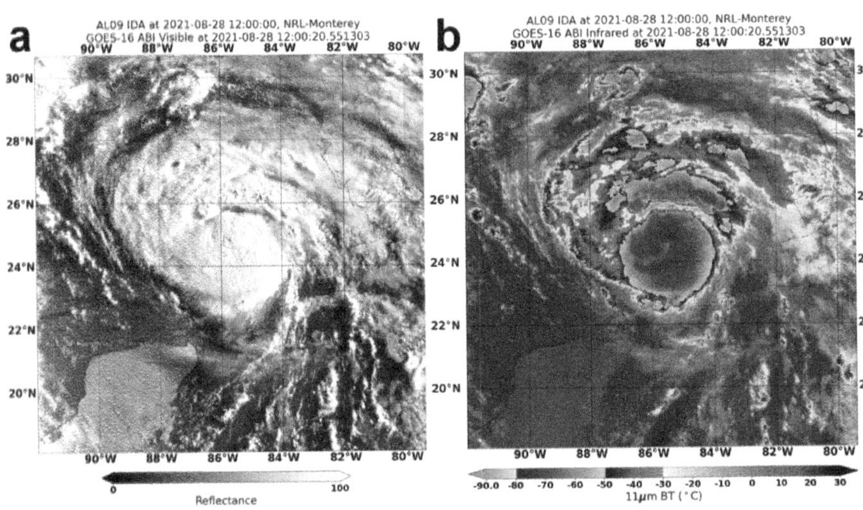

Figure 6.1. *Visible (a) and infrared (b) images of tropical cyclone (Hurricane) Ida taken by geostationary satellite GOES-16 on August 28, 2021, at 12:00 UTC (source: United States Naval Research Laboratory*[1]

1. https://www.nrlmry.navy.mil/tc_pages/tc_home.html.

In 1975, an important advancement was the publication by Vernon Dvorak of an objective method of analyzing satellite imagery by recognizing specific shapes to quantify the intensity of cyclones and predict their evolution (Figure 6.2). This technique has been improved several times (Dvorak 1984; Olander and Velden 2007) to make better use of observations and reduce the role human judgment plays in favor of automated numerical methods. It is currently an essential tool for monitoring cyclones at all points in the tropical belt.

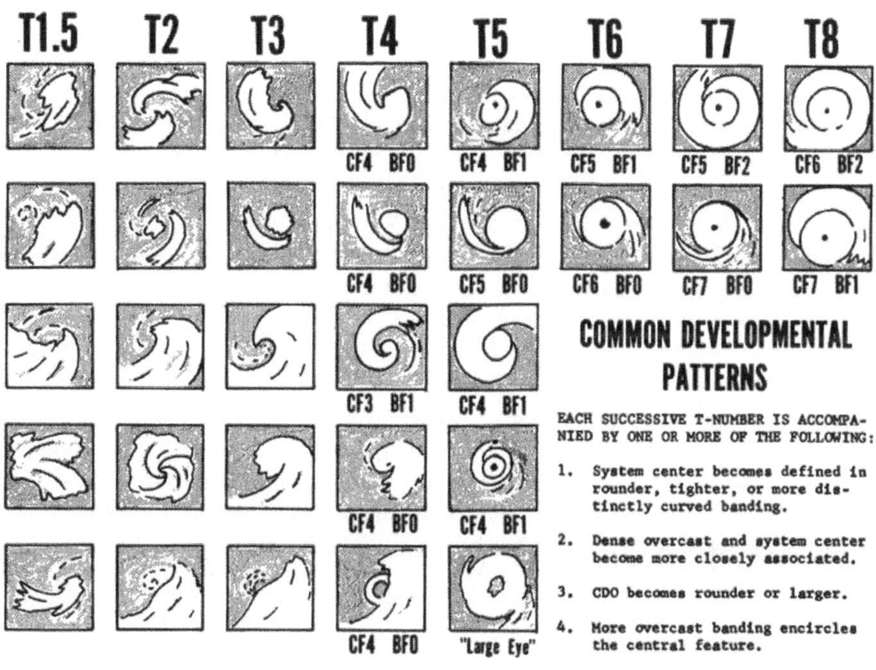

Figure 6.2. *Characteristic cloud structures for classifying tropical cyclones from intensity levels T1–T8 using the Dvorak technique (source: Dvorak (1973))*

Tracking cloud structures on visible and infrared images taken by satellites in geostationary orbit also provides information on the strength and direction of winds associated with tropical cyclones. For accuracy, quick scan modes (< 5 minutes) can be activated on specific areas. Algorithms combining these "satellite winds" and the recognition of characteristic cloud structures provide an estimate of the intensity of a cyclone and the radial distribution of the maximum wind up to 200 km from the center in the absence of in situ measurements (Kossin et al. 2007). Similarly, the distribution of wind shear[2] – an inhibitory parameter of cyclonic development – that can thus be deduced in the nearby environment is useful for predicting evolution.

Methods based on cloud top temperature (derived from thermal infrared measurements) also provide estimates of precipitation or built-in rainfall rates associated with cyclones. These values are subject to significant uncertainty but are useful in the absence of other more accurate measurements.

Measurements in "water vapor" channels (6–7 μm wavelengths) provide information on humidity in the middle and upper troposphere and the presence of dry areas of altitude unfavorable to cyclonic activity. Large zones of dry dust-laden air from the Sahara sometimes cross the tropical Atlantic Ocean from east to west. They are evidenced by the differential analysis of thermal infrared channels at 10.3 and 12.3 μm, sensitive to the presence of such air masses in the lower and middle troposphere (Dunion and Velden 2004). Measurements of aerosol content by space lidar (CALIPSO/CALIOP since 2006) are more accurate but less frequent. Cyclones caught in such air masses usually weaken and then struggle to regain intensity when they leave.

Modern Sun-synchronous orbit weather satellites offer better spatial resolution than geostationary ones, but at the expense of lower temporal resolution. For tropical cyclones, images with kilometer or hectometric resolution in visible and infrared domains can identify fine structures within the eye, spiral bands or cirrus cloud cover. Some provide insights into the development and evolution of a cyclone, and help refine short-term analysis and forecasts.

6.3. Microwave imaging

Microwave radiometers on low-orbit weather satellites measure the brightness temperature of the atmosphere at frequencies between a few GHz and a few hundred GHz. The first observations were made with the SMMR instrument on board the Seasat satellite in 1978 and Nimbus-7 from 1978 to 1987, then from 1987 with SSM/I aboard the DMSP satellites of the U.S. Department of Defense. Since then, many microwave imagers have flown and are now flying on several satellites (e.g. in 2021: DMSP/SSMIS, Aqua/AMSR-E, GCOM/AMSR-2, GPM Core Observatory/GMI, Feng-Yun 3/MWRI, Meteor M2/MTVZA-GY).

Channels used for tropical cyclone imagery (e.g. 37 and 85–91 GHz) correspond to windows for the transmission of surface thermal radiation by atmospheric gases.

2. Wind shear: the variation in wind strength or direction with altitude.

Microwave radiation interacts with clouds and liquid or icy precipitation by absorption–re-emission or scattering (see Chapter 9 of Volume 1). At 37 GHz, the ocean surface has a low emissivity[3] (≈0.4–0.5), so it appears "cold" in microwave imaging. Rain drops that absorb and re-emit surface radiation have a higher emissivity (≈0.9) and appear "warmer". Diffusion is negligible for small cloud droplets (with diameters from 1 to 100 μm); it becomes more important for precipitating droplets. At higher frequencies, diffusion by ice particles dominates that by raindrops. A high concentration of ice then results in brightness temperatures at 85–89 GHz "colder" than that associated with the ocean surface.

Tropical cyclones are characterized by large precipitating zones, more or less organized in spiral bands around the eye (Figure 6.3). The 37 GHz imagery gives an overview of rainfall areas with a horizontal resolution of about 30 km and helps determine the position of the cyclone center. This information is often difficult to obtain in visible or infrared because of the frequent presence of a thick veil of cirrus clouds in the upper troposphere. In the eyewall[4] and within spiral bands, regions of strong convection peak at very high altitude (15–18 km) with temperatures of –50 to –80°C. They contain large ice particles over thicknesses of several kilometers that result in relatively cold brightness temperatures at 85–91 GHz. At these frequencies, a spatial resolution of about 10 km gives a fairly accurate picture of the most active regions.

As several satellites in low orbit are now equipped with imaging microwave radiometers, successive observations of the same cyclone provide a more or less complete temporal monitoring of the organization of precipitating bands depending on the position of the cyclone in relation to the orbits of satellites. Forecasters can thus identify phases of rapid change in intensity, which are difficult to represent by numerical models. The "eyewall replacement cycle" is thus marked by the gradual organization of precipitating bands into an external eyewall that leads to the tightening, then the disappearance of the old eyewall (weakening phase), until only the well-defined external eyewall remains (intensification phase).

3. Emissivity: the ratio between the power of thermal radiation emitted by a surface element at a given temperature, and that which a black body would emit at the same temperature.

4. Eyewall: a roughly circular ring of strong rain–cloud convection around the calm central area (the eye) of a cyclone. It is the region with the strongest surface winds and the most intense rainfall.

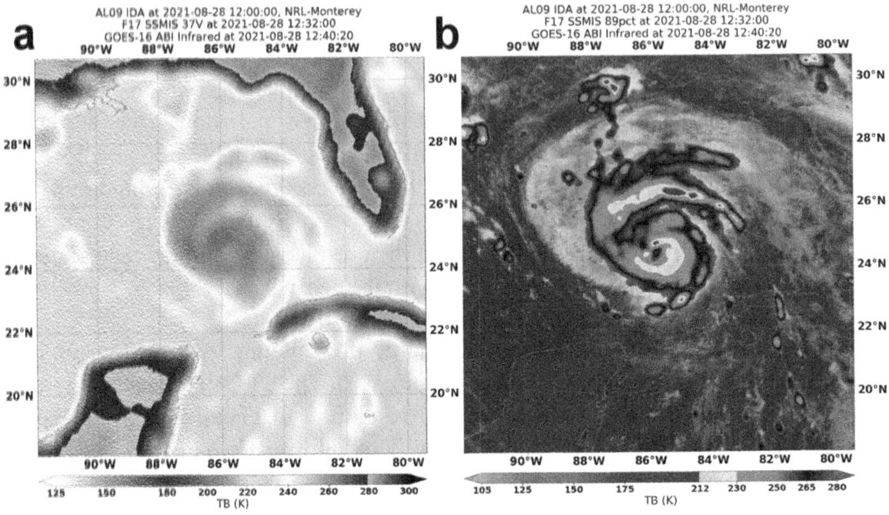

Figure 6.3. *Images of brightness temperatures in the 37 GHz (a) and 89 GHz (b) microwave channels of tropical cyclone Ida taken by the polar-orbiting satellite DMSP F-17 on August 28, 2021, at 12:32 UTC (source: United States Naval Research Laboratory*[5]

Microwave measurements at 37 and 85–91 GHz, in conjunction with geostationary infrared imagery that gives the altitude of cloud tops, are used for quantitative estimation of rainfall (precipitation and cumulative rates) (see Chapter 15). This makes it possible to identify areas at risk of flooding. More accurate algorithms for precipitation restoration use complementary measurements by weather radars in Ku (12–18 GHz) or W (75–110 GHz) band on board satellites (TRMM/PR 1997–2015, GPM Core Observatory/DPM since 2014, CloudSat/CPR 2006–2020). Due to orbital constraints, not all cyclones can be properly observed by a space-based radar, but where possible the information provided on the vertical structure of precipitation helps to refine the intensity analysis and can be compared with the results of the most accurate numerical models. Similarly, space-based information on the number and frequency of lightning strikes is complementary information on the internal activity of cyclones. The deep convective developments ("Hot Towers") identified in this way are considered an essential factor in the initial organization of cyclones and, at maturity, in the initiation of intensification phases. But their exact role is still debated, and their numerical representation remains imperfect.

5. https://www.nrlmry.navy.mil/tc_pages/tc_home.html.

6.4. Microwave sounding

The MSU instrument, with four channels between 50 and 60 GHz, on board the American weather satellite TIROS-N in 1978 was the first microwave space sounder of temperature in the troposphere and lower stratosphere. But its horizontal resolution of about 250 km was far too wide for studying tropical cyclones. The improved accuracy of its successor AMSU from 1998 made it much more useful. AMSU-A, with 15 channels between 23.8 and 89 GHz, gives six to eight characteristic temperature levels in the troposphere, with a horizontal resolution of 45 km at nadir. AMSU-B, with five channels between 89 and 183 GHz, gives three to four characteristic levels of humidity in the troposphere, with a horizontal resolution of 15 km at nadir. AMSU-A and B have been carried on many satellites in polar orbit, and newer microwave sounders (NOAA and MetOp/MHS since 2005, Suomi-NPP/ATMS since 2011, Megha-Tropics/SAPHIR since 2011) continue these missions.

When obtained at the center of tropical cyclones, vertical temperature and humidity profiles derived from microwave soundings help characterize the "warm core" of disturbances (Kidder et al. 2000; Figure 6.4). The intensity and extension of the central temperature anomaly determine the minimum surface pressure by the "hydrostatic equilibrium"[6] and the radial and vertical distributions of wind by the "thermal wind equilibrium"[7]. This information complements that which can be deduced from the analysis of visible, infrared and microwave cloud structures.

Nevertheless, the thermodynamic information deduced from microwave soundings is tainted by uncertainty due, on the one hand, to the contamination of the microwave signal by the attenuation due to liquid water and by the diffusion due to ice, and, on the other hand, to the horizontal resolution of a few tens of kilometers. Thus, this information can only be really useful for cyclones well positioned with respect to the satellite orbit and possessing a sufficiently wide eye without cloud cover at altitude.

6. Hydrostatic equilibrium: the steady state of the atmosphere when the force of gravity (directed downwards) is counterbalanced by the pressure force due to a decrease in pressure with altitude.

7. Thermal wind equilibrium: the steady state of the atmosphere when a relationship is established between vertical wind shear and horizontal temperature variation in the direction perpendicular to the wind.

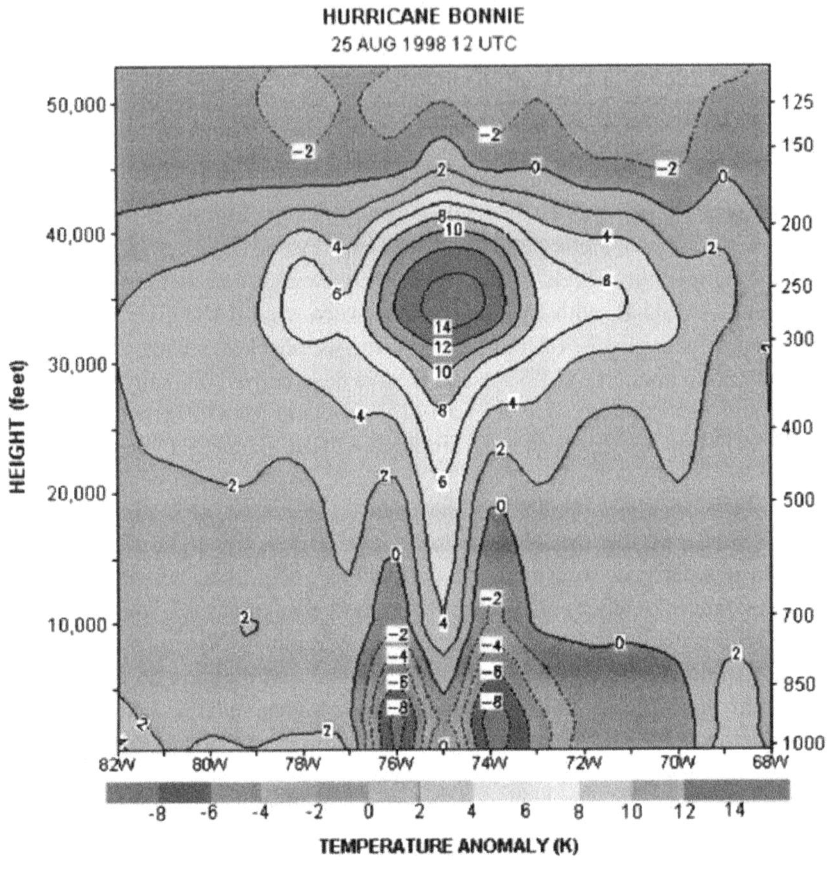

Figure 6.4. *Vertical cross-section of the temperature anomaly associated with the tropical cyclone (hurricane) Bonnie on August 25, 1998 at 12:00 UTC (source: Kidder et al. (2000))*

6.5. Surface wind measurements

In the presence of wind, the microwave emission of the ocean surface depends on its roughness, due to waves and the presence of foam and spray. Empirical relationships link surface observations by a C-band radiometer (4–8 GHz) or X-band radiometer (8–12 GHz) with wind blowing there. Comparisons (Zhang et al. 2018) between in situ observations and winds restored from measurements by imaging radiometers (DMSP/SSMIS, Aqua/AMSR-E, GCOM/AMSR-2) and by the CORIOLIS/WINDSAT polarimetric radiometer show a good correlation for intensities of less than about 20 m s^{-1} environ. Beyond that, the saturated microwave signal and the retrieved wind become not very significant. Application to tropical cyclones is difficult because of this limitation: pollution of the C- or X-band microwave signal by precipitation, and the low spatial resolution (a few tens of kilometers).

The L-band measurement (1–2 GHz) is less affected by hydrometeors. Such polarimetric space radiometers, originally designed to observe soil moisture and ocean salinity on board SMOS since 2009 and SMAP since 2015, provide estimates of strong winds (> 30 m s^{-1}), in good agreement with independent measurements (Reul et al. 2012; Meissner et al. 2017). In addition, polarimetric measurements make it possible to restore wind direction. However, spatial resolution (40 km for SMOS, 25 km for SMAP) is not precise enough to reveal the fine distribution of winds and their maximum intensity in tropical cyclones.

Scatterometers are radars that measure the cross-section of backscattered electromagnetic waves by the ocean surface. Ku-band spatial observations at an incidence of 30–60° (QuikSCAT 1999–2018, Oceansat-2/SCAT since 2009) have good sensitivity to surface wind velocity. Measurements in at least two different azimuths are necessary to determine its direction but ambiguities nevertheless require human or automatic corrections. Scatterometer observations of tropical cyclones have been a notable advance in identifying the center of circulation and the distribution of surface wind (up to about 25 m s^{-1}) with a horizontal resolution of 10–25 km (Chavas and Emanuel 2010). Comparisons between estimated winds and in situ measurements by aircraft, averaged at the same spatial resolution, indicate mean errors and bias in clear air of a few m s^{-1} in speed and 10–20° in direction, and higher values in the presence of precipitation. C-band scatterometers on European satellites (AMI on ERS-1 1991–2000 and ERS-2 1995–2011, ASCAT on MetOp-A 2006–2021, MetOp-B since 2012, and MetOp-C since 2018) are slightly less affected by precipitation and saturate for slightly stronger winds. Scatterometers provide prediction centers with valuable information on tropical cyclones occurring in basins not covered by in situ airborne measurements (North Atlantic and Northeast Pacific). However, rain contamination of the signal, poor performance for winds above 30 m s^{-1} and low spatial and temporal coverage due to orbital constraints somewhat limit the operational use of scatterometer measurements (Figure 6.5).

C-band synthetic aperture radars (SARs) with polarization diversity and multiple modes of operation (on Envisat 2002–2012, Radarsat-1 1995–2013, Radarsat-2 since 2008, Sentinel-1A since 2014, Sentinel-1B since 2016) produce wind fields at kilometric resolution with relatively wide coverage (about 500 km x 500 km) and perform best in strong winds (Li et al. 2013; Jackson et al. 2021; Figure 6.6). SAR data gives fairly accurate information on the most active central region of cyclones: maximum wind intensity and radius, asymmetry, extension of winds to 17, 25 and 33 m s^{-1} (34, 50 and 64 kts). However, SAR data adapted to tropical cyclones can only be acquired for relatively short durations due to available power and data storage constraints, and observation areas need to be scheduled a few days in advance. Also, Sentinel-1 and Radarsat-2 satellites can only provide incomplete coverage of overall cyclonic activity. However, the operational value of these measures encourages space agencies to complement surveillance with new satellites equipped with SAR.

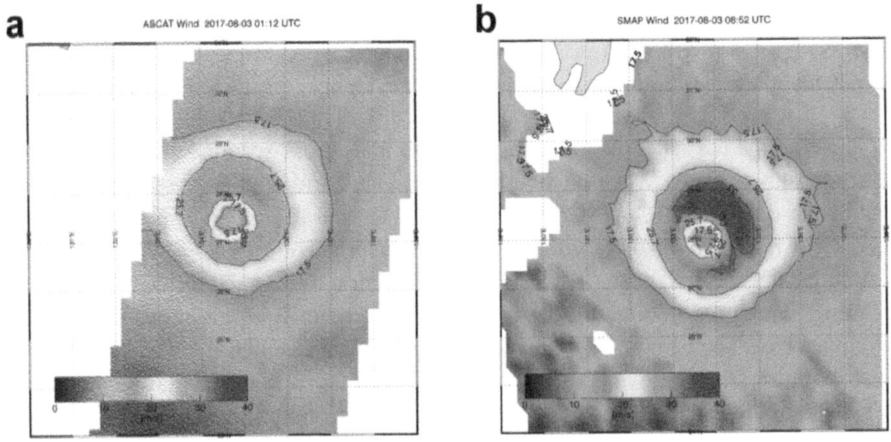

Figure 6.5. *Surface winds for the tropical cyclone (Super Typhoon) Noru on August 3, 2017, deduced from measurements by (a) the ASCAT scatterometer at 01:12 UTC and (b) the SMAP radiometer at 08:52 UTC (source: Kidder et al. (2000); Ricciardulli and Meissner (2017)*[8]

CYGNSS is a set of eight NASA micro-satellites, in orbit since 2016, using variations in the L-band GPS signal reflected by the ocean surface to determine wind intensity in four different directions with a horizontal resolution of 25 km (Ruf et al. 2016). For winds of less than about 20 m s^{-1}, the accuracy is a few m s^{-1}. CYGNSS has several advantages: the orbital displacements of the CYGNSS and GPS constellations make it possible to, in a few hours, almost completely monitor the tropical belt where cyclones spend most of their life; the L-band signal is hardly affected by precipitation; and passive receivers limit energy consumption and

8. https://www.remss.com/blog/smap-winds-super-typhoon-noru/.

instrumental fragility. Spatial and temporal resolutions make CYGNSS complementary to passive and active microwave surface wind observations. Despite limitations in wind intensity, the frequency and coverage of CYGNSS data make them interesting for assimilation in large- and medium-scale numerical weather forecasting models.

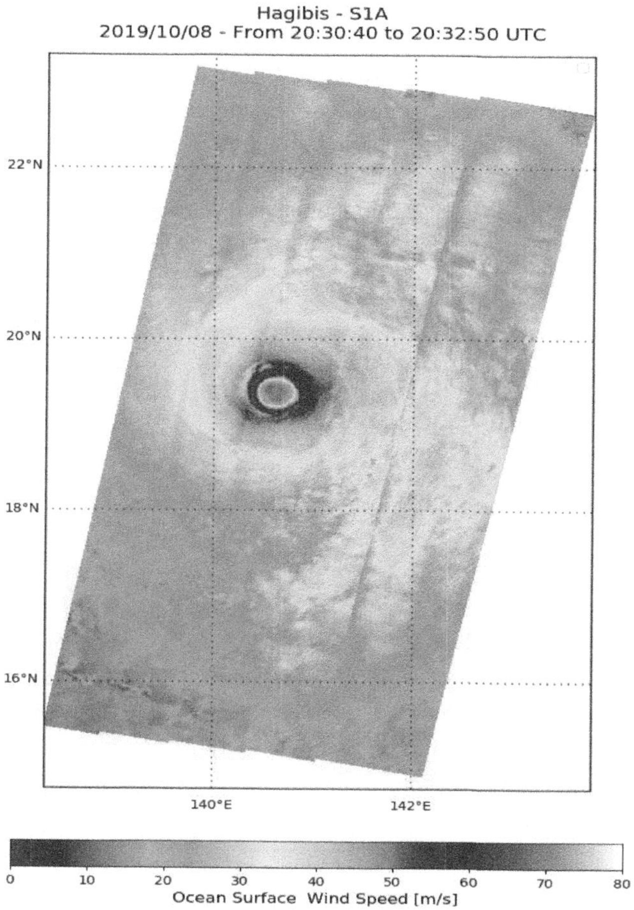

Figure 6.6. *Surface winds for tropical cyclone (typhoon) Hagibis on October 8, 2019, from 20:30 to 20:32 UTC, derived from measurements by Sentinel-1 synthetic aperture radar (source: ESA[9])*

9. https://www.esa.int/ESA_Multimedia/Images/2019/10/Ocean_surface_wind_speed_of_Typhoon_Hagibis_derived_from_Copernicus_Sentinel-1.

6.6. Ocean parameters

Sea surface temperature (SST) is an important parameter in the activity of tropical cyclones, atmospheric disturbances that draw their energy from heat accumulated by ocean surfaces during summer. Satellite observations of the thermal radiation emitted by the surface make it possible to map SST with good (kilometric) spatial and temporal resolutions. Infrared radiometric measurements (approximately 3.7 μm and/or 10 μm) are the oldest and most accurate, but they are contaminated by the presence of clouds and by atmospheric diffusion by water vapor and aerosols (Legeckis 1975). More recent microwave measurements (around 7 GHz) are less affected by clouds and the atmospheric correction is simpler, with poorer spatial resolution (> 10 km) and lower accuracy (Wang and Vesecky 1999). In the vicinity of tropical cyclones, the influence of wind on surface roughness and the presence of precipitation must be taken into account by means of measurements in other microwave channels.

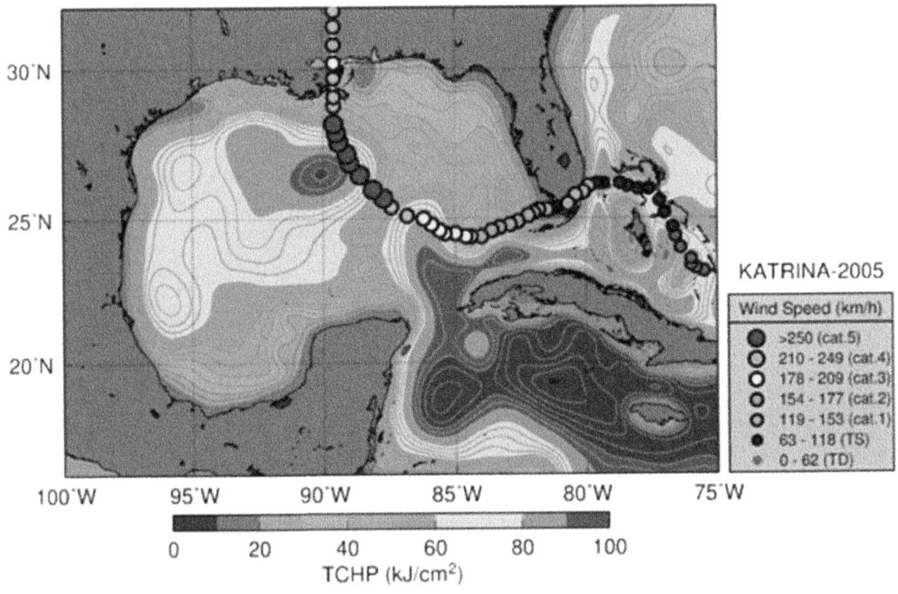

Figure 6.7. *Trajectory and intensity of the tropical cyclone (Hurricane) Katrina over the Gulf of Mexico on August 20, 2005, and distribution of the Tropical Cyclone Heat Potential derived from altimetric measurements (source: NOAA/AOML/PHOD*[10]*)*

10. https://www.aoml.noaa.gov/phod/gom/figures/Katrina_zoom.png.

More significant than surface temperature, the energy available for tropical cyclones can be quantified as the integral of the temperature from the surface to the depth of the 26°C isotherm. High values are often associated with intensification phases (Figure 6.7). Low values attenuate ocean–atmosphere fluxes of sensitive and latent heat, and can cause a reduction in intensity. Changes in ocean surface height measured by in-orbit altimeter radars can be analyzed as tracers of this thermal content of the surface ocean (Shay et al. 2000). TCHP (Tropical Cyclone Heat Potential) maps are created daily for different cyclone basins from spatial altimetry and SST data to identify areas in which ocean conditions can cause rapid changes in intensity, which are difficult to predict without this information.

Tropical cyclones also generate powerful waves, which are dangerous for navigation and coastal regions. Cyclonic swells can also spread hundreds of kilometers away. Waves, which sometimes exceed 10 meters, are measured by altimeter radar when a satellite equipped with them overflights the concerned zone. These data are assimilated into near real-time in coupled ocean–wave–atmosphere models. However, the use of these observations is more difficult for low-intensity cyclones because the wave spectrum they generate is less intense and more complex.

6.7. Climatology of cyclones

In a context of proven global warming since the beginning of the 20th century, and the prospects of an amplification of this phenomenon during the 21st century, it is important to know the influence this can have on cyclonic activity. The IBTrACS archive (Knapp et al. 2010) gathers in open access standardized information on tropical cyclones observed throughout the world since 1900. Nevertheless, inhomogeneities between analytical techniques and methodological changes over time induce uncertainties in possible statistics and trends that could be deduced.

During the 1980s, the Specialized Regional Meteorological Centers and the Tropical Cyclone Warning Centers – responsible for cyclone monitoring and forecasting on the various basins – all converted to the Dvorak technique (see section 6.2) to estimate the intensity of cyclones from the objective analysis of images in the visible and infrared channels taken by meteorological satellites in low and geostationary orbits. The Dvorak method has remained relatively stable over the decades, and statistical results obtained from this information over the past 35 years are reliable enough to identify trends (Kishtawal et al. 2012).

A first conclusion concerns the influence of natural climate variability. On a multidecadal scale, cyclonic activity in the North Tropical Atlantic is strongly modulated by the Atlantic multidecadal oscillation. Hurricanes were less frequent during the cold period from 1965 to 1998, and more frequent during the recent warm period since 1995. On a multi-year scale, the ENSO (El Niño Southern Oscillation) has a major influence on cyclones in the Atlantic and North Pacific. During the El Niño phases (e.g. 1997 or 2015), cyclones are less numerous than the average on the Atlantic but more numerous on the Pacific. A reverse situation is observed during the La Niña phases (e.g. 1998 or 2008).

Trends over longer periods also emerge from these series. Thus, the overall number of tropical storms and cyclones has remained almost stable in recent decades (or even in the past century, if IBTrACS is to be reckoned with). On the other hand, the proportion of intense events – cyclones of category 3 or more, with maximum winds greater than 180 km/h – has increased by about 5% per decade. The latitude at which cyclones reach their maximum intensity also appears to have increased, consistent with the observed expansion of the tropical meteorological domain (Kossin et al. 2014). More recently, aggregate precipitation data derived from the TRMM and GPM space missions show an average increase in cyclonic rainfall of about 1.3 per cent per year for the 1998–2016 period, more pronounced in the North Atlantic and the Northwest Pacific (Guzman and Jiang 2021). This effect affects all categories of cyclones and is consistent with the overall trend of warming of ocean surface temperatures and increasing atmospheric humidity.

6.8. Conclusion

Observation by meteorological satellites is the only way to track cyclonic activity across the planet. Visible and infrared imagery from the geostationary orbit reveals the movement and evolution of tropical cyclones, but frequent cloud cover limits the determination of the center of circulation and the estimation of intensity. Active and passive microwave measurements from satellites in low orbit provide more precise information on the distribution and intensity of precipitation. However, it remains difficult to obtain reliable information in the most active central regions.

Satellite data available in real time are assimilated into weather prediction models with a very positive impact. In particular, this has contributed to a substantial reduction in track errors in previous decades. The slightest improvement in intensity forecasts, in terms of maximum and distribution of winds and rainfall, encourages further analysis of passive and active microwave satellite data and their assimilation into numerical weather prediction models. Analysis methods should also cover situations that are currently difficult to deal with: tropical and extra-tropical transitions, structural changes at landfall, cyclonic activity within large oceanic monsoon depressions, subtropical or hybrid phenomena, etc.

In order to ensure at least a constant quality of cyclonic monitoring, it is essential to maintain and develop space observation, in particular in microwave from low orbit. The development of new, more accurate sensors, with better spatial and temporal resolutions, and wide dissemination of data, should be encouraged. Some projects envisage using micro-satellites, possibly in constellations, for this purpose but this must not be to the detriment of the quality of measurements or of the temporal continuity. It is also essential that the parameters retrieved from satellite data are validated by in situ measurements, which are, however, difficult in cyclonic conditions. The dissemination of sophisticated spatial data analysis techniques to operational cyclone monitoring centers in different basins must also be encouraged. In this context, the coordination offered by the Tropical Cyclone Programme of the World Meteorological Organization is important.

6.9. References

Chavas, D.R. and Emanuel, K.A. (2010). A QuikSCAT climatology of tropical cyclone size. *Geophys. Res. Lett.*, 37, L18816. doi:10.1029/2010GL044558.

Dunion, J.P. and Velden, C.S. (2004). The impact of the Saharan air layer on Atlantic tropical cyclone activity. *Bull. Amer. Meteor. Soc.*, 85, 352–365. doi:10.1175/BAMS-85-3-353.

Dvorak, V.F. (1973). A technique for the analysis and forecasting of tropical cyclone intensities from satellite pictures [Online]. Available at: https://repository.library.noaa.gov/view/noaa/18546.

Dvorak, V.F. (1975). Tropical cyclone intensity analysis and forecasting from satellite imagery. *Mon. Wea. Rev.*, 103, 420–430. doi: 10.1175/1520-0493(1975)103<0420:TCIAAF>2.0.CO;2.

Dvorak, V.F. (1984). Tropical cyclone intensity analysis using satellite data. NOAA technical report NESDIS 11, NTIS, Springfield.

Guzman, O. and Jiang, H. (2021). Global increase in tropical cyclone rain rate. *Nat. Commun.*, 12, 544. doi: 10.1038/s41467-021-25685-2.

Jackson, C.R., Ruff, T.W., Knaff, J.A., Mouche, A., Sampson, C.R. (2021). Chasing cyclones from space. *Eos*, 102. doi:10.1029/2021EO159148.

Katsaros, K.B. and Mitnik, L. (2014). Microwave instruments for observing tropical cyclones. In *Typhoon Impact and Crisis Management*, Tang, D. and Sui, G. (ed.). *Adv. in Natural Technol. Hazards Res.*, 40, 201–226. doi: 10.1007/978-3-642-40695-9_2.

Kidder, S.Q., Goldberg, M.D., Zehr, R.M., DeMaria, M., Purdom, J.F.W., Velden, C.S., Grody, N.C., Kusselson, S.J. (2000). Satellite analysis of tropical cyclones using the advanced microwave sounding unit (AMSU). *Bull. Amer. Meteor. Soc.*, 81, 1241–1260. doi:10.1175/1520-0477(2000)081%3C1241:SAOTCU%3E2.3.CO;2.

Knapp, K.R., Kruk, M.C., Levinson, D.H., Diamond, H.J., Neumann, C.J. (2010). International best track archive for climate stewardship (IBTrACS): Unifying tropical cyclone best track data. *Bull. Amer. Meteor. Soc.*, 91, 363–376. doi: 10.1175/2009BAMS2755.1.

Kossin, J.P., Knaff, J.A., Berger, H.I., Herndon, D.C., Cram, T.A., Velden, C.S., Murnane, R.J., Hawkins, J.D. (2007). Estimating hurricane wind structure in the absence of aircraft reconnaissance. *Wea. Forecasting*, 22, 89–101. doi: 10.1175/WAF985.1.

Kossin, J.P., Emanuel, K.A., Vecchi, G.A. (2014). The poleward migration of the location of tropical cyclones maximum intensity. *Nature*, 509, 349–352. doi: 10.1038/nature 13278.

Legeckis, R. (1975). Application of synchronous meteorological satellite data to the study of time dependent sea surface temperature changes along the boundary of the Gulf Stream. *Geophys. Res. Lett.*, 2, 435–438. doi: 10.1029/GL002i010p00435.

Li, X., Zhang, J.A., Yang, X., Pichel, W.G., DeMaria, M., Long, D., Li, Z. (2013). Tropical cyclone morphology from spaceborne synthetic aperture radar. *Bull. Amer. Meteor. Soc.*, 94, 215–230. doi: 10.1175/BAMS-D-11-00211.1.

Meissner, T., Ricciardulli, L., Wentz, F.J. (2017). Capability of the SMAP mission to measure ocean surface winds in storms. *Bull. Amer. Meteor. Soc.*, 98, 1662–1677. doi: 10.1175/BAMS-D-16-0052.1.

Olander, T. and Velden, C.S. (2007). The advanced Dvorak technique (ADT) – Continued development of an objective scheme to estimate TC intensity using geostationary IR satellite imagery. *Wea. Forecasting*, 22, 287–298. doi: 10.1175/WAF975.1.

Reul, N., Tenerelli, J., Chapron, B., Vandemark, D., Quilfen, Y., Kerr, Y. (2012). SMOS satellite L-band radiometer: A new capability for ocean surface remote sensing in hurricanes. *J. Geophys. Res.*, 117, C02006. doi: 10.1029/2011JC007474.

Ruf, C.S., Atlas, R., Chang, P.S., Clarizia, M.P., Garrison, J.L., Gleason, S., Katzberg, S.J., Jelenak, Z., Johnson, J.T., Majumdar, S.J. et al. (2016). New ocean winds satellite mission to probe hurricanes and tropical convection. *Bull. Amer. Meteor. Soc.*, 97, 385–395. doi: 10.1175/BAMS-D-14-00218.1.

Shay, L.K., Goni, G.J., Black, P.G. (2000). Effects of a warm oceanic feature on Hurricane Opal. *Mon. Wea. Rev.*, 128, 366–383.

Velden, C. and Hawkins, J. (2010). Satellite observations of tropical cyclones, In *Global Perspectives on Tropical Cyclones*, Chan, J.C.L. and Kepert, J.D. (ed.). World Scientific, Singapore. doi: 10.1142/9789814293488_0007.

Velden, C. and Herndon, D. (2020). A consensus approach for estimating tropical cyclone intensity from meteorological satellites: SATCON. *Wea. Forecasting*, 35, 1645–1662. doi: 10.1175/WAF-D-20-0015.1.

Wang, N.-Y. and Vesecky, J.F. (1999). Sea surface temperature estimation using active/passive microwave remote sensing, *IEEE 1999 International Geoscience and Remote Sensing Symposium*, 971–973, 2. doi: 10.1109/IGARSS.1999.774502.

Yang, S. and Cossuth, J. (2016). Satellite remote sensing of tropical cyclones. In *Recent Developments in Tropical Cyclone Dynamics, Prediction, and Detection*, Lupo, A. (ed.). InTechOpen, Rijeka. doi : 10.5772/64114.

Zhang, L, Shi, H., Wang, Z., Liu, H., Lin, X., Liao, Q. (2018). Comparison of wind speeds from spaceborne microwave radiometers with in situ observations and ECMWF data over the global ocean. *Remote Sens.*, 10, 425. doi: 10.3390/rs10030425.

PART 2

Atmospheric Composition

Introduction to Part 2

Air Composition and the Contribution from Satellite Observations

Thierry PHULPIN[1] and Claude CAMY-PEYRET[2]

[1] *Retired from CNES and Météo-France, Toulouse, France*

[2] *IPSL, Paris, France*

I2.1. Air composition

The Earth's atmosphere is composed of gases and suspended particles. Nitrogen (~78%) and oxygen (~21%) are the major gaseous constituents. Water in its three phases is the most important minor constituent in the lower part of the atmosphere (troposphere). It is this ubiquitous constituent of the atmosphere of our blue planet (as seen from space) that determines the Earth's climate and affects all weather phenomena. We can recall that the first images from the geostationary orbit revolutionized our global vision of the atmosphere and led to the rapid development of operational satellite meteorology. The air also contains many other chemical species, rare gases (argon, neon, helium, etc.) and gases in much lower abundance called "trace gases" (carbon dioxide, methane, carbon monoxide, nitrogen oxides, sulfur dioxide, etc.). Some of these gases play a major role for life on Earth: some are toxic if they are inhaled, and are said to be pollutants; others by their radiative properties (absorption/emission of infrared radiation and transparency of the atmosphere in the visible range) are responsible for the greenhouse effect of the atmosphere. Finally, other gases are involved in protecting the planet from ultraviolet (or shorter wavelength) solar radiation, since UV is most dangerous for the different forms of life on the surface. It is established that in the history of planet

Satellites for Atmospheric Sciences 2,
coordinated by Thierry PHULPIN, Didier RENAUT, Hervé ROQUET and Claude CAMY-PEYRET.

Earth the presence of molecular oxygen results from the appearance of the first blue algae in the primitive ocean while the protective ozone layer (which can only be formed from a certain concentration of molecular oxygen O_2) did not yet exist. Some gases are naturally present; others are emitted by plants and living creatures (especially humans). As a result, their spatial distribution is highly variable. As some species may react with each other or be involved in many chemical reactions, their atmospheric lifetime can be highly variable.

In Part 2, we will focus more specifically on gaseous species and particles that can be detected by satellite instruments and that contribute to air quality, the greenhouse effect and the chemistry of the ozone layer.

12.2. The ozone hole

The first application of the use of satellites in the field of atmospheric composition was focused on the ozone hole, the first evidence of which dates back to the late 1970s. Low Earth orbit (LEO) polar satellites flying over the poles every 100 minutes are extremely well suited for continuous monitoring of the evolution of the total ozone column or partial columns thanks to hyperspectral sounders in the UV or IR range. For a more precise picture of the vertical distribution of ozone and monitoring of ozone depletion phenomena involving chlorinated or brominated compounds, limb viewing instruments have proved valuable and have helped to refine understanding of the process. Chapter 11 provides an exhaustive account of the contribution of satellite observations to this problem.

12.3. Air quality

Air quality is a growing concern of humankind, and the ability of space sensors to detect pollutants is an important asset in comprehensively observing, understanding, modeling and predicting changes in atmospheric composition.

Chapters 7–9 deal with the detection of reactive species present near the surface and in the lower atmosphere as well as dust, which significantly deteriorates air quality. So far, few space missions have been dedicated to the determination of air quality from space, since operational monitoring requires frequent observations (and therefore permanent views by geostationary satellites or frequent overflights by satellite constellations), good spectral and radiometric resolution (to detect and distinguish the spectral signatures of the different chemical species) as well as good spatial resolution to unambiguously distinguish the polluted areas and the sources of pollutants. All of these requirements are difficult to meet simultaneously for satellite instruments, but technological advances have now made it possible to meet the objectives of global monitoring of air quality from outer space and many experimental as well as operational missions have emerged or are planned.

I2.4. The greenhouse effect

Greenhouse gases such as water vapor or carbon dioxide are dealt with in Part 3. However, reactive species that are also involved in air pollution such as ozone, which behaves like a greenhouse gas (when it is present in the lower troposphere) and methane, are dealt with in Chapters 7 and 8.

I2.5. Dust and its impact

The particles emitted by fires and the sand dust lifted by wind and transported over long distances create risks for humans or their activities. Methods of detection and tracking from outer space are described in Chapters 9 and 10. The same applies to volcanic ash released at high altitude during explosive eruptions, which can pose a real danger to airliners.

Numerous studies have been carried out (Prata 2008) on methods for detecting and monitoring ash and sulfur dioxide, SO_2, often emitted simultaneously by volcanoes (and for the dangerous outcome of its transformation into sulfuric acid and sulfate particles). These dangers have justified the establishment by the International Civil Aviation Organization (ICAO) of a worldwide network of nine specialized centers (VAACs, centers for monitoring volcanic activity), which provide us with constant information on the behavior and/or eruptions of the many active volcanoes.

I2.6. Stratosphere

It is not possible in this part to cover all aspects of the Earth's atmospheric environment as observed from space, and the subjects introduced above have focused on the lower atmosphere and the air we breathe. But we have emphasized the importance of the ozone layer, which protects us from ultraviolet radiation and associated formation and destruction reactions involving species such as nitrous oxide N_2O and chlorofluorobrominated compounds that are inactive in the troposphere but "activated" by photodissociation when they reach the stratosphere. Similarly, CH_4 methane, which is a fairly long-lived species in the troposphere

(~10 years) and as such contributes strongly to the greenhouse effect, participates in chemical reactions in the stratosphere. These aspects are dealt with specifically in Chapter 11.

12.7. References

Prata, A.J. (2008). Satellite detection of hazardous volcanic clouds and the risk to global air traffic. *Nat Hazards*, 74(1). doi:10.1007/s11069-008-9273-z.

7

Reactive Tropospheric Chemistry

Sarah SAFIEDDINE and Camille VIATTE
LATMOS/IPSL, Sorbonne Université, UVSQ, CNRS, Paris, France

7.1. Introduction

The global observation system for reactive species in the atmosphere, that is, species that can react in the atmosphere within a time frame of seconds to weeks or months, has improved over the past decade. This is mainly due to both availabilities of new satellite-based observations and to the development of in situ observations from the ground and from commercial aircrafts. For example, the newTROPOspheric Monitoring Instrument (TROPOMI) on board the Sentinel 5 precursor (S5P) satellite has measured since 2017 main pollutants absorbing in the UV–vis part of the spectrum, such as NO_2, SO_2, HCHO and CO, and also O_3 and CH_4. In the thermal infrared, the Infrared Atmospheric Sounding Interferometer (IASI) series of instruments launched on board the MetOp satellites in 2006 measures more than 30 species in the troposphere and the stratosphere. Several satellite instruments dedicated to GHGs have also been launched in recent years such as OCO-2 and GOSAT. In parallel, the ground-based observational system was also significantly consolidated by the establishment of long-term regional organizations. These are usually supported by the WMO's Global Atmospheric Watch or other widely recognized international networks (NDACC, AERONET). The In-service Aircraft for a Global Observing System (IAGOS) Aerosol Package on board commercial aircrafts provides tropospheric profiles of CO_2, CH_4, CO, NO_2,

Satellites for Atmospheric Sciences 2,
coordinated by Thierry PHULPIN, Didier RENAUT, Hervé ROQUET and Claude CAMY-PEYRET.

aerosol particle size distribution and the number concentrations, and ozone mole fractions. With this, atmospheric transport models are able to better simulate the atmosphere, and as such enhance our understanding of the state, dynamics and transport of different pollutants across the globe.

7.2. Methane

Methane (CH_4) is the second most important anthropogenic greenhouse gas after CO_2, with a 100-year warming potential 28 times larger than CO_2 according to the IPCC. Since the pre-industrial era, CH_4 concentrations have more than doubled. Worldwide, CH_4 emissions increased by 17% between 1990 and 2012, compared to a 53% increase in CO_2 (European Commission 2018). CH_4 oxidation (Figure 7.1) is a major source of reactive carbon in the atmosphere. It is also a precursor of ozone and organic aerosol, both presenting a public health threat (Chapter 7).

About 60% of the current global methane is emitted by sources like agriculture, landfills and wastewater, and the production and pipeline transport of fossil fuels, while the rest is from natural sources. Globally, CH_4 emissions and concentrations are still increasing, raising concerns for air quality and climate change. Climate policies such as the Kyoto protocol and the Paris agreement are important means to reduce CH_4 emissions.

To date, a series of satellite sensors are able to detect and monitor CH_4 concentrations, such as AIRS (Atmospheric Infrared Sounder instrument), IASI, SCIAMACHY (Scanning Imaging Absorption spectrometer for Atmospheric Chartography instrument), GOSAT (Greenhouse Gases Observing Satellite) and TROPOMI.

7.3. Reactive organic species

Reactive organic species are typically those with an atom of carbon or more, and an atom of hydrogen or more. This notation is used to differentiate reactive carbon from carbon monoxide (CO) and carbon dioxide (CO_2), which are less reactive and at the end of the reactive organic carbon (ROC) cascade. ROC is the fuel of tropospheric chemistry, not only leading to ozone and organic aerosol formation but also providing a chemical source of CO_2 estimated to be approximately 13% of the total present-day source of CO_2 (Safieddine et al. 2017).

Non-methane reactive organic carbon (ROC) is the sum of atmospheric non-methane volatile organic compounds (NMVOCs) and primary and secondary organic aerosol (OA). NMVOCs drive the oxidative chemistry of the atmosphere, affecting the hydroxyl radical (OH) concentrations and methane lifetime. They also serve as precursors to ozone and aerosol formation, both of which are deleterious to human health and affect the climate.

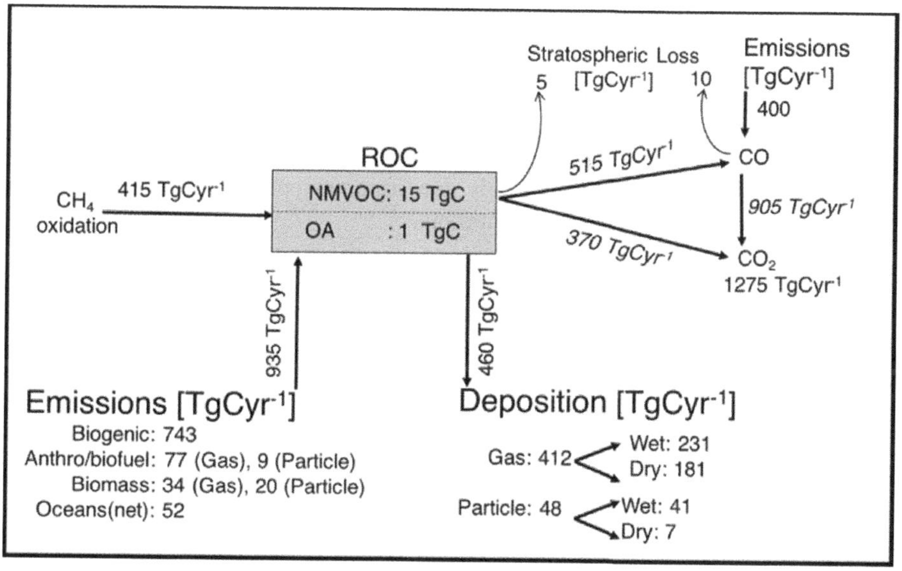

Figure 7.1. *The global reactive organic carbon (ROC) tropospheric budget (source: Safieddine et al. (2017))*

Gaseous NMVOCs dominate the ROC burden, as Figure 7.1 shows. They include isoprene, aliphatic, aromatic and oxygenated hydrocarbons (e.g. aldehydes, ketones, alcohols and organic acids), and organic nitrates and sulfates. It is estimated that of the organic compounds likely present in the atmosphere (10^4–10^5 species), only a fraction of these have been identified and the total pool of NMVOCs in the troposphere is poorly characterized and evaluated.

Organic aerosols participate in heterogeneous chemical reactions and can impact cloud formation. Primary organic aerosol (POA) is directly emitted from sources, and secondary organic aerosol (SOA) is formed when NMVOC oxidation products undergo gas-to-particle transfer. To date, we still do not have a complete understanding of the formation and loss of OA in the atmosphere, including the oxidative chemistry of NMVOCs that serve as precursors to SOA.

7.3.1. *Isoprene*

Isoprene (C_5H_8, 1-methyl-1,3-butadiene) makes up the largest fraction of NMVOCs emitted into the atmosphere. Its main source is biogenic: it is produced and emitted by vegetation. The largest contributor to the "biogenic emissions" shown in Figure 7.1 is isoprene (approximately 80). These emissions are temperature-dependent, and the largest sources of isoprene on the globe are in the Amazonia and equatorial forests. It is very reactive in the atmosphere and as such has a very short lifetime, and affects the oxidation capacity of the global atmosphere since its oxidation consumes the OH radical.

Isoprene plays a key role in tropospheric chemistry and climate, as a precursor to both ozone and aerosols. Thus, not only does isoprene emission influence climate, but also climate influences isoprene emission.

7.3.2. *Other non-methane volatile organic compounds*

Due to the chemical complexity and variety of VOC species, limited global measurements and the lack of mass closure in both ambient measurements and models, uncertainties regarding the life cycle of VOCs in the atmosphere are significant. Figure 7.2 shows satellite observations of different VOCs as seen from space. We are now able to map different short- and long-lived VOCs as more sophisticated retrieval methods are available, including those using artificial intelligence (Franco et al. 2018). Figure 7.2 shows different volatile organic compounds as they are detected from space by IASI. We note how many of the VOCs shown have high values around the Tropics, suggesting that vegetation or fires are a major source of VOCs.

– Methanol is one of the most abundant non-methane organic gases in the atmosphere. It is highly reactive in the troposphere and is an important source of CO and formaldehyde. The largest source of methanol is biogenic; that is, from terrestrial plant growth, much like formic acid and acetic acid. This is clearly seen in the figure, with high concentrations of these VOCs around the equator in Africa, and over the Amazon rainforest in South America.

– Acetone is also an abundant oxygenated VOC in the atmosphere. It is either directly emitted from a wide variety of natural and anthropogenic sources or produced from the photochemical oxidation of a number of precursors. The role of the ocean is complicated; acetone is produced in the ocean from the photolysis of colored, dissolved organic materials or from biological processes but is also removed through microbial uptake. Major sinks of acetone in the atmosphere include photolysis, oxidation by OH and dry deposition. Acetone has a long global atmospheric lifetime, of approximately a few weeks, allowing for its long-range transport and contributing to the production of peroxyacetyl nitrate (PAN, also shown in Figure 7.2) in the transported plumes.

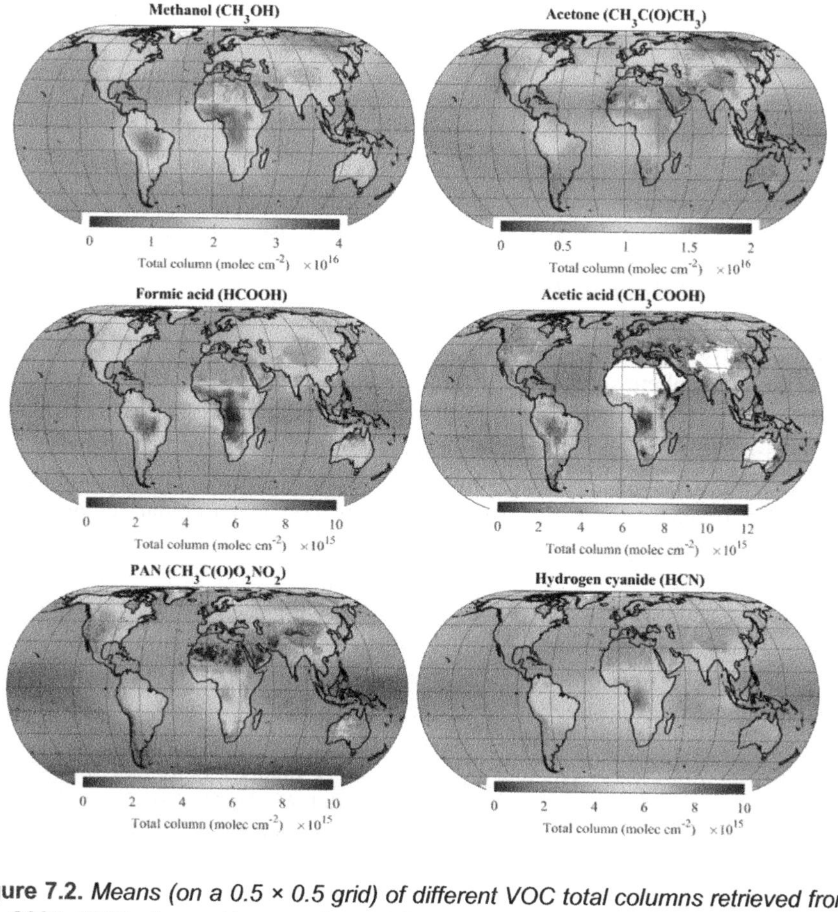

Figure 7.2. *Means (on a 0.5 × 0.5 grid) of different VOC total columns retrieved from the 2007–2020 observations performed by the IASI infrared sensor on board the MetOp-A satellite. Acetic acid data over deserts are not displayed due to the presence of major measurement interferences. For a color version of this figure, see www.iste.co.uk/phulpin/satellites2.zip*

– Formic and acetic acids are the most abundant carboxylic acids in the Earth's atmosphere and contribute to the acidity of rainwater and clouds. They exhibit similar abundance, distributions and seasonality, pointing to major common sources from biomass burning, fossil fuel, soil and vegetation.

– Peroxyacetyl nitrate (PAN) is the only compound shown in Figure 7.2 that is mainly formed (and not emitted) in the atmosphere through the oxidation of NMVOCs. It is the principal tropospheric reservoir for nitrogen oxide radicals ($NO_x = NO + NO_2$) and enables the transport and release of NO_x globally with major implications for the global distributions of ozone and OH, the main tropospheric oxidants.

– Hydrogen cyanide has a variety of sources: tobacco smoke, combustion of fuel, fires and industrial processes (e.g. combustion of synthetic polymers, wool and silk). Cyanide ions are also generated naturally during biogenic processes of plant bacteria and fungi.

7.4. Reactive inorganic species

SO_2 is a reactive pollutant with a lifetime of a few days in the troposphere. It enters the atmosphere as a result of both natural (volcanoes) and anthropogenic emissions (fossil fuel combustion). Coal burning itself accounts for more than 50% of human-made annual emissions. With the obsolescence of the coal industry, SO_2 concentrations decreased drastically in many regions of the world, such as in Europe and the US (Constantin et al. 2020) but remain high in particular regions of the world such as China and Russia (Bauduin et al. 2014).

In the atmosphere, NO_2 is emitted by both anthropogenic activities, such as transportation, energy production and other industrial activities, and natural processes such as soil microbial processes, wildfires and lightning. NO_2 is a short-term lived species affecting human respiratory health.

Global monitoring of NO_2 and SO_2 began more than 20 years ago with GOME (Global Ozone Monitoring Experiment on board ERS-2), GOME-2 (on Envisat), OMI (on board AURA) and TROPOMI (TROPOspheric Monitoring Instrument) and is foreseen to continue in the future.

With its high spatial resolution, NO_2 plumes from individual ships can be clearly observed with TROPOMI demonstrating exciting possibilities for satellite observations in support of environmental regulation. In addition, the impact of the Covid-19 outbreak on NO_2 pollution has been assessed over many regions of the world using TROPOMI. Figure 7.3 shows the decline of NO_2 concentrations over China during the 2020 Covid-19 lockdown.

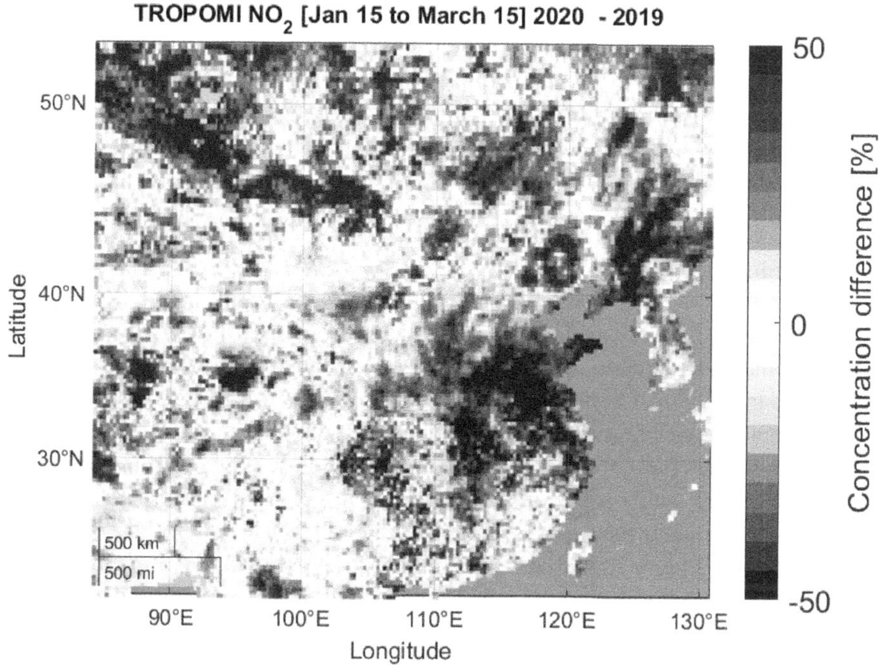

Figure 7.3. *NO_2 concentration differences (2020–2019) derived from TROPOMI over China during the lockdown (January 15 to March 15). For a color version of this figure, see www.iste.co.uk/phulpin/satellites2.zip*

NH_3 is also an important reactive pollutant involved in air quality as it is a precursor of particulate matter which is very harmful to human health. Global monitoring of NH_3 is now possible but challenging due to its short lifetime (and as such to a low "spectral" signature as detected by sensors on board satellites). At present, three missions are able to measure it: AIRS, the Cross-track Infrared Sounder (CrIS), and IASI.

Recently, more than 250 NH_3 hotspots have been identified from space using nine years of IASI data (Van Damme et al. 2018). This study reveals a global underestimation of NH_3 hotspot emissions compared to state-of-art emissions inventory. Atmospheric chemical transport models have difficulty representing both NH_3 and $PM_{2.5}$ distributions due to the challenge of reproducing NH_3 temporal variability.

Satellite observations can therefore be used to restore the daily ammonia emissions during the fertilizer application period and thus the model's simulation ability to predict ammonia and particulate matter concentrations during the pollution peaks. Such an application of inverse methods is expected to extend in the future towards new pollutants and finer spatial resolution using advanced satellite measurements.

7.5. Conclusion

This chapter lists a variety of methods that can be used today to measure and detect methane and many of the volatile organic compounds (VOCs) that are ubiquitous in the global atmosphere. We also show how, while each of the VOCs is found in trace concentrations in the atmosphere, their large number (up to 10^5) makes them the fuel of tropospheric chemistry. They are the precursors of particulate matter and ozone, and their end point of oxidation is CO and CO_2 (Figure 7.1). As such they play an underrated role in the climate that is hard to quantify today. Future model development, as well as satellite instruments and inversions will help to fill the gap in our understanding of these various and complex species.

7.6. Acknowledgment

The authors would like to thank Bruno Franco (Université Libre de Bruxelles) for Figure 7.2.

7.7. References

Bauduin, S., Clarisse, L., Clerbaux, C., Hurtmans, D., Coheur, P.-F. (2014). IASI observations of sulfur dioxide (SO2) in the boundary layer of Norilsk. *Journal of Geophysical Research: Atmospheres*, 119, 4253–4263.

Constantin, D.-E., Bocăneala, C., Voiculescu, M., Roşu, A., Merlaud, A., Van Roozendael, M., Georgescu, P.L. (2020). Evolution of SO2 and NOx emissions from several large combustion plants in Europe during 2005–2015. *International Journal of Environmental Research and Public Health*, 17(10), 3630.

Franco, B., Clarisse, L., Stavrakou, T., Müller, J.-F., Van Damme, M., Whitburn, S., Hadji-Lazaro, J., Hurtmans, D., Taraborrelli, D., Clerbaux, C. et al. (2018). A general framework for global retrievals of trace gases from IASI: Application to methanol, formic acid, and PAN. *Journal of Geophysical Research: Atmospheres*, 123, 13,963–13,984. https://doi.org/10.1029/2018JD029633.

Safieddine, S.A., Heald, C.L., Henderson, B.H. (2017). The global nonmethane reactive organic carbon budget: A modeling perspective. *Geophysical Research Letters*, 44, 3897–3906. https://doi.org/10.1002/2017GL072602.

The European Commission (2018). Global trends of methane emissions and their impacts on ozone concentrations [Online]. Available at: http://op.europa.eu/en/publication-detail/-/publication/c40e6fc4-dbf9-11e8-afb3-01aa75ed71a1/language-en [Accessed 23 February 2021].

Van Damme, M., Clarisse, L., Whitburn, S., Hadji-Lazaro, J., Hurtmans, D., Clerbaux, C., Coheur, P.-F. (2018). Industrial and agricultural ammonia point sources exposed. *Nature*, 564(7734), 99–103. https://doi.org/10.1038/s41586-018-0747-1.

8

Major Pollutants: Ozone and Fine Particulate Matter

Juan Cuesta and Gaëlle Dufour
LISA – UMR7583, Créteil, France

8.1. Introduction

Anthropogenic changes in atmospheric composition in the lower atmosphere have a direct impact on regional and global air quality, climate, public health and the environment. Among air pollutants, atmospheric aerosols and ground-level ozone are fundamental players in air quality and the climate system. At present, these two air pollutants have been identified as the most harmful to the health of populations and ecosystems in Europe (European Environment Agency 2011) and worldwide (Molina and Molina 2004). Exposure to particulate matter and ozone can produce and severely exacerbate respiratory diseases, leading to increased premature mortality and damage to buildings and historic heritage objects. Particles can also cause cardiovascular dysfunction or cancer. Ozone is an oxidizing agent, which in high concentrations damages plants and reduces the growth of forests and crops as well as their carbon dioxide uptake. Both pollutants are among those that produce the most significant radiative forcings, with ozone contributing to global warming through the greenhouse effect and particulate matter through cooling. In addition to adverse health effects, air pollution affects the economy. Recent studies estimate

---

For a color version of all the figures in this chapter, see www.iste.co.uk/phulpin/satellites2.zip.

Satellites for Atmospheric Sciences 2,
coordinated by Thierry Phulpin, Didier Renaut, Hervé Roquet and Claude Camy-Peyret.

that several tens of billions of euros (health impact, absenteeism, etc.) could be saved in Europe if pollution were reduced to below the recommended thresholds[1].

Advances in space-borne remote sensing techniques have contributed significantly to the regional observation of major air pollutants, such as aerosols and tropospheric ozone. The main satellite approaches for observing them are described below.

8.2. Tropospheric ozone

8.2.1. *Beginnings of satellite-based tropospheric ozone observations*

Ozone, both a pollutant in the boundary layer and a greenhouse gas in the upper troposphere, plays a central role in tropospheric chemistry by being a source of hydroxyl radicals, and thus controlling the oxidizing capacity of the atmosphere. The variability of ozone sources and pathways associated with a sufficiently long lifetime for long-range transport makes tropospheric ozone particularly variable spatially and temporally (Monks et al. 2015). Monitoring it from the local to the global scale is essential in order to better understand its impact in terms of air quality, pollution transport and radiative forcing. With the advent of satellite observations, a new perspective has opened up for monitoring the tropospheric ozone in order to supplement surface, lidar and sounding balloon observations by offering the potential of a global vision of the distribution of tropospheric ozone, in connection with the first developments of tropospheric chemistry transport models. In the late 1980s, Fishman and Larsen (1987) were the first to estimate satellite tropospheric ozone columns by the so-called "residue" method, that is, by subtracting the stratospheric O_3 column (derived from the SAGE limb sounder) from the total atmospheric column derived from TOMS. The difficulty and limitation of this method are due to the difference in sampling and resolution between limb and nadir observations. Only stratospheric columns derived from climatologies could be effectively used. These methods have been refined progressively, in particular by assimilating the observations of limb sounders (HALOE, MLS, etc.) in atmospheric models in order to retrieve stratospheric ozone columns better resolved in time (Ziemke et al. 1998) or by using the TOMS and SBUV sounders in synergy (Fishman et al. 2003). Subsequently, some methods use convective clouds of high altitude (tropopause) that are highly reflective to determine the stratospheric column of ozone above clouds, which is then subtracted from the total ozone columns for clear pixels observed near cloudy scenes (Ziemke et al. 1998), with the limitation of being applicable only in regions impacted by convection, mainly the tropics. These methods, developed for the first generation of satellites, are still applied to the

1. http://www.aphekom.org.

current generation of satellites, notably by combining OMI and MLS to derive global tropospheric ozone climatologies (Ziemke et al. 2011). Figure 8.1 shows a comparison of the tropospheric ozone columns obtained during the summer with the first generation of satellites (Tropospheric Ozone Residual or TOR) and the current generation (Ozone Monitoring Instrument/Microwave Limb sounder or OMI/MLS).

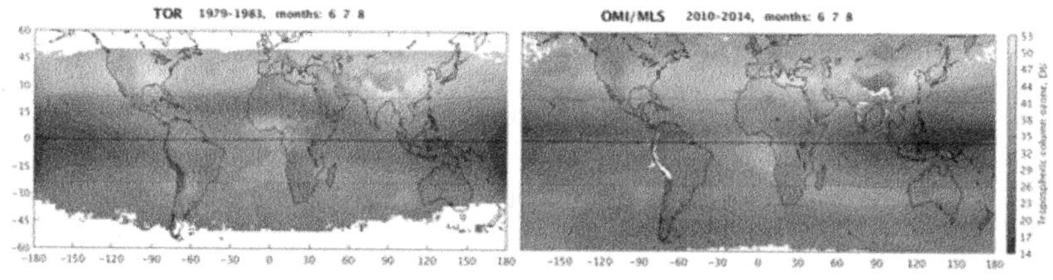

Figure 8.1. *Average tropospheric ozone columns in Dobson units (DU) for the months of June to August obtained by the TOR product over the 1979–1983 period (left) and by the OMI/MLS product for the 2010–2014 period (right) (source: Gaudel et al. (2018))*

8.2.2. *Current capabilities for tropospheric ozone monitoring*

The improvement of the performance, in particular spectral performance (transition from measuring a few channels to measuring the full spectrum for UV sounders) of the second generation of satellite instruments and the development of inversion methods applied to atmospheric sounding (Rodgers 2000) have made it possible to envisage the direct determination of tropospheric ozone without going through a differential method. The first direct inversion of tropospheric ozone was performed with the GOME instrument (Chance et al. 1997; Munro et al. 1998). The spectral capacities of this instrument (and its successors, SCIAMACHY, OMI and GOME-2) make it possible to solve spectral structures of the Huggins band (between 305 and 340 nm), which depend on the temperature and therefore on the altitude below the ozone maximum. Ozone profiles, including tropospheric profiles, are then retrieved. The degrees of freedom (amount of independent information in the retrieved profile, see Chapter 10 of Volume 1) in the troposphere vary between 0.5 and 1.2/1.5 depending on the instruments with vertical resolutions of the order of 10–14 km (Liu et al. 2005; Liu et al. 2010). This makes it possible to retrieve tropospheric columns of ozone partially independent of the stratospheric column according to the latitude.

From the mid-1990s onwards, infrared sounders dedicated in part to tropospheric sounding made it possible to obtain tropospheric columns (between the surface and 12 km of altitude) and ozone profiles from the IMG instrument (Clerbaux et al.

2003; Coheur et al. 2005), followed by TES (Bowman et al. 2006; Worden et al. 2007) and IASI (Eremenko et al. 2008; Boynard et al. 2009). The spectral and radiometric performances of these instruments ensure optimum sensitivity from the free troposphere to the lower stratosphere with a vertical resolution of the order of 6–8 km (Coheur et al. 2005) and degrees of freedom varying from 1 to 2 or even a little more depending on the spectral resolution of the instruments and the retrieval methods used (Worden et al. 2004; Dufour et al. 2012).

For these instruments operating in the ultraviolet and infrared range, the primary purpose of the measurement of tropospheric ozone is the study of its budget, of long-range transport of its radiative impact on a global scale. In recent years, the international community has been mobilizing to provide an overview of tropospheric ozone and its evolution within the framework of the TOAR (Tropospheric Ozone Assessment Report[2]). The main existing satellite for UV and IR sounders has been integrated into this study and provides an overall view of the ozone distribution (Figure 8.2). While the distribution of ozone and the inferred budgets of different products are relatively consistent, disagreements have arisen on trends derived from the different products with UV sounders giving overall recent positive trends and IR sounders giving negative trends, thus leaving open questions about the use of satellite observations to calculate tropospheric ozone trends.

Smaller-scale applications for monitoring pollution and the problem of air quality are also envisaged for these sounders. For such applications, it is necessary for the signal observed to be sensitive to the lowest atmospheric layers and to be able to extract as best as possible the information (even poor) of these layers from the total signal received by the instruments. UV sounders are relatively sensitive to low atmospheric layers but with a coarse vertical resolution (of the order of 10 km), which does not make it possible to decorrelate strongly the information coming from the lower, middle and upper troposphere. Ozone profiles recently retrieved from OMI, however, show good sensitivity in the low layers in the most intense source regions such as China (Hayashida et al. 2015; Shen et al. 2019). For high spectral resolution infrared sounders, the vertical resolution reached is better (6–8 km), offering better potential for separating different tropospheric layers (Dufour et al. 2012). The sensitivity of infrared instruments is strongly dependent on the thermal contrast of the sampled scene. The greater the thermal contrast, the better the ability to dissociate the different layers of the atmosphere, especially in the lower troposphere. As ozone pollution generally occurs during summer anticyclonic periods, the thermal contrast is often quite high with a well-developed boundary layer. These warm surface conditions make possible the observation of pollution cases and the identification of processes controlling the distribution of ozone in the

2. https://igacproject.org/activities/TOAR.

lower troposphere using infrared sounders (Dufour et al. 2010, 2015; Safieddine et al. 2013; Doche et al. 2014). The evolution of atmospheric retrieval methods has also contributed to improving the amount of information from lower layers that can be extracted from infrared spectra (Kulawik et al. 2006; Eremenko et al. 2008) with the possibility of extracting partial tropospheric columns, the lowest of which with a maximum sensitivity between 3 and 4 km (Dufour et al. 2012).

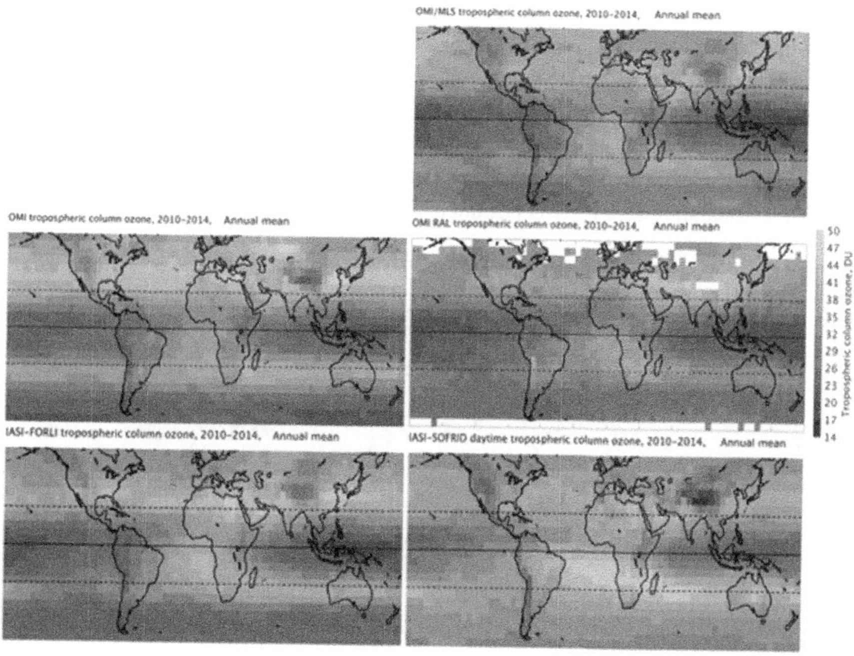

Figure 8.2. *Annual tropospheric ozone column in Dobson units (DU) from OMI/MLS satellite products (top right), OMI-SAO middle left), OMI-RAL (middle right), IASI-FORLI (bottom left) and IASI-SOFRID (bottom right) for the period 2010–2014 (adapted from Figure 10 of Gaudel et al. (2018))*

8.2.3. *Multi-wavelength synergy for ozone pollution monitoring*

An essential route for improving the sensitivity of satellite products to near-surface ozone concentrations and their vertical resolution is multi-spectral coupling. The synergy gain of different sounders and multiple spectral regions (ultraviolet, infrared, visible more recently) has been demonstrated several times theoretically on synthetic observations over the last 20 years. The development and application of such methods coupling infrared and ultraviolet on real data are recent

(Cuesta et al. 2013; Fu et al. 2013). These observations show improved sensitivity to ozone pollution in the lowest atmospheric layers (maximum sensitivity between 2 and 3 km) and improved ability to estimate ozone distribution in the troposphere (Cuesta et al. 2013). Figure 8.3 illustrates the overall distribution of ozone in the very low troposphere (LMT: lowermost troposphere) below 3 km altitude resulting from the synergy of IASI measurements in the infrared and GOME-2 measurements in the ultraviolet (distributed by the AERIS data division). Observations of near-surface ozone from the latter approach show good agreement with in situ surface measurements for pollution episodes (Cuesta et al. 2018) and seasonal averages (Gaudel et al. 2018). In addition, new methods try to use satellite measurements in the Chappuis band (maximum between 500 and 650 nm) in the visible range, which potentially offers the possibility of retrieving ozone concentrations closer to the surface. This concept is currently being studied for several missions such as TEMPO (Chance et al. 2013) and GEMS (Kim et al. 2020).

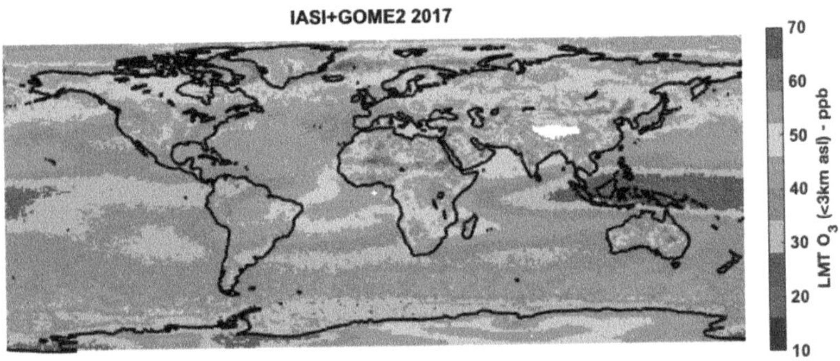

Figure 8.3. *Observations of ozone in the very low troposphere (LMT: lowermost Troposphere) in average concentration over the year 2017 deduced from the IASI+GOME2 multi-spectral approach (source: AERIS[3])*

8.3. Pollution aerosols

The health impact of atmospheric aerosols is directly related to their abundance (in terms of mass concentrations) in the air we breathe, that is, close to the surface. This quantity is typically referred to as "PM", an acronym for particle matter (European Environment Agency 2011). Similarly, the size of particles determines the level of their penetration into the human body and therefore the respiratory diseases that they can cause or exacerbate. The toxicity of the particles also depends on their chemical composition, for example carbonaceous particles are more harmful

3. www.aeris-data.fr.

to health than inorganic particles. From a health point of view, it is therefore essential to characterize aerosols in terms of their concentrations, as is the case with gaseous pollutants, and also their size and composition.

Satellite remote sensing approaches for atmospheric aerosols allow for the retrieval of their abundance and for properties to be quantified as a function of optical properties, since these methods are based on the analysis of the interactions between particles and radiation. The optical quantity widely used as an indicator of the horizontal distribution of the particles is the optical thickness of the aerosols (aerosol optical depth or AOD) integrated over the entire atmospheric column. This property is obtained with passive remote sensing methods, which analyze natural radiation, either solar or terrestrial, and is used in some cases as an indicator of the variability of particle concentration (Al Saadi et al. 2005), although the links between these two quantities are complex. On the contrary, the vertical distribution of the aerosols is mainly studied thanks to active sounders, based on a pulsed laser, which measure vertical profiles of the backscatter coefficient of the aerosols. The variation of optical properties as a function of wavelength and the changes in the polarization of the returned signal induced by particles are often used as indicators of the size of the particles and help to define "families" or types of particles, such as urban, desert, continental origin and biomass fires. The following sections present methods typically used for satellite remote sensing of pollution aerosols, but are also used for other types of aerosols (see Chapter 9).

8.3.1. *Optical thickness of pollution aerosols*

Unlike most trace gases that absorb radiation in very specific spectral bands, associated with specific transitions of molecules that absorb photons, aerosols absorb and scatter photons over wider spectral bands (several μm). The interactions between aerosols and radiation often vary gradually with wavelength, with the exception of absorption bands in the thermal infrared. As a result, aerosol remote sensing approaches can use satellite sensors that measure radiation over relatively wide bands, such as Moderate Resolution Imaging Spectroradiometer (MODIS) and Visible Infrared Imaging Radiometer Suite (VIIRS), but with fine spatial resolution (pixels of several hundred meters) to better distinguish the presence of clouds (which also have a spectrally broad radiative impact) and to more accurately map the horizontal distribution of aerosols. Moreover, sensors with high spectral resolutions (typically less than 1 nm), such as the Ozone Monitoring Instrument (OMI), can also be used but, in general, their spatial resolution is coarser (pixels more than 10 km wide). The new TROPOMI (TROPOspheric Monitoring Instrument) sensor measures the spectrum from the UV to the near IR with a spatial resolution of less than 5 km.

A wide variety of AOD products from many sensors, and based on a wide variety of methods, are currently available (Sogacheva et al. 2020). An important aspect that differentiates these methods is how the signature of aerosols is identified in the radiation measurements compared to that of the surface. AOD products widely used by the international community are derived from the MODIS sensors on Terra and Aqua satellites. In this case, the methods that estimate AOD are referred to as "Dark Target" (Remer et al. 2005) and "Deep Blue" (Sayer et al. 2013). The Dark Target approach retrieves AOD over sea and land (except in bright areas such as deserts) using the low opacity of most near-infrared aerosols (at wavelengths of 2–3 μm) to differentiate them from dark surfaces. The same approach is applicable to measurements of other satellites with similar spectral channels, such as those of VIIRS, Himawari, EPIC, etc. These AOD measurements are illustrated in Figure 8.4a–b for MODIS and VIIRS, for a typical episode of particulate pollution in Europe. On the non-cloudy areas, both sensors show fairly similar AOD distributions, and are able to clearly identify the most polluted areas (northern Italy, France and Germany) and the relative abundance of particles in terms of AOD integrated on the vertical profile. The Deep Blue method provides AOD on land, on all types of land surface (including deserts), based on the contrast in the "deep" blue (412 nm) of the radiation scattered by the aerosols (brighter) compared to that reflected by the surfaces (darker).

The variation in AOD according to wavelength indicates the type of aerosol, as the AOD of fine particles such as those from pollution or from fires is clearly higher for short wavelengths than it is for long wavelengths. Conversely, AOD is almost constant for larger aerosols such as desert dust or volcanic ash. The variation in AOD is commonly expressed by the Angström exponent (Figure 8.4c) where values close to zero (green) and around ~2 (blue) suggest the presence of coarse and fine particles, respectively. Taking into account the geographical location and the spectral variation of the AOD between several wavelengths allows for a finer classification (Figure 8.4d).

Other spatial sensors measure the radiation reaching a pixel at the surface from several viewing angles. These are sounders like MISR on Terra, ATSR-2 on ERS or SLSTR on Sentinel 3, whereas others can also measure the degree of polarization of light like POLDER and PARASOL. These measurements offer an enhanced contrast, making it possible to differentiate more precisely the signatures of the aerosols with respect to the surface. They make it possible to retrieve additional properties such as the altitude of the aerosol layers by stereoscopy with MISR (Nelson et al. 2013) or the simple scattering albedo and discrimination of the AOD according to the size of the aerosols with PARASOL (Dubovik et al. 2021).

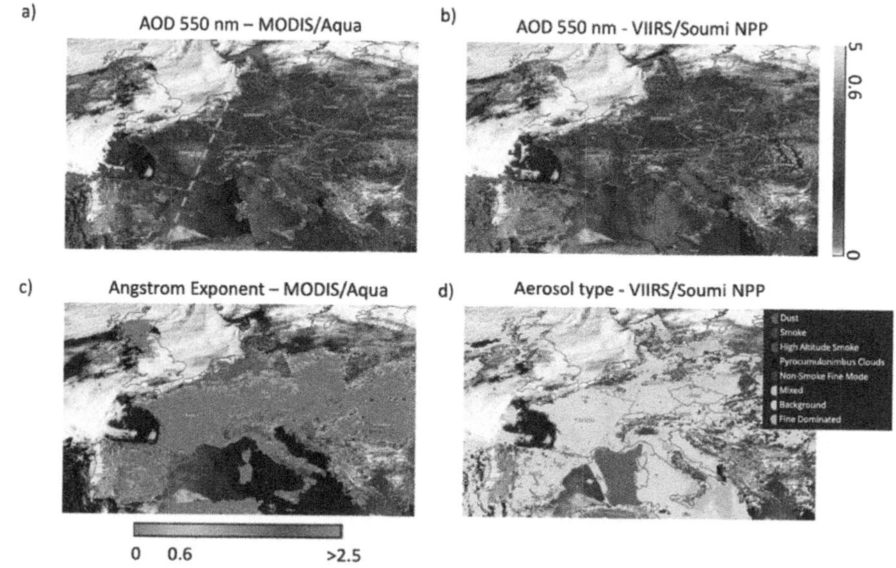

Figure 8.4. *Observations of the AOD at 550 nm retrieved from the measurements of the sensor (a) MODIS on the AQUA satellite and (b) VIIRS on Suomi NPP, on June 26, 2019. Restitutions of (c) the Angström exhibitor with MODIS and (d) the type of aerosols of VIIRS are indicators of the origins of aerosols (source: NASA Worldview[4])*

8.3.2. *Altitude of pollution aerosols*

The relationship between AOD and surface particle concentration is complex, in part because of the high spatial and temporal variability in the vertical distribution of aerosols. The more direct approach for observing the vertical aerosol profile from space is based on active laser probing of aerosol layers with, for example, the CALIOP lidar on board the CALIPSO satellite (Winker et al. 2009). This instrument measures the radiation reflected by the particles as a function of distance with a vertical resolution of 30 m. The abundance of particles is quantified by using the intensity of the backscattered energy which is also attenuated by the atmosphere during its propagation on the forward and backward path of the laser pulse between the instrument and the aerosol layer. These CALIOP measurements are illustrated in Figure 8.5 for the same pollution episode observed by MODIS and VIIRS. Pollution particles are observed as a dense layer between the surface and about 1 km of altitude (between 50 and 55°N in Figure 8.5a), and also surrounded by multiple layers of elevated aerosols located south of them. An estimate of aerosol types based

4. https://worldview.earthdata.nasa.gov.

on the degree of depolarization and spectral variation of the backscattered signal identifies desert dust south of 47°N (Figure 8.5b) and a mixture of aerosols, indicated as "polluted dust" to the north.

Lidar backscatter measurements are very useful for distinguishing aerosol layers and their relative abundance. They are mainly qualitative indicators because they alone do not make it possible to avoid atmospheric absorption by molecules. A retrieval method is necessary to deduce, for example, the vertical extinction profiles of the aerosols. However, these are based on assumptions about aerosol properties, which can induce significant biases. To avoid these hypotheses, new LIDAR space missions with additional measurement channels are being developed, such as the high-resolution ATLID LIDAR on the EarthCare satellite[5].

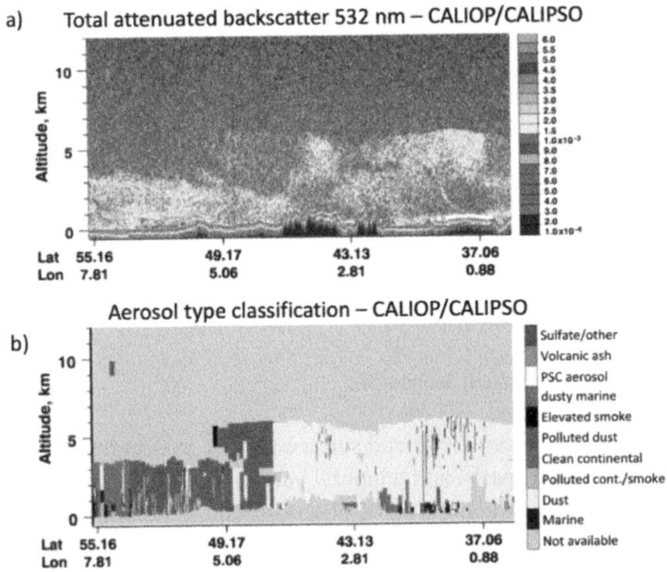

Figure 8.5. *Transects of the profiles of (a) attenuated backscattering of aerosols observed by CALIOP on board CALIPSO and (b) types of aerosols deduced from CALIOP measurements, on June 26, 2019. The location of the transect is indicated in Figure 8.4a by a dotted magenta line (source: CALIPSO[6])*

In addition, passive satellite measurements also provide information on the aerosol altitude. Basically, an average altitude of the aerosol layers is deduced from thermal infrared radiation, as measured by IASI (Kylling et al. 2017) or using

5. https://earth.esa.int/eogateway/missions/earthcare.

6. https://www-calipso.larc.nasa.gov.

molecular oxygen absorption bands (Sanders et al. 2017; Chimot et al. 2018; Nanda et al. 2020). Similarly, the passive retrieval of vertical aerosol profiles is also possible from IASI but for desert dust (see Chapter 9). Very recently, vertical profiles of fine aerosols emitted by wildfires have been observed for the first time from passive measurements using TROPOMI high spectral resolution reflectance measurements at the oxygen A-band near 760 nm (Lemmouchi et al. 2022).

8.4. References

Al-Saadi, J., Szykman, J., Pierce, R.B., Kittaka, C., Neil, D., Chu, D.A., Fishman, J. (2005). Improving national air quality forecasts with satellite aerosol observations. *Bull. Am. Meteorol. Soc.*, 86(9), 1249–1262.

Bowman, K.W. et al. (2006). Tropospheric emission spectrometer: Retrieval method and error analysis. *IEEE Trans. Geosci. Remote Sens.*, 44(5), 1297–1307.

Boynard, A., Clerbaux, C., Coheur, P.-F., Hurtmans, D., Turquety, S., George, M., Hadji-Lazaro, J., Keim, C., Meyer-Arnek, J. (2009). Measurements of total and tropospheric ozone from IASI: Comparison with correlative satellite, ground-based and ozonesonde observations. *Atmos. Chem. Phys.*, 9(16), 6255–6271.

Chance, K.V., Burrows, J.P., Perner, D., Schneider, W. (1997). Satellite measurements of atmospheric ozone profiles, including tropospheric ozone, from ultraviolet/visible measurements in the nadir geometry: A potential method to retrieve tropospheric ozone. *J. Quant. Spectrosc. Rad. Transfer*, 57(4), 467–476.

Chance, K., Liu, X., Suleiman, R.M., Flittner, D.E., Al-Saadi, J., Janz, S.J. (2013). Tropospheric emissions: Monitoring of pollution (TEMPO). In *International Society for Optics and Photonics*, Butler, J.J., Xiong, X.J., Gu, X. (eds). 88660D, International Society for Optical Engineering.

Chimot, J. et al. (2017). An exploratory study on the aerosol height retrieval from OMI measurements of the 477 nm O_2–O_2 spectral band using a neural network approach. *Atmos. Meas. Tech.*, 10(3), 783–809.

Clerbaux, C., Hadji-Lazaro, J., Turquety, S., Mégie, G., Coheur, P.-F. (2003). Trace gas measurements from infrared satellite for chemistry and climate applications. *Atmos. Chem. Phys.*, 3(5), 1495–1508.

Coheur, P.-F., Barret, B., Turquety, S., Hurtmans, D., Hadji-Lazaro, J., Clerbaux, C. (2005). Retrieval and characterization of ozone vertical profiles from a thermal infrared nadir sounder. *J. Geophys. Res.*, 110(D24), D24303.

Cuesta, J., Eremenko, M., Liu, X., Dufour, G., Cai, Z., Höpfner, M., Von Clarmann, T., Sellitto, P., Foret, G., Gaubert, B. et al. (2013). Satellite observation of lowermost tropospheric ozone by multispectral synergism of IASI thermal infrared and GOME-2 ultraviolet measurements over Europe. *Atmos. Chem. Phys.*, 13(19), 9675–9693.

Cuesta, J., Kanaya, Y., Takigawa, M., Dufour, G., Eremenko, M., Foret, G., Miyazaki, K., Beekmann, M. (2018). Transboundary ozone pollution across East Asia: Daily evolution and photochemical production analysed by IASI + GOME2 multispectral satellite observations and models. *Atmos. Chem. Phys.*, 18, 9499–9525. https://doi.org/10.5194/acp-18-9499-2018.

Doche, C., Dufour, G., Foret, G., Eremenko, M., Cuesta, J., Beekmann, M., Kalabokas, P. (2014). Summertime tropospheric ozone variability over the Mediterranean basin observed with IASI. *Atmos. Chem. Phys.*, 14(9), 13021–13058.

Dubovik, O. et al. (2021). Multi-term LSM for applying multiple a priori constraints in problems of atmospheric remote sensing: GRASP algorithm, concept and applications. *Front. Remote Sens.*, October 19.

Dufour, G., Eremenko, M., Orphal, J., Flaud, J.-M. (2010). IASI observations of seasonal and day-to-day variations of tropospheric ozone over three highly populated areas of China: Beijing, Shanghai, and Hong Kong. *Atmos. Chem. Phys.*, 10(8), 3787–3801.

Dufour, G., Eremenko, M., Griesfeller, A., Barret, B., Leflochmoën, E., Clerbaux, C., Hadji-Lazaro, J., Coheur, P.F., Hurtmans, D. (2012). Validation of three different scientific ozone products retrieved from IASI spectra using ozonesondes. *Atmos. Meas. Tech.*, 5(3), 611–630.

Dufour, G., Eremenko, M., Cuesta, J., Doche, C., Foret, G., Beekmann, M., Cheiney, A., Wang, Y., Cai, Z., Liu, Y. et al. (2015). Springtime daily variations in lower-tropospheric ozone over east Asia: The role of cyclonic activity and pollution as observed from space with IASI. *Atmos. Chem. Phys.*, 15(18), 10839–10856.

Eremenko, M., Dufour, G., Foret, G., Keim, C., Orphal, J., Beekmann, M., Bergametti, G., Flaud, J.-M. (2008). Tropospheric ozone distributions over Europe during the heat wave in July 2007 observed from infrared nadir spectra recorded by IASI. *Geophys. Res. Lett.*, 35(18), 0–4.

European Environment Agency (2011). Air quality in Europe – 2011 technical report. Publications Office of the European Union, ISSN 1725–2237. doi:10. 2800/83213.

Fishman, J. and Larsen, J.C. (1987). Distribution of total ozone and stratospheric ozone in the tropics: Implications for the distribution of tropospheric ozone. *J. Geophys. Res.*, 92(D6), 6627–6627.

Fishman, J., Wozniak, A.E., Creilson, J.K. (2003). Global distribution of tropospheric ozone from satellite measurements using the empirically corrected tropospheric ozone residual technique: Identification of the regional aspects of air pollution. *Atmos. Chem. Phys.*, 3, 893–907.

Fu, D., Worden, J.R., Liu, X., Kulawik, S.S., Bowman, K.W., Natraj, V. (2013). Characterization of ozone profiles derived from Aura TES and OMI radiances. *Atmos. Chem. Phys.*, 13(6), 3445–3462.

Gaudel, A. et al. (2018). Tropospheric ozone assessment report: Present-day distribution and trends of tropospheric ozone relevant to climate and global atmospheric chemistry model evaluation [Online]. Available at: http://www.elementascience.org/articles/10.1525/elementa.291/.

Hayashida, S., Liu, X., Ono, A., Yang, K., Chance, K. (2015). Observation of ozone enhancement in the lower troposphere over East Asia from a space-borne ultraviolet spectrometer. *Atmos. Chem. Phys.*, 15(17), 9865–9881.

Kim, J., Jeong, U., Ahn, M.H., Kim, J.H., Park, R.J., Lee, H., Choi, Y. (2020). New era of air quality monitoring from space: Geostationary environment monitoring spectrometer (GEMS). *Bull. Am. Meteorol. Soc.*, 101(1), E1–E22.

Kulawik, S.S., Osterman, G., Jones, D.B.A., Bowman, K.W. (2006). Calculation of altitude-dependent Tikhonov constraints for TES nadir retrievals. *IEEE Trans. Geosci. Remote Sens.*, 44(5), 1334–1342.

Kylling A., Vandenbussche, S., Capelle, V., Cuesta, J., Klüser, L., Lelli, L., Popp, T., Stebel, K., Veefkind, P. (2018). Comparison of dust-layer heights from active and passive satellite sensors. *Atmos. Meas. Tech.*, 11, 2911–2936. https://doi.org/10.5194/amt-11-2911-2018.

Landgraf, J. and Hasekamp, O.P. (2007). Retrieval of tropospheric ozone: The synergistic use of thermal infrared emission and ultraviolet reflectivity measurements from space. *J. Geophys. Res.*, 112(D8), D08310.

Lemmouchi, F., Cuesta, J., Eremenko, M., Derognat, C., Siour, G., Dufour, G., Sellitto, P., Turquety, S., Tran, D., Liu, X. et al. (2022). Three-dimensional distribution of biomass burning aerosols from Australian wildfires observed by TROPOMI satellite observations. *Remote Sens.*, 14(11), 2582. https://doi.org/10.3390/rs14112582.

Liu, X. et al. (2005). Ozone profile and tropospheric ozone retrievals from the global ozone monitoring experiment: Algorithm description and validation. *J. Geophys. Res.*, 110(D20), D20307.

Liu, X. et al. (2010). Ozone profile retrievals from the ozone monitoring instrument. *Atmos. Chem. Phys.*, 10(5), 2521–2537.

Molina, M.J. and Molina, L.T. (2004). Megacities and atmospheric pollution, *J. Air Waste Manag. Assoc.*, 54(6), 644–680.

Monks, P.S. et al. (2015). Tropospheric ozone and its precursors from the urban to the global scale from air quality to short-lived climate forcer. *Atmos. Chem. Phys.*, 15(15), 8889–8973.

Munro, R., Siddans, R., Reburn, W.J., Kerridge, B.J. (1998). Direct measurement of tropospheric ozone distributions from space. *Nature*, 392(6672), 168–171.

Nanda, S., Graaf, M.D., Veefkind, J.P., Sneep, M., Linden, M.T., Sun, J., Levelt, P.F. (2020). A first comparison of TROPOMI aerosol layer height (ALH) to CALIOP data. *Atmos. Meas. Tech.*, 13(6), 3043–3059.

Nelson, D.L., Garay, M.J., Kahn, R.A., Dunst, B.A. (2013). Stereoscopic height and wind retrievals for aerosol plumes with the MISR INteractive eXplorer (MINX). *Remote Sens.*, 5(9), 4593–4628.

Remer, L.A. et al. (2005). The MODIS aerosol algorithm, products, and validation, *J. Atmos. Sci.*, 62(4), 947–973.

Rodgers, C.D. (2000). *Inverse Methods for Atmospheric Sounding: Theory and Practice 2.* World Scientific, Singapore.

Safieddine, S., Clerbaux, C., George, M., Hadji-Lazaro, J., Hurtmans, D., Coheur, P.-F., Wespes, C., Loyola, D., Valks, P., Hao, N. (2013). Tropospheric ozone and nitrogen dioxide measurements in urban and rural regions as seen by IASI and GOME-2. *J. Geophys. Res.*, 118(18), 10,555–10,566.

Sanders, A.F.J., De Haan, J.F., Sneep, M., Apituley, A., Stammes, P., Vieitez, M.O., Veefkind, J.P. (2015). Evaluation of the operational aerosol layer height retrieval algorithm for Sentinel-5 Precursor: Application to O_2 A band observations from GOME-2A. *Atmos. Meas. Tech.*, 8(11), 4947–4977.

Sayer, A.M. et al. (2013). Validation and uncertainty estimates for MODIS Collection 6 Deep Blue aerosol data. *J. Geophys. Res. Atmos.*, 118, 7864–7872.

Shen, L., Jacob, D.J., Liu, X., Huang, G., Li, K., Liao, H., Wang, T. (2019). An evaluation of the ability of the ozone monitoring instrument (OMI) to observe boundary layer ozone pollution across China: Application to 2005–2017 ozone trends. *Atmos. Chem. Phys.*, 19(9), 6551–6560.

Sogacheva, L. et al. (2020). Merging regional and global aerosol optical depth records from major available satellite products. *Atmos. Chem. Phys.*, 20(4), 2031–2056.

Winker, D.M., Vaughan, M.A., Omar, A., Hu, Y., Powell, K.A., Liu, Z., Hunt, W.H., Young, S.A. (2009). Overview of the CALIPSO mission and CALIOP data processing algorithms. *J. Atmos. Ocean. Technol.*, 26(11), 2310–2323.

Worden, J. et al. (2004). Predicted errors of tropospheric emission spectrometer nadir retrievals from spectral window selection. *J. Geophys. Res.*, 109(D9), D09308.

Worden, H.M. et al. (2007a). Comparisons of tropospheric emission spectrometer (TES) ozone profiles to ozonesondes: Methods and initial results. *J. Geophys. Res.*, 112(D3), D03309.

Worden, J., Liu, X., Bowman, K., Chance, K., Beer, R., Eldering, A., Gunson, M., Worden, H. (2007b). Improved tropospheric ozone profile retrievals using OMI and TES radiances. *Geophys. Res. Lett.*, 34(1), L01809.

Ziemke, J.R., Chandra, S., Bhartia, P.K. (1998). Two new methods for deriving tropospheric column ozone from TOMS measurements: Assimilated UARS MLS/HALOE and convective-cloud differential techniques. *J. Geophys. Res.*, 103(D17), 22115–22127.

Ziemke, J.R. et al. (2011). A global climatology of tropospheric and stratospheric ozone derived from Aura OMI and MLS measurements. *Atmos. Chem. Phys.*, 11(17), 9237–9251.

9

Desert Dust

Juan CUESTA
LISA – UMR7583, Créteil, France

9.1. Introduction

Massive amounts of dust are lifted into the atmosphere over the vast tropical desert region, commonly known as the "Dust Belt", comprising the Sahara, the Near and Middle East, all the way to Mongolia and northern China. In terms of mass concentration, desert dust is one of the most abundant aerosol species in the atmosphere. It is a major player in the climate and terrestrial system (Knippertz and Stuut 2014). Given their lifetime in the atmosphere of a few days to a week, these dusts can be carried by winds for several thousand kilometers, arriving at mid-latitudes and even to the poles. Their multiple environmental impacts are therefore significant from the regional to the intercontinental scales.

The life cycle of desert dust and its many environmental impacts (Mahowald et al. 2001) are closely related to its three-dimensional (3D) spatial distribution. Once lifted from the surface above source regions, dust is vertically mixed by turbulence and carried by the wind to remote areas. It directly alters the Earth's energy balance, by absorbing and scattering solar radiation and interacting with telluric radiation. As it absorbs light, dust heats the atmospheric layers in which it is present, thus altering the stability of the atmosphere, its circulation and weather conditions. Dust can also cause changes in cloud properties and chemical processes in the atmosphere near the altitude at which it is transported. Close to the surface, desert dust can severely degrade air quality and drastically reduce visibility,

---

For a color version of all the figures in this chapter, see www.iste.co.uk/phulpin/satellites2.zip.

Satellites for Atmospheric Sciences 2,
coordinated by Thierry PHULPIN, Didier RENAUT, Hervé ROQUET and Claude CAMY-PEYRET.

affecting living conditions in downwind populated desert regions (Wang et al. 2012). As they move, minerals are deposited over oceans and continents, providing nutrients that can be integrated by ecosystems. Similarly, their deposition modifies the color of snow and ice surfaces, even over the poles, and thus, the intensity of light absorbed and reflected by these last ones. Quantifying these effects is a major challenge, particularly because of the sporadic nature of dust emissions, the meteorological phenomena that control their transport, and the great variability of their optical or mineralogical properties. The observation of the geographical extent of dusts and their vertical distribution is therefore essential to assess their (major) impact on the environment.

Currently, satellite observations play a fundamental role in describing the horizontal distribution of dust plumes and their transport in the atmosphere. However, space-based observations of dust over deserts have required specific developments to distinguish their spectral signatures from those of the surface. In fact, the reflectivity of dust in the visible spectrum is very similar to that of the desert surfaces from which it is blown. As a result, the classical "Dark Target" type of approaches (see Chapter 8) used with MODIS or VIIRS measurements to quantify the aerosol optical depth (AOD) are not applicable over deserts. This chapter presents a synthesis of satellite approaches dedicated to the observation of desert aerosols from space. These methods of remote sensing of desert dust allow its detection and its quantification in terms of optical depth in the UV and infrared spectral domains. It is also the case for the retrieval of their vertical profile using active sounding and spectral analyses in thermal infrared radiation (TIR). This final one also enables analysis of the 3D distribution of desert dust.

9.2. Qualitative satellite detection of desert dust

The first approaches proposed to identify the location of dust plumes over deserts (bright surfaces) are based on semi-quantitative indicators, obtained in the UV range using observations from TOMS (Prospero et al. 2002), OMI, GOME and GOME-2. The most commonly used aerosol index (AI) is calculated as follows (Torres et al. 2007):

$$AI = -100\, log_{10}\left[\frac{I_{354}^{obs}}{I_{354}^{no\ aero}\, R_{354}^{surf}}\right]$$

Here, I^{obs}_{354} is the reflectance measured by the satellite at 354 nm, $I^{no\ aero}_{354}$ is calculated assuming the absence of aerosols in the atmospheric column, and R^{surf}_{354} is the surface reflectance (assumed to be Lambertian). AI is the difference between the reflectance measured in the presence of dust and that calculated in its absence. This index has been used in numerous studies to map the main source areas where dust is most commonly detected (Prospero et al. 2002). The index also identifies the horizontal extent of dust plumes during horizontal transport in the atmosphere, as illustrated in Figure 9.1a for a major event, that is, a dust storm over Central Sahara on June 18, 2011.

Other approaches for detecting dust in the thermal infrared range have also been proposed from observations in several spectral channels of the AVHRR and SEVIRI instruments (Schepanski et al. 2012). These imagers provide luminance measurements (we will also use the term radiance) in various channels (12 for SEVIRI) with good spatial resolution (3 km) and high repeatability (15 minutes for SEVIRI). An indicator (called "thermal anomaly") is shown in Figure 9.1b with a false-color image generated from the differences between three channels in the TIR, after subtracting the surface contribution (roughly estimated as the 10-day average value).

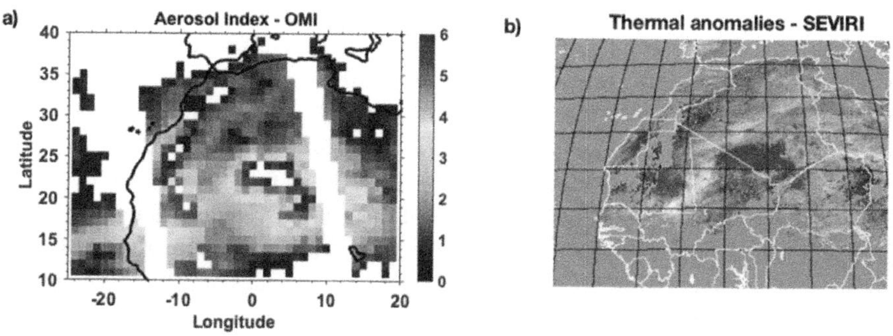

Figure 9.1. *Horizontal distribution of Saharan dust on June 18, 2011, observed in terms of (a) aerosol indexes AI in the UV of OMI*[1] *and (b) "thermal anomalies" derived from SEVIRI*[2]*. Desert dust is indicated in red*

Limitations of indices such as AI in the UV and thermal anomalies in the TIR are that their values depend not only on the amount of dust and its optical depth, but also on other factors influencing radiances reaching the instruments such as the height of the dust in the atmosphere, humidity, surface properties, etc. Therefore, these indices are only used for depicting the spatio-temporal occurrence of dust.

1. https://giovanni.gsfc.nasa.gov.
2. http://nascube.univ-lille1.fr.

9.3. Satellite observation of the optical depth of desert dust

Methods to quantitatively determine the optical depth of dust from space are typically based on calculations of radiative transfer in the atmosphere. The latter simulate the radiance measured by sensors by explicitly accounting for the interactions of radiation with dust, with other atmospheric compounds (molecules) and with the surface. Assumptions must be made about dust properties such as size distribution and complex refractive index. Similarly, inversion methods must be used to derive aerosol optical depths from radiances. This can be done by finding the solution in a so-called multi-dimensional look-up table or by iteratively minimizing a cost function. Constraints on the solution make it possible to obtain solutions that are compatible with the natural variability of dust properties, as done by the optimal estimation approach (Rodgers 2000). In addition to these classical methods based on physical principles, the optical depth of dust can also be inferred from statistical approaches such as artificial neural networks or principal component analysis.

Physical approaches are implemented, for example for optical depth products from the "Deep Blue" method applied to MODIS UV and blue measurements (Sayer et al. 2013) over the desert and by the "Dark Target" method elsewhere (see Chapter 8). OMI's "multi-wavelength" retrieval (Torres et al. 2007) and MISR's "multi-angular" method (Diner et al. 2001) are illustrated in Figure 9.2. These three methods use different strategies to distinguish the contributions of eolian dust and the desert surface from the measured radiances. "Deep Blue" uses the UV reflectivity of the surface, which is somewhat lower than that of dust in the visible and near infrared. The multi-wavelength method is applicable to reflectance spectra measured by OMI and uses the spectral variation of UV and visible reflectances at 19 wavelengths to distinguish dust from the surface. This multi-wavelength method is also used in TIR with radiance spectra measured by IASI and AIRS sounders. The MISR method uses the angular dependence of radiation scattering, which differs significantly between particles and the surface.

The accuracy of these methods is closely related to the assumptions made for the retrieval, mainly on the optical properties of the dust and the surface. This explains the differences between the optical depth of dust derived, for example by MODIS, OMI or MISR, and the reference measurements made by the AERONET ground-based sun photometer. Figure 9.2 shows that the horizontal structures of the dust plumes retrieved by MODIS and OMI are similar. However, there is an underestimation of the optical depth of the central plume for MODIS and a better agreement with the photometers for OMI. The spatial coverage of MISR is significantly smaller, with a swath about 10 times narrower than that of the other two sensors.

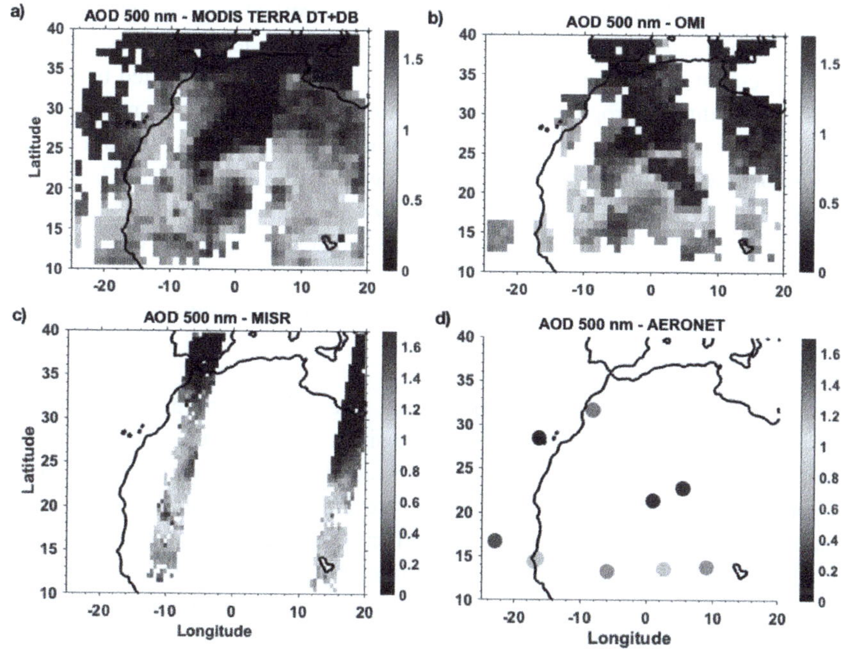

Figure 9.2. *Horizontal distribution of the optical depth of Saharan dust on June 18, 2011 derived by (a) MODIS Deep Blue "DB" over the desert and Dark Target "DT" over the sea, (b) OMI with the multi-wavelength method, (c) MISR with a multi-angular approach and (d) ground-based sun photometers of the AERONET network*[3]. *Satellite products are accessible through NASA's GIOVANNI portal*[4]

9.4. Vertical profiles of desert dust by spaceborne lidar

As mentioned in Chapter 8, the vertical distribution of aerosols in the atmosphere can be measured with high spatial resolution thanks to the capability of CALIOP, a nadir-pointing spaceborne lidar onboard the CALIPSO satellite since 2006 (Winker et al. 2009). Aerosol plumes within the very fine footprint of the lidar (90 m wide) are sampled along the CALIPSO orbit. Horizontal transport of aerosol plumes is difficult to track since the satellite tracks are separated by about 2,000 km in longitude on a given day (during daytime or nighttime) at mid-latitudes. Nevertheless, the three-dimensional (3D) distribution of aerosols can be inferred from CALIOP on seasonal and interannual scales by averaging numerous transects.

---

3. https://aeronet.gsfc.nasa.gov.
4. https://giovanni.gsfc.nasa.gov.

The CALIOP lidar measures vertical atmospheric reflectivity profiles (at 532 nm and 1,064 nm) that characterize the vertical structure of aerosol layers. However, this measurement is qualitative because the laser radiation is attenuated by particles and molecules during propagation in the atmosphere. An inversion method of the lidar measurements must be used to withdraw the effects of atmospheric transmission and quantitatively derive the vertical profiles of particles. The classical approach called "Fernald" (Fernald et al. 1972) or "Klett" assumes a constant vertical profile of the ratio of extinction and backscatter of particles. This hypothesis assumes that the intrinsic properties of the particles, such as size and composition, are homogenous in the atmosphere and that the light is scattered only once (i.e. simple scattering conditions).

In the case of desert dust, an additional phenomenon takes place: multiple scattering. It is related to the relatively large size of these particles with respect to the wavelength, as a significant part of the radiation is scattered in the forward direction. This radiation can then be scattered several times along the propagation of the laser. To account for this effect, an effective value of the ratio between the extinction and backscatter of dust is estimated by using a multiple scattering coefficient η. According to Monte Carlo simulations (Young et al. 2008), the multiple scattering coefficient varies with altitude as it is stronger for dust layers closer to the lidar and weaker for those further away. This method was used to retrieve extinction profiles of desert dust above the Sahara on June 17, 2011 (Figure 9.3), even for relatively dense layers (thus inducing increased multiple scattering).

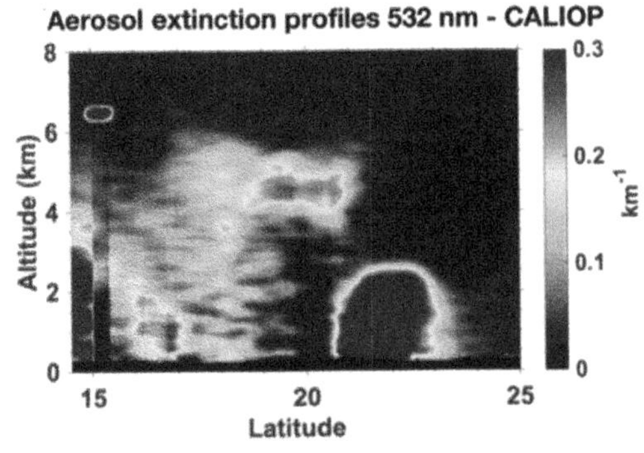

Figure 9.3. *Transect of vertical profiles of desert dust extinction coefficients retrieved from active measurements from the CALIOP lidar on board the CALIPSO satellite on June 17, 2011, over the Sahara (see the satellite track on Figure 9.5b and 9.5c)*

9.5. 3D distribution of desert dust by infrared spectrometer

Satellite observations are essential to characterize the regional and continental distribution of aerosols. However, classical satellite methods are uniquely capable of observing the spatial distribution of aerosols in 2D. On the one hand, the horizontal distribution of dust is estimated in terms of optical depth using satellite sensors such as MODIS. On the other hand, the vertical distribution of aerosols in the atmosphere can be obtained with high vertical resolution thanks to the CALIOP space lidar. However, only the aerosol plumes below the path of the lidar are sampled.

A satellite approach, called AEROIASI, has been developed to observe desert dust plumes in 3D on a twice-daily basis (Cuesta et al. 2015). This method is based on the use of hyperspectral measurements in the thermal infrared of the IASI satellite sensor. A set of spectral microwindows were selected to simultaneously derive the vertical distribution of dust, its optical depth and the surface temperature from each cloud-free measurement of IASI, both over land and over sea. AEROIASI retrieves vertical profiles of extinction coefficients at 10 μm of desert dust for most IFOVs (Instantaneous Field Of View) of IASI, using an auto-adaptive Tikhonov–Phillips type of regularization. This flexible tool is designed for retrieving the 3D distribution of aerosols in a large variety of atmospheric and surface conditions.

Figure 9.4 illustrates the spectral effects of desert dust on IASI measurements. The spectral signatures are significantly different for AOD changes (red), the vertical distribution of dust (green) and surface temperatures (yellow). Considering three parts of the spectrum at intervals of 800–910, 910–980 and 1,070–1,200 cm^{-1}, the spectral slope of the radiation changes has different signs for the three variables (Figure 9.4b). For a given AOD value, colder dust layers (located at high altitudes) emit less radiation in the TIR. This reduction in thermal emission is more significant at 800 and at 1,070 cm^{-1}, but less at 910 and at 1,200 cm^{-1}. These modulations induce changes in the spectral slope. An increase in AOD leads to a greater reduction of infrared radiation values at 910–980 cm^{-1}, where bands of intense absorption of radiation by dust exist (Di Biagio et al. 2017). On the contrary, the spectral signatures associated with surface temperature changes are different from those associated with dust, only leading to a change in the slope at 1,050 cm^{-1}. In addition, the spectral signature of many absorption lines of molecular species (i.e. peaks in the spectra) is also different for the four atmospheric scenes.

The vertical distribution of dust retrieved from passive IASI measurements (Figure 9.5a) shows a remarkable agreement with the vertical profile transect of the CALIOP space lidar (Figure 9.3). The same dust layers with similar vertical extent are found in both measurements. Active and passive observations clearly show the same two dust plumes reaching the surface, extending to 5–6 km and 2.5 km of altitude, and a dust layer 3.5–5.5 km high. This comparison illustrates the potential of IASI measurements for analyzing the 3D distribution of desert dust, as shown in the example of Figure 9.5b.

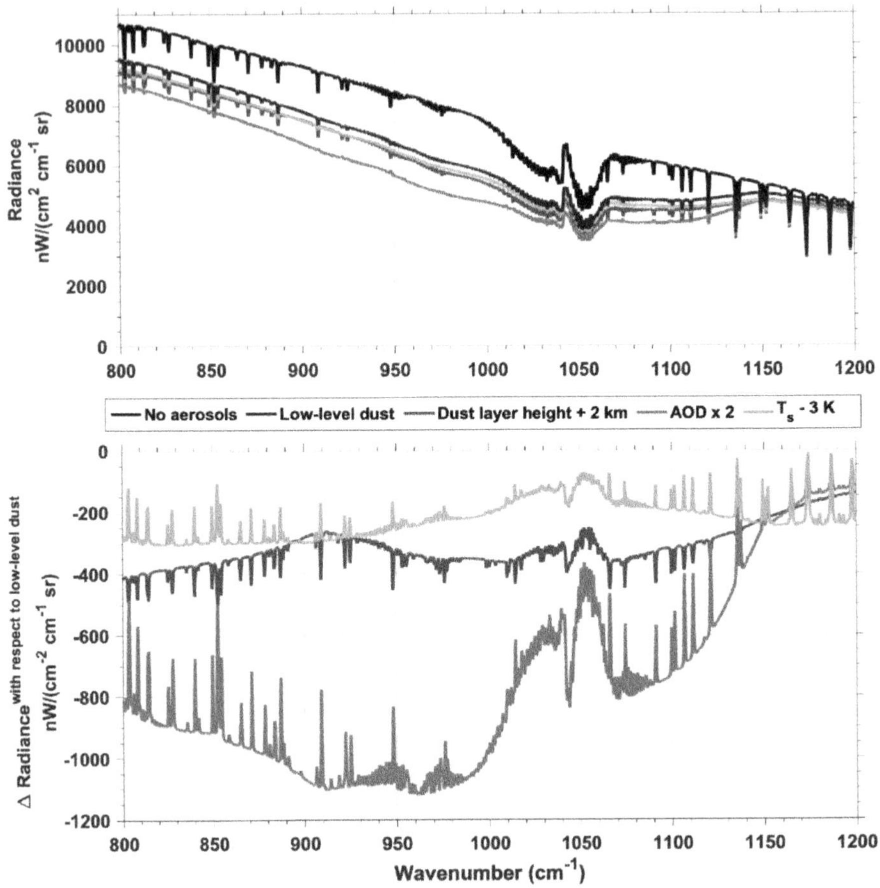

Figure 9.4. *(Upper panel) IASI spectra simulated in the thermal infrared for four atmospheric scenes: no aerosols (black), a dust layer between the surface and 2 km of altitude (blue), the same layer of dust but 2 km higher (green), with an AOD twice as large (red) or with a surface temperature reduced by 3 K (yellow). (Lower panel) Difference between IASI spectra when the dust layer is 2 km higher (green), AOD twice as high (red) and surface temperature 3 K colder (yellow) than when desert dust is closer to the surface*

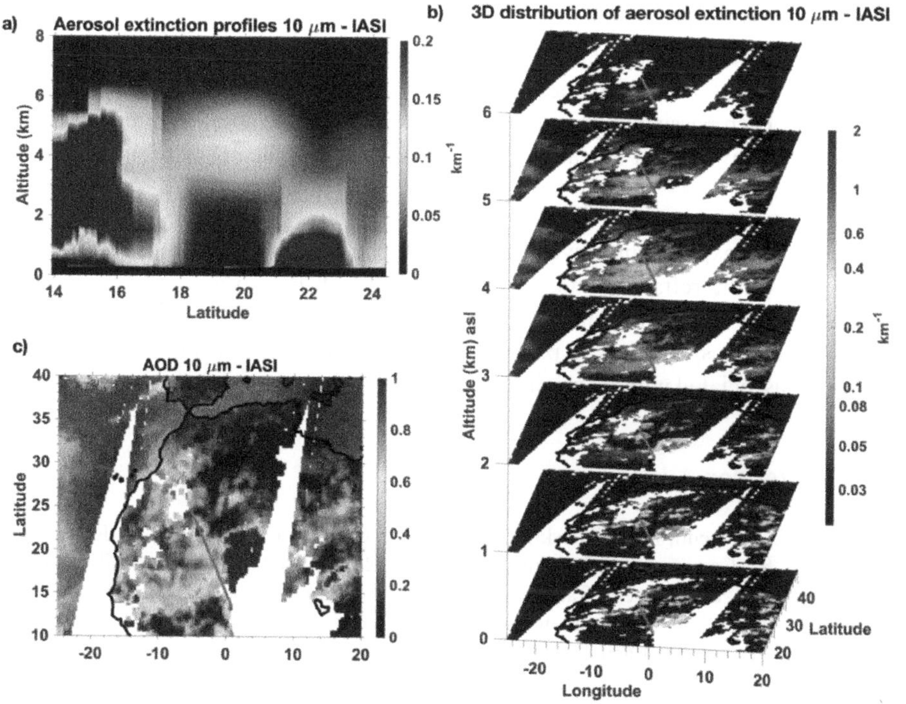

Figure 9.5. *Distribution of dust over the Sahara on June 17, 2011, retrieved from passive IASI measurements with the AEROIASI method, in terms of (a) a transect of vertical dust extinction profiles (co-located in the lidar transect of Figure 9.3), (b) the horizontal distribution of dust optical depth and (c) the 3D distribution of dust extinction. The transect in panel (a) is obtained by taking the average of the profiles retrieved for the IASI IFOVs at ± 1° around the CALIOP track (indicated by magenta lines in (b) and (c))*

9.6. Conclusion

Satellite observations of desert dust have significantly evolved over the past few decades. Initially, dust was detected only qualitatively, with indices depending on optical quantities that allowed for the identification of its source regions and horizontal transport patterns. Thanks to specific approaches to better identify dust aerosols from the radiative signature of the surface, the horizontal distribution of dust could be quantified in terms of AOD.

Since 2007, the vertical distribution of dust has been studied using measurements from the CALIOP active sounder. However, this is done in a semi-quantitative way, since assumptions about the optical properties of dust are necessary and only for north–south transects separated daily by about 2,000 km in longitude. In recent years, the use of IASI TIR spectra has allowed for the retrieval of vertical dust profiles for each passive measurement of this sensor (in the absence of clouds). The quasi–global coverage of the latter enables observation of the 3D distribution of dust twice a day.

Laboratory experiments have allowed for high-resolution spectral measurements in the TIR of the refractive indices of desert dust (Di Biagio et al. 2017). We can clearly identify distinct signatures in absorption according to the deserts from which the dust originates, opening up promising means for remote sensing of specific dust properties associated with its origin.

9.7. References

Biagio, C.D., Formenti, P., Balkanski, Y., Caponi, L., Cazaunau, M., Pangui, E., Journet, E., Nowak, S., Caquineau, S., Andreae, M. et al. (2017). Global scale variability of the mineral dust long-wave refractive index: A new dataset of in situ measurements for climate modeling and remote sensing. *Atmos. Chem. Phys.*, 17, 1901–1929. https://doi.org/10.5194/acp-17-1901-2017.

Cuesta, J., Eremenko, M., Flamant, C., Dufour, G., Laurent, B., Bergametti, G., Hopfner, M., Orphal, J., Zhou, D. (2015). Three-dimensional distribution of a major desert dust outbreak over East Asia in March 2008 derived from IASI satellite observations. *J. Geophys. Res.*, 120, 7099–7127.

Diner, D.J., Abdou, W.A., Conel, J.E., Crean, K.A., Gaitley, B.J., Helmlinger, M.C., Kahn, R.A., Martonchik, J.V., Pilorz, S.H. (2001). MISR aerosol retrievals over southern Africa during the SAFARI-2000 dry season campaign. *Geophys. Res. Lett.*, 28, 3127–3130.

Fernald, F.G., Herman, B.M., Reagan, J.A. (1972). Determination of aerosol height distributions by lidar. *J. Appl. Meteorol.*, 11, 482–489.

Knippertz, P. and Stuut, J.B.W. (2014). *Mineral Dust*. Springer, Dordrecht.

Mahowald, N.M., Kloster, S., Engelstaedter, S., Moore, J.K., Mukhopadhyay, S., McConnell, J.R., Albani, S., Doney, S.C., Bhattacharya, A., Curran, M.A.J., et al. (2001). Observed 20th century desert dust variability: Impact on climate and biogeochemistry. *Atmos. Chem. Phys.*, 10(22), 10875–10893.

Prospero, J.M., Ginoux, P., Torres, O., Nicholson, S.E., Gill, T.E. (2002). Environmental characterization of global sources of atmospheric soil dust identified with the NIMBUS 7 total ozone mapping spectro-meter (TOMS) absorbing aerosol product. *Rev. Geophys.*, 40(1), 1002. doi:10.1029/2000RG000095.

Rodgers, C.D. (2000). *Inverse Methods for Atmospheric Sounding: Theory and Practice*. World Scientific Publishing Company, London.

Sayer, A.M., Hsu, N.C., Bettenhausen, C., Jeong, M.-J. (2013). Validation and uncertainty estimates for MODIS Collection 6 Deep Blue aerosol data. *J. Geophys. Res.*, 14, 7864–7872.

Schepanski, K., Tegen, I., Macke, A. (2012). Comparison of satellite based observations of Saharan dust source areas. *Remote Sens. Env.*, 123, 90–97.

Torres, O., Tanskanen, A., Veihelmann, B., Ahn, C., Braak, R., Bhartia, P.K., Veefkind, P., Levelt, P. (2007). Aerosols and surface UV products from ozone monitoring instrument observations: An overview. *J. Geophys. Res.*, 112(D24).

Young, S., Winker, D., Vaughan, M., Hu, Y., Kuehn, R. (2008). Extinction retrieval algorithms, CALIOP algorithm theoretical basis document Part 4 [Online]. Available at: http://www-calipso.larc.nasa.gov/resources/pdfs/PC-SCI-202_Part4_v1.0.pdf.

Wang, K., Zhang, Y., Nenes, A., Fountoukis, C. (2012). Implementation of dust emission and chemistry into the community multiscale air quality modeling system and initial application to an Asian dust storm episode. *Atmos. Chem. Phys.*, 12, 10209–10237. doi:10.5194/acp-12-10209-2012.

Winker, D.M., Vaughan, M.A., Omar, A., Hu, Y., Powell, K.A., Liu, Z., Hunt, W.H., Young, S.A. (2009). Overview of the CALIPSO mission and CALIOP data processing algorithms. *J. Atmos. Ocean. Tech.*, 26(11), 2310–2323.

10

Species Emitted by Fires

Camille VIATTE[1] **and Pasquale SELLITTO**[2]
[1] *LATMOS/IPSL, Sorbonne Université, UVSQ, CNRS, Paris, France*
[2] *LISA – UMR7583, Créteil, France*

10.1. Introduction

Fires play an important role in the global emission budget of greenhouse and reactive gases, as well as particulate matter, in the atmosphere (Table 10.1). They infer large perturbations to atmospheric composition and reactivity, and affect climate and human health. As feedback, fire frequency and intensity are sensitive to climate change and climate variability, as well as land use practices, and more severe fires are expected in the future in many parts of the world (IPCC 2022).

Emissions from biomass burning are known to be a source of greenhouse gases such as carbon dioxide (CO_2) and methane (CH_4), and thus directly affect climate. Beyond that, biomass burning is a considerable source of chemically active gases such as carbon monoxide (CO), volatile organic compounds (VOC), nitric oxide (NO_x) and ammonia (NH_3). These emissions affect the oxidation capacity of the troposphere by reacting with OH radicals, leading to the formation of ozone and other photo oxidants.

Biomass burning is the main global source of aerosol like black carbon (BC) and primary organic aerosol (POA). In addition, through atmospheric chemical and microphysical processes, some of the gaseous emissions of biomass burning can undergo gas-to-particle conversion and be an important atmospheric source of secondary organic and sulfate aerosols. Particulate matter emitted by fires in the lower troposphere and the boundary layer can have a negative impact on air quality

Satellites for Atmospheric Sciences 2,
coordinated by Thierry PHULPIN, Didier RENAUT, Hervé ROQUET and Claude CAMY-PEYRET.

and affect ecosystems and human health. Through scattering and absorption of shortwave and longwave radiation, these aerosols can have an important direct impact on the climate system. In addition, biomass burning aerosols can modify cloud occurrence and properties, thus producing an indirect impact on climate. Through pyro-convection and the peculiar optical properties of the very absorbing BC particles that dominate fire-generated plumes, some fires can be lofted at very high altitudes, that is, up to the stratosphere, thus impacting on the stratospheric composition and climate at the global scale. Peculiar phenomena can result from the radiative–dynamics interaction linked to the strong absorption of sunlight and the resulting localized radiative heating of the absorbing aerosols emitted by fires, like the recently discovered smoke-charged anticyclonic/rising vortices observed during the record-breaking Australian bushfire in 2019–2020 (Khaykin et al. 2021).

Fire emissions depend critically on the type of the burned vegetation. Since the 1980s, numerous studies based on field measurement campaigns and laboratory experiments have provided information on biomass burning emissions for different vegetation types around the world (Andreae et al. 1988; Akagi et al. 2011; Andreae 2019). The so-called Emission Factor (EF) quantifies the mass of a specific gas or aerosol emitted per dry biomass burned for many different biomes. To estimate the impact of fires on climate and atmospheric chemistry, global fire emission inventories need to be accurately estimated using satellite-based instruments which provide simultaneous observations of different gaseous species and particles types. Recently, NO_2 and CO TROPOMI observations have been used to estimate combustion efficiency, making it possible to distinguish flaming from smoldering combustion (van der Velde et al. 2021).

Several satellite systems have proven effective as fire monitoring tools and provide suitable information required to better understand the temporal and spatial distribution of biomass burning products.

	Total ($Tg.a^{-1}$)
CO_2	13,900
CO	820
Total VOCs	62
CH_4	50
NO_x	19
NH_3	10
Total particulate matter	85

Table 10.1. *Global emission ($Tg.a^{-1}$) of selected biomass burning species (source: Andreae (2019))*

10.2. Biomass burning gases

10.2.1. *Greenhouses gases*

Fires contribute to about 20% of global GHG emissions. CO_2 emitted by fires will be mostly taken up by vegetation for regrowth, whereas parts of GHGs will remain in the atmosphere for years and potentially even up to centuries. GHG fire monitoring is therefore challenging but we are still able to monitor CO_2 and CH_4 using SCIAMACHY, GOSAT missions, OCO missions (Guo et al. 2019) and future missions like MERLIN (CH_4), MicroCarb (CO_2) and ASCENDS, among others.

Extreme Australian fires released more than 800 million tons of CO_2 from November 2019 to January 2020, which is roughly equivalent to the entire annual emissions of Germany (van der Velde et al. 2021). We may fear that fires would be a major contributor to global warming, but 80% of this carbon dioxide has been absorbed by ocean algal bloom (Tang et al. 2021).

10.2.2. *Carbon monoxide (CO)*

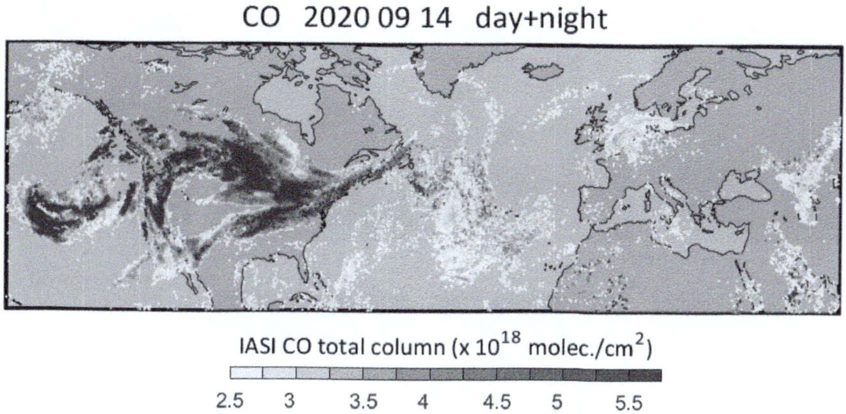

Figure 10.1. *CO concentrations derived from IASI during California fires occurring September 14, 2020. Courtesy of Maya George (LATMOS). For a color version of this figure, see www.iste.co.uk/phulpin/satellites2.zip*

CO is the second most important species emitted by fires after CO_2. With its relatively long lifetime in the atmosphere (1–2 months), CO is an excellent tracer of fire plume and pollutant transport. Thus, satellite observations of CO began in the 1980s with MAPS on-board Challenger and continue nowadays with the MOPITT instrument since 2000, the IASI missions since 2007 and TROPOMI since 2017, among others.

As an example, in August and September 2020, the California fire plumes that crossed the Atlantic Ocean were monitored using IASI CO observations from MetOp-A, -B and -C (Figure 10.1). These devastating fires emitted large amounts of CO which circulated the planet.

10.2.3. *Volatile organic compounds (VOCs)*

During massive wildfires that erupted in Amazonia in summer 2019, substantial amounts of VOCs were emitted directly by the fires as well as being rapidly produced in plumes via the degradation of short-lived gas precursors. The IASI instruments enable monitoring of various VOCs, such as formic acid, acetic acid, PAN, ammonia, methanol, acetylene, hydrogen cyanide and carbon monoxide (Figure 10.2).

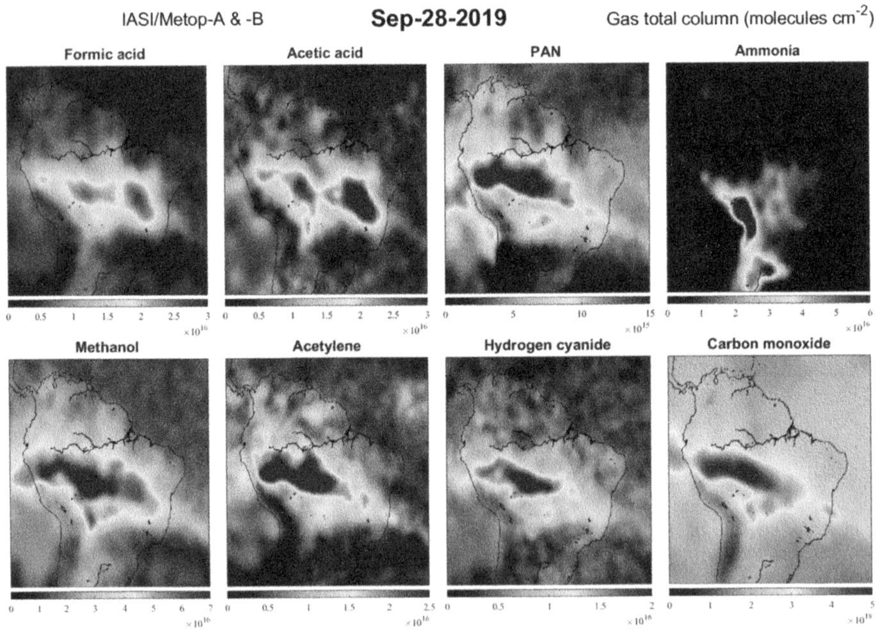

Figure 10.2. *Vertical abundance (total column) of gas fire tracers over the Amazon Basin, September 28, 2019, derived from the infrared observations of the IASI satellite sounder on board the MetOp-A and -B platforms (source: Franco et al. (2020)). For a color version of this figure, see www.iste.co.uk/phulpin/satellites2.zip*

10.2.4. *Ammonia (NH3)*

Ammonia is a short-lived species emitted by fires (Paulot et al. 2017) with roughly 60–70% originating from fires in Africa. Recently, spaceborne instruments, such as IASI (Whitburn et al. 2016), TES, (Luo et al. 2015) and CrIS (Shephard et al. 2020) have been used to quantify the enhancement of NH_3 over fires. A recent study found that NH_3 concentrations increased in West Africa as a result of biomass burning intensification in the region (Hickman et al. 2021).

10.2.5. *Nitrous acid (HONO)*

Nitrous acid (HONO) is a precursor of OH, which controls the degradation of greenhouse gases, contributes to photochemical smog and ozone production, and influences air quality. Using recent TROPOMI measurements, it has been shown that previous assessments underestimate pyrogenic HONO emissions by a factor of 2–4 across all ecosystem types (Theys et al. 2021). Using a combination of CrIS and IASI observations, the dispersion of HONO plumes from the Australian bushfires in 2019/20 has been tracked daily (Dufour et al. 2021).

10.3. Biomass burning aerosols

10.3.1. *AOD observations with nadir-viewing instruments*

Broad-band nadir-viewing instruments can measure the aerosol optical depth (AOD) and, in the case of multiple channels, can also derive the spectral variability of the AOD (possibly in terms of the compact Angström exponent); these two parameters can help to identify aerosols from biomass burning based on their optical properties. The MODIS sensor has been used to study fire aerosol perturbations based on spatio-temporal variability of its AOD observations (Yang et al. 2021). Exploiting the information on spectral and morphological variability of observations taken at different overpass time intervals and associated with deposited aerosols and ground charcoal, MODIS can provide information on burned area (Giglio et al. 2018). In addition to the generic AOD, GOES also provides a product for smoke aerosol detection, based on its known spectral variability of optical properties.

High-spectral resolution nadir instruments operating in the UV/VIS have long been used to detect absorbing aerosols, including those from fires, by means of the ultraviolet aerosol index, UVAI. The S5P TROPOMI instrument has recently been used to exploit its very high spatial resolution, in combination with an updated version of the UVAI algorithm, to detect and follow fire plumes (Torres et al. 2020).

The high spectral resolution of IR nadir-viewing instruments such as, for example, IASI (Clarisse et al. 2013), has been exploited to characterize the specific spectral signature of smoke aerosols using high-resolution spectral fitting methods.

10.3.2. *Extinction observations with limb-viewing instruments*

In the case of upper-tropospheric and stratospheric injections of fire plumes, aerosol can be observed with limb-viewing instruments. The SAGE series has been in operation since 1979; the most recent of the series, SAGE-III/ISS is mounted to the International Space Station (ISS). SAGE-III measures the aerosol extinction coefficient at multiple UV/VIS wavelengths in the upper-troposphere and the stratosphere at high signal-to-noise ratio, thanks to its solar occultation geometry of observations. The limb profiler of the OMPS mission (OMPS-LP) measures scattered solar UV/VIS radiation, which provides aerosol extinction observations in the upper-troposphere and stratosphere with a lower signal-to-noise ratio than the occultation systems like SAGE, albeit with better spatio-temporal coverage. The OMPS-LP provides multiple wavelength aerosol information (Taha et al. 2021). An example of using the aerosol extinction of limb instruments and its spectral variability to discriminate fire aerosol from anthropogenic aerosols has been discussed (Kloss et al. 2019) for the British Columbia fire of 2017.

10.3.3. *Lidar profiles observations*

Space lidars provide unique information on the altitude of aerosol layers and their vertical structure at very high vertical resolution. While its primary observed parameter is aerosol backscattering, the CALIOP space lidar can also measure the depolarization ratio and the so-called color ratio, that is, a parameter associated with the spectral variability of the lidar signal; using these pieces of information, a bulk categorization of aerosol layers, based on the aerosol type, can be obtained. In the operational aerosol sub-type classification of CALIOP, smoke is one of the possible aerosol types (Kim et al. 2018).

10.4. Fire detection systems from space

Active fire detection, fire radiative power and other fire detection parameters are provided from space observations by multiple instruments. Using algorithms based on the detection of thermal anomalies, AVHRR, MODIS and VIIRS can detect active fires at the time of overpass (Giglio et al. 2018). The Sentinel SLSTR, based on spectral channel use, provides active fire detection and fire radiative power products. The series ATSR-2/AATSR has been used to aggregate a World Fire Atlas covering the period 1995–2012.

10.5. Conclusion

– Satellite observations detect and provide objective information on global patterns of fire emissions and occurrences.

– Recent improvements in spatial, temporal and spectral resolution of satellite remote sensing systems reduce past uncertainties and allow for unprecedented fire plume monitoring and emission estimation.

– Planned future missions and constellations of geostationary satellites will allow long-term trends and improved observational coverage of fire emissions, and their feedbacks with global warming.

10.6. Acknowledgments

The authors thank Maya George (LATMOS) for Figure 10.1 and Bruno Franco (Université Libre de Bruxelles) for Figure 10.2.

10.7. References

Andreae, M.O. (2019). Emission of trace gases and aerosols from biomass burning – An updated assessment. *Atmos. Chem. Phys.*, 19, 8523–8546. https://doi.org/10.5194/acp-19-8523-2019.

Clarisse, L., Coheur, P.-F., Prata, F., Hadji-Lazaro, J., Hurtmans, D., Clerbaux, C. (2013). A unified approach to infrared aerosol remote sensing and type specification. *Atmos. Chem. Phys.*, 13, 2195–2221. https://doi.org/10.5194/acp-13-2195-2013.

Dufour, G., Eremenko, M., Siour, G., Sellitto, P., Cuesta, J., Perrin, A., Beekmann, M. (2022). 24 h evolution of an exceptional HONO plume emitted by the record-breaking 2019/2020 Australian wildfire tracked from space, *Atmosphere*, 13, 1485, https://doi.org/10.3390/atmos13091485.

Franco, B., Clarisse, L., Hadji-Lazaro, J., Hurtmans, D., Lecomte, G., Turquety, S., Clerbaux, C., Coheur, P.-F. (2020). Large VOC enhancements in recent massive wildfires observed from space [Online]. Available at: https://doi.org/10.5194/egusphere-egu2020-20243.

Giglio, L., Boschetti, L., Roy, D.P., Humber, M.L., Justice, C.O. (2018). The collection 6 MODIS burned area mapping algorithm and product. *Remote Sens. Environ.*, 217, 72–85.

Guo, M., Li, J., Wen, L., Huang, S. (2019). Estimation of CO_2 emissions from wildfires using OCO-2 data. *Atmosphere*, 10, 581. https://doi.org/10.3390/atmos10100581.

Hickman, J.E., Andela, N., Dammers, E., Clarisse, L., Coheur, P.-F., Van Damme, M., Di Vittorio, C.A., Ossohou, M., Galy-Lacaux, C., Tsigaridis, K. et al. (2021). Changes in biomass burning, wetland extent, or agriculture drive atmospheric NH_3 trends in select African regions. *Atmos. Chem. Phys.*, 21, 16277–16291. https://doi.org/10.5194/acp-21-16277-2021.

IPCC (2022). Intergovernmental Panel on Climate Change, impacts, adaptation and vulnerability [Online]. Available at: https://report.ipcc.ch/ar6wg2/pdf/IPCC_AR6_WGII_SummaryForPolicymakers.pdf.

Khaykin, S., Legras, B., Bucci, S., Sellitto, P., Isaksen, L., Tencé, F., Bekki, S., Bourassa, A., Rieger, L., Zawada, D. et al. (2020). The 2019/20 Australian wildfires generated a persistent smoke-charged vortex rising up to 35 km altitude. *Commun. Earth Environ.*, 1, 22. https://doi.org/10.1038/s43247-020-00022-5.

Kim, M.-H., Omar, A.H., Tackett, J.L., Vaughan, M.A., Winker, D.M., Trepte, C.R., Hu, Y., Liu, Z., Poole, L.R., Pitts, M.C. et al. (2018). The CALIPSO version 4 automated aerosol classification and lidar ratio selection algorithm. *Atmos. Meas. Tech.*, 11, 6107–6135. https://doi.org/10.5194/amt-11-6107-2018.

Kloss, C., Berthet, G., Sellitto, P., Ploeger, F., Bucci, S., Khaykin, S., Jégou, F., Taha, G., Thomason, L.W., Barret, B. et al. (2019). Transport of the 2017 Canadian wildfire plume to the tropics via the Asian monsoon circulation. *Atmos. Chem. Phys.*, 19, 13547–13567. https://doi.org/10.5194/acp-19-13547-2019.

Luo, M., Shephard, M.W., Cady-Pereira, K.E., Henze, D.K., Zhu, L., Bash, J.O., Pinder, R.W., Capps, S.L., Walker, J.T., Jones, M.R. et al. (2015). Satellite observations of tropospheric ammonia and carbon monoxide: Global distributions, regional correlations and comparisons to model simulations. *Atmos. Environ.*, 106, 262–277.

Paulot, F., Paynter, D., Ginoux, P., Naik, V., Whitburn, S., Van Damme, M., Clarisse, L., Coheur, P.-F., Horowitz, L.W. (2017). Gas-aerosol partitioning of ammonia in biomass burning plumes: Implications for the interpretation of spaceborne observations of ammonia and the radiative forcing of ammonium nitrate. *Geophys. Res. Lett.*, 44, 8084–8093. doi:10.1002/2017GL074215.

Shephard, M.W., Dammers, E., Cady-Pereira, K.E., Kharol, S.K., Thompson, J., Gainariu-Matz, Y., Zhang, J., McLinden, C.A., Kovachik, A., Moran, M. et al. (2020). Ammonia measurements from space with the cross-track infrared sounder: Characteristics and applications. *Atmos. Chem. Phys.*, 20, 2277–2302. https://doi.org/10.5194/acp-20-2277-2020.

Taha, G., Loughman, R., Zhu, T., Thomason, L., Kar, J., Rieger, L., Bourassa, A. (2021a). OMPS LP version 2.0 multi-wavelength aerosol extinction coefficient retrieval algorithm, *Atmos. Meas. Tech.*, 14, 1015–1036. https://doi.org/10.5194/amt-14-1015-2021.

Tang, W., Llort, J., Weis, J., Perron, M.M.G., Basart, S., Li, Z., Sathyendranath, S., Jackson, T., Rodriguez, E.S., Proemse, B.C. et al. (2021b). Widespread phytoplankton blooms triggered by 2019–2020 Australian wildfires. *Nature*, 597(7876), 370–375. https://doi.org/10.1038/s41586-021-03805-8.

Theys, N., Volkamer, R., Müller, J.-F., Zarzana, K.J., Kille, N., Clarisse, L., Smedt, I.D., Finkenzeller, L.H., Hendrick, F., Koenig, T.K. et al. (2020). Global nitrous acid emissions and levels of regional oxidants enhanced by wildfires. *Nat. Geosci.*, 13(10), 681–686. https://doi.org/10.1038/s41561-020-0637-7.

Torres, O., Jethva, H., Ahn, C., Jaross, G., Loyola, D.G. (2020). TROPOMI aerosol products: Evaluation and observations of synoptic-scale carbonaceous aerosol plumes during 2018–2020. *Atmos. Meas. Tech.*, 13, 6789–6806. https://doi.org/10.5194/amt-13-6789-2020.

van de Velde, I.R., van der Werf, G.R., Houweling, S., Maasakkers, J.D., Borsdorff, T., Landgraf, J., Tol, P., van Kempen, T.A., van Hees, R., Hoogeveen, R. et al. (2021). Vast CO_2 release from Australian fires in 2019–2020 constrained by satellite. *Nature*, 597, 366–369. https://doi.org/10.1038/s41586-021-03712-y.

Whitburn, S., Van Damme, M., Clarisse, L., Turquety, S., Clerbaux, C., Coheur, P.-F. (2016). Doubling of annual ammonia emissions from the peat fires in Indonesia during the 2015 El Niño. *Geophys. Res. Lett.*, 43, 11,007–11,014. doi:10.1002/2016GL070620.

Yang, X., Zhao, C., Yang, Y., Yan, X., Fan, H. (2021). Statistical aerosol properties associated with fire events from 2002 to 2019 and a case analysis in 2019 over Australia. *Atmos. Chem. Phys.*, 21, 3833–3853. https://doi.org/10.5194/acp-21-3833-2021.

11

Stratospheric Chemistry

Claude CAMY-PEYRET[1] **and Sarah SAFIEDDINE**[2]
[1] *IPSL, Paris, France*
[2] *LATMOS/IPSL, Sorbonne Université, UVSQ, CNRS, Paris, France*

11.1. Introduction

The stratosphere is the region of the atmosphere that lies above the tropopause, the altitude where the temperature ceases to decrease with altitude to be constant or starts to grow with altitude (between 7 km at the poles and 17 km at equator, depending on seasons). The driving factor for the presence of this so-called dynamic barrier is mostly related to the concentration of ozone (O_3), in the so-called "ozone layer", which increases with altitude. The term barrier is also suitable since it reduces the exchange of molecules and aerosols between the troposphere and the stratosphere: the cold trap separating the two is very efficient to reduce the injection of water vapor, H_2O, from the troposphere where it is abundant to the much drier stratosphere (because gas phase H_2O molecules are converted to ice particles that tend to fall down). This is not the case for non-condensable chemical species (CO_2, CH_4, N_2O, other greenhouse gases). For this reason, and because of the importance of ozone chemistry, we will first cover the topic of stratospheric ozone chemistry and, in a second step, discuss the chemical reactions related to other stratospheric species.

11.2. Stratospheric ozone chemistry

Ozone in the troposphere is a greenhouse gas that is dangerous to human health and vegetation growth. In the stratosphere, the ozone layer contains about 90% of the atmospheric ozone. It protects humans and other living things by absorbing

---

For a color version of all the figures in this chapter, see www.iste.co.uk/phulpin/satellites2.zip.

Satellites for Atmospheric Sciences 2,
coordinated by Thierry PHULPIN, Didier RENAUT, Hervé ROQUET and Claude CAMY-PEYRET.

UV-B radiation. Ozone is mainly produced in tropical latitudes by the photolysis of molecular oxygen into two oxygen atoms, followed by the reaction of the photolysis product with O_2 to form ozone. In the stratosphere, ozone is converted back into molecular oxygen by various reactions. The combination of ozone production and destruction is called the Chapman Cycle.

Ozone destruction can happen effectively through catalytic cycles driven by certain radicals (e.g. NO, NO_2, OH, HO_2, Cl, ClO, BrO). Due to the anthropogenic emissions of halogen or bromine containing species, the ozone chemistry is strongly perturbed. Large amounts of these chemicals, such as chlorofluorocarbons (CFCs), have been used as refrigerant air in air conditioning systems, in spray bottles and in foam production. These compounds are extremely stable in the troposphere and enter the stratosphere through upward transport, mainly in the tropics with a delay of a few years. The connection between the accumulation of these substances and ozone loss was first described in 1974 by Molina and Rowland; both received the Nobel Prize in Chemistry in 1995 because of their discovery and stratospheric ozone research. The discovery of the consequences of ozone-depleting substances prompted the establishment of the Montreal Protocol of 1987 and its amendments, thanks to which many ozone-depleting substances have been declared illegal.

Today, the consequences of ozone-depleting substances (ODS) such as those containing chlorine or bromine on the ozone layer are routinely measured and monitored. In the following, we will go through the main reactions responsible for the depletion of polar ozone, where most of the ODS end up and are trapped by the cold air masses in the so-called polar vortex.

11.2.1. *Polar ozone depletion*

In winter, stratospheric polar circulation is dominated in high latitudes by a strong and persistent cyclonic flow that is commonly referred to as the polar vortex. Within that, extreme cold temperatures are induced. The Arctic stratosphere is generally less isolated than its Antarctic counterpart, as the presence of land masses and nearby mountain ranges disrupts weather conditions more than in the Southern Hemisphere. This explains why the polar vortex in the Northern hemisphere is generally weaker and more disturbed than in the Southern hemisphere where the temperatures are much lower. The Antarctic has a very stable polar vortex, and as such the most severe losses of stratospheric ozone have been observed over Antarctic in austral spring since the late 1980s.

In the stratosphere, different **catalytic reactions** lead to the destruction of ozone, for example cycles involving chlorine radicals are effective at 40 km as such:

$$\begin{aligned} &ClO + O \rightarrow Cl + O_2 \\ &Cl + O_3 \rightarrow ClO + O_2 \\ Net:\ &O + O_3 \rightarrow 2O_2 \end{aligned} \qquad [11.1]$$

CATALYTIC REACTIONS.– A reaction in which a reactant increases the rate of reaction without being consumed in the process. Thus, a small amount of the catalyst, for example ClO in equation [11.1], can have a large effect; it can deplete ozone over and over again.

With the presence of very cold icy clouds, called Polar Stratospheric Clouds (PSCs), whose temperature is below 195 K (−78°C), heterogeneous reactions take place. They convert ODS reservoirs such as HCl and $ClONO_2$ to active substances which destroy ozone such as in equation [11.1]. As an example, and for chlorine and bromine ODS, the most important heterogeneous reactions on the surface of PSCs are:

$$\begin{aligned} &ClONO_2 + HCl \rightarrow Cl_2 + HNO_3 \\ &ClONO_2 + H_2O \rightarrow HOCl + HNO_3 \\ &HOCl + HCl \rightarrow H_2O + Cl_2 \\ &HOBr + HCl \rightarrow H_2O + ClBr \end{aligned} \qquad [11.2]$$

The ensemble of equation [11.2] happens during winter. With the first sunrays of austral spring, Cl_2 and $HOCl$ are photolyzed upon exposure to sunlight (even at low sun) to Cl atoms, which then induce ozone destruction as described in equation [11.1] (second equation).

Although here we only show the example involving chlorine, over the years, scientists have discovered many ozone-depleting substances with different depleting potential. The Montreal protocol and its amendments cover over 200 individual substances with high ozone-depleting potential, including CFCs, halons, carbon tetrachloride (CTC), 1,1,1-trichloroethane (TCA), hydrochlorofluorocarbons (HCFCs), hydrobromofluorocarbons (HBFCs), bromochloromethane (BCM) and methyl bromide (MB), among others.

11.2.2. *Antarctic ozone distribution*

During the austral winter, the polar vortex with cold air masses forms over Antarctica. It is generally very stable and concentric with the South Pole terminator. The polar vortex shape and strength and the variation of temperatures around the vortex in Antarctic winter/spring controls the overall ozone hole size and distribution. Figure 11.1 shows the inter-annual variability of the total column ozone (TCO) in Dobson units (DU) calculated from satellite measurements.

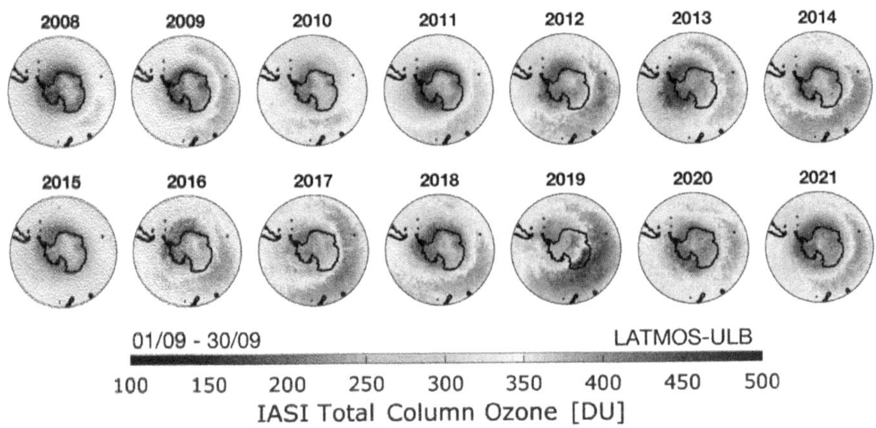

Figure 11.1. *Total column ozone (TCO) in Dobson Units from the IASI instrument during the period 2008–2021*

Sometimes, as in 2019, we have a small ozone hole (high ozone values in Figure 11.1), but it is not due to the recovery of the ozone hole, but rather specific meteorological conditions. In 2019, for example, stratospheric temperature were anomalously high, with an increase in temperature > 30° in the [60–90]°S latitude range when compared to previous years (Safieddine et al. 2020). Dynamical parameters suggest a shift in the location of the polar jet vortex. This led to air masses mixing, to reduced polar stratospheric cloud formation detected, and as such to lower ozone depletion. One year later, in September 2020, the weather made the polar vortex over Antarctica very stable, resulting in a significant depletion of ozone. The full recovery of the pre-industrial ozone level is not expected before the middle of the 21st century, mainly due to the very long lifetime of ODS.

11.2.3. *Arctic ozone distribution*

In the Arctic, spring weather conditions are generally not favorable to a significant reduction in ozone and therefore to the creation of a "hole". However, in 2020, the Arctic polar vortex was exceptionally strong and persistent. Temperatures in the Arctic stratosphere were low enough for several months in early 2020, allowing for the formation of polar stratospheric clouds, causing significant ozone losses over the Arctic, as Figure 11.2 shows.

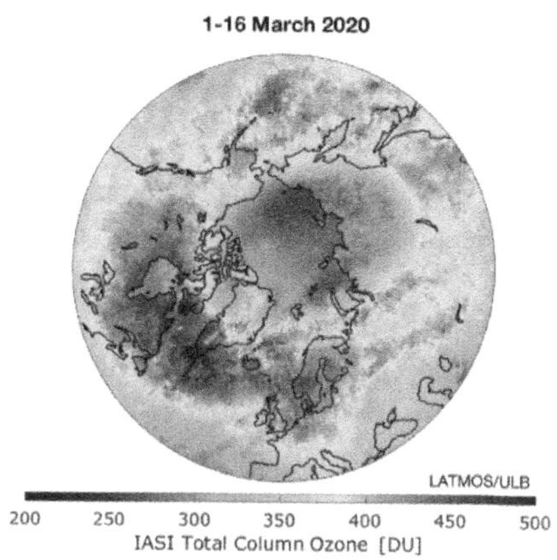

Figure 11.2. *Total column ozone in Dobson units from the IASI instrument during the anomalously cold spring of 2020, which led to the formation of an Arctic ozone hole*

11.3. Stratospheric chemistry of other species

Two major types of chemical processes determine the composition of the stratosphere and its variations when the altitude is increasing up to the mesopause (~60 km) that we will not consider here. First, because the oxygen atom $O(^1D)$ is present in increasing concentration, reactions of source species (the ones having passed the tropopause through troposphere–stratosphere exchange) with this very reactive excited atom are destroying the sources species and progressively converting them into other more stable reservoir species. Second, because of the increasing UV flux, filtered less by the ozone layer as the altitude is increasing, the source species themselves are directly broken (photo-dissociated) into simpler, more reactive species inducing a new series of chemical cycles that determine the partitioning between the different chemical families.

PHOTO-DISSOCIATION PROCESSES.– Most chemical species begin to photo-dissociate when the solar UV flux $J(\lambda, z)$ for wavelength $\lambda = c/\nu$ at the altitude z becomes significant. The corresponding reaction between the incident photon (bearing the energy $h\nu$) and the active molecule A is of the type:

$A + h\upsilon \rightarrow B + C$, where B and C are the photo dissociation products

The photo dissociation rate for the reaction is such that:

$$-\frac{d[A]}{dt} = \frac{d[B]}{dt} = \frac{d[C]}{dt} = J_A(\lambda, z)[A]$$

In this equation J_A is expressed in s^{-1} and can be written as the product $J_A(\lambda,z) = \phi(\lambda,z)\sigma_A(\lambda)$ of the local solar photon flux (expressed in photon cm^{-2} s^{-1}) times the photo dissociation cross-section of species A (in cm^2/molecule). The final dimension of J is then [photon cm^{-2} s^{-1}] × [cm^2/molecule] = [photon/molecule] [s^{-1}] and is indeed a rate (inverse of time) of the photo-dissociation reaction where one photon is absorbed and one molecule is destroyed (generating the secondary photo-dissociation products). If there are several photo-dissociation paths (generating species D and E in addition to A and B), a quantum yield coefficient (often noted φ) is used as a multiplicative factor for each photo-dissociation channel.

A very extensive and up-to-date version of the tables and information on cross-section and quantum yield (with bibliographic sources) is provided by the NASA Panel for Data Evaluation of JPL[1].

11.3.1. *Chemistry of the stratosphere and models*

It is out of the scope of the present chapter to describe in detail all the ingredients needed for the simulation of chemical reactions occurring in the stratosphere (the same concepts apply in the troposphere, as discussed in previous chapters). We refer to Brasseur and Salomon (2005) for a rather complete monograph on the chemistry (and physics) of the middle atmosphere (defined as the stratosphere and mesosphere). A well-maintained source of photochemical and chemical reaction

1. See: https://jpldataeval.jpl.nasa.gov/pdf/NASA-JPL%20Evaluation%2019-5.pdf and the relevant section 4, “Photochemical data”. The same report also provides the recommended set of reaction rate coefficients (see section 1 for bimolecular reactions and section 2 for termolecular reactions).

information for the atmosphere was mentioned just above. Wider information (more extensive for the troposphere) is provided by the Chemical Kinetic Data Evaluation for Atmospheric Chemistry[2] and the IUPAC Kinetics Database[3].

A strong impetus for stratospheric modeling came in the early 1970s by the possible impact of a fleet of high flying jet aircrafts releasing nitrogen oxides ($NO_x = NO + NO_2$) in the region of the upper troposphere/lower stratosphere (UT/LS). Initial photochemical models of the stratosphere were 1D (altitude and time), but were soon replaced by more realistic 2D models (altitude, latitude and time) to more realistically assess the expected impact of these high flying aircrafts on the ozone layer. The fact that, aside from the French/UK Concorde, the number of such aircrafts was limited and that the impact of the released chlorine compounds (CFC, HCFC, etc.) was observed as a much more urgent concern (the Montreal Protocol of 1987) resulted in slower development of stratospheric models up to the 1990s. Then, full-fledged 3D stratospheric models were developed for UARS (the Upper Atmospheric Research Satellite of NASA) with the limb viewing instruments HALOE, CLAES and MLS, as well as the European ENVISAT satellite of ESA with the limb-viewing instruments MIPAS, GOMOS and SCIAMACHY, all providing an extensive and long-lasting set of measured (remotely sensed) stratospheric concentration profiles of a huge number of chemical species (and stratospheric aerosol information).

Since stratospheric dynamics is driving chemical and physical processes as well as transport of chemical species, the state-of-the-art stratospheric 3D models are currently using meteorological fields as a basis for the underlying thermodynamics and wind fields and, in this sense, they are called chemical transport models. We mention here three examples of such models:

– the GEOS-Chem model[4] (Bey et al. 2001) is widely used in the US community and is driven by the meteorological input from the Goddard Earth Observing System (GEOS) of the NASA Global Modeling and Assimilation Office[5];

– the REPROBUS (Reactive Processes Ruling the Ozone Budget in the Stratosphere) model (Lefèvre et al. 1998) is driven by the ECMWF/IFS (Integrated Forecast System) and is considering 55 chemical species involving 147 reactions, including heterogeneous reactions on liquid aerosols and polar stratospheric clouds[6];

2. https://jpldataeval.jpl.nasa.gov/pdf/NASA-JPL%20Evaluation%2019-5.pdf.

3. http://publications.iupac.org/ci/2012/3405/pp2_2011-049-1-100.html.

4. https://acmg.seas.harvard.edu/geos_chem.

5. https://gmao.gsfc.nasa.gov.

6. http://espri.aeris-data.fr/reprobus.

– the INCA (INteraction with Chemistry and Aerosols) is a chemistry and aerosol model (Hauglustaine et al. 2004) coupled to the *Laboratoire de Météorologie Dynamique* (LMD) General Circulation Model, LMDz. It can also be run in a nudged mode (i.e. loosely driven), relaxing to ECMWF winds and temperature[7].

11.3.2. *Radical processes and cycles for the major families*

The major chemical families involved in stratospheric chemistry are the oxygen O_x family (= $O_3 + O + O(^1D)$, i.e. the species related to the ozone layer chemistry as discussed in the corresponding section), the hydrogen family HO_x (= H + OH + HO_2), the nitrogen family NO_x (= NO + NO_2), the various halogen families including the chlorine family ClO_x (= Cl + ClO + Cl_2O_2) and the bromine family BrO_x (= Br + BrO + BrCl). The corresponding reactive species, are involved in catalytic cycles that can destroy ozone, but these cycles can terminate (temporarily) through the formation of the so-called reservoir species, leading to the definition of extended "families" like $O_y = O_x + HO_y = HO_x + H_2O_2 + \ldots$, $NO_y = NO_x + HNO_3 + N_2O_5 + \ldots$, $ClO_y = ClO_x + ClONO_2 + \ldots$, $BrO_y = BrO_x + BrONO_2 + \ldots$. In these definitions of chemical families in which the reactive species (fast chemistry) appear first, only the major reservoir species are listed, but many more chemicals are considered in the most sophisticated models (55 are involved in REPROBUS).

The reactive species are initially produced by photodissociation of source species coming from the troposphere through the UT/LS and typical examples are the CFCs generating Cl atoms, the HCFCs generating Cl and F atoms, the halons generating Br atoms, nitrous oxide N_2O generating nitrogen oxides and so on.

The fate of the complicated set of stratospheric reactions between all of these species are the most stable reservoir species like HCl, HF, NO, CO up to a given altitude where they will themselves be destroyed into their atomic constituents. But this is only at altitudes higher than the mesopause.

Very long lived species like CF_4 (and PFCs or perfluorocarbons) as well as SF_6 (i.e. very stable chemically) released at the surface by human activities are playing a role as greenhouse gases in the troposphere but are not easily destroyed into the stratosphere except at high altitude (electron capture).

The family of carbon species is usually not considered in the stratosphere since the photodissociation of CO_2 only pertains to altitudes above 80 km in the

7. https://inca.lsce.ipsl.fr/index_UK.html.

mesosphere/thermosphere region not considered here. However, CO, CH_4 and related species H_2CO are to be considered as such.

11.3.3. *The example of methane in the stratosphere*

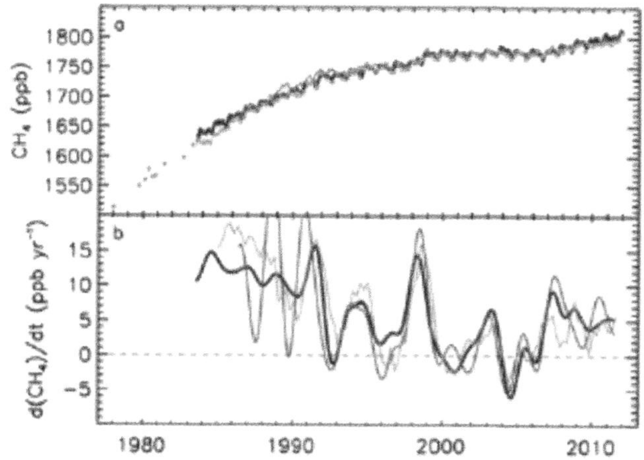

Figure 11.3. *(a) Globally averaged CH_4 dry air mole fractions and (b) instantaneous growth rate for globally averaged atmospheric CH_4 (source: IPCC (2013))*

Methane is released at the surface by natural and anthropogenic processes and the tropospheric trend of CH_4 is clearly observed (see Figure 11.3). Even though methane, like many other organic volatile compounds (VOCs), has a sink trough reaction with the OH radical, its relatively long tropospheric lifetime (8–12 years) makes it a potent greenhouse gas in the troposphere. But as for the other long lived species, CH_4 is transported to the stratosphere (mostly by convection in the tropics) and is then more efficiently destroyed than in the troposphere.

Above the tropopause, in addition to the reaction with OH (as in the troposphere):

$$CH_4 + OH \rightarrow CH_3 + H_2O \quad [11.3]$$

the following reactions with the atoms $O(^1D)$, Cl and F are also a sink for CH_4 in the stratosphere:

$$CH_4 + O(^1D) \rightarrow H_2CO + H_2 \quad [11.4]$$

$$CH_4 + O(^1D) \rightarrow CH_3 + OH \quad [11.5]$$

$$CH_4 + Cl \rightarrow CH_3 + HCl \quad [11.6]$$

$$CH_4 + F \rightarrow CH_3 + HF \quad [11.7]$$

The fate of the very short lived methyl radical CH_3 (like in the troposphere) involves many subsequent reactions that we will not discuss here, but the final degradation products of CH_4 will be temporary reservoir species like H_2CO and those that are longer lived such as CO, HCl, HF and H_2. The direct methane photodissociation is only occurring in the vacuum UV (VUV) and needs to be considered only in the Lyman-α region around 121.6 nm at high altitudes (thermosphere), where the solar flux is not strongly attenuated. All of the above reactions are sinks for CH_4 explaining its global atmospheric lifetime between 9.1 and 11.8 years as from the most recent IPCC report[8], but due to the increasing concentration of methane in the troposphere, the concentration of this species is also increasing in the stratosphere as will be discussed in section 11.4. Like other species (CO2, H2O, CFC, PFC) with an increasing trend in the stratosphere, methane will contribute to a cooling of the stratosphere, hence positive or negative feedback for chemical reaction rates that have a significant temperature dependence, leading to what is called chemistry–climate feedbacks.

11.4. Satellite measurements of trace species in the stratosphere

Passive nadir sounders like IASI (see its contribution to tropospheric chemistry in Chapters 5 and 6 and to the chemistry of the ozone layer in section 11.2 of this chapter) have only limited capability to probe the stratospheric composition of the atmosphere with good enough vertical resolution. In a complementary manner, limb sounders in various regions of the electromagnetic spectrum have been providing and continue to provide (although with only a limited number of currently operational instruments) very important information on the vertical distribution (profiles) of a large number of chemical species.

A very complete assessment of the available current data sets is provided in the SPARC[9] Report No. 8. Table 11.1 generated from the information contained in this report covers (in a simplified manner) the instruments that have provided extensive climatologies of the chemical composition of the stratosphere. Since we have taken CH_4 as an example, Figure 11.4, from the SPARC report, shows the comparative

8. https://www.ipcc.ch/report/ar6/wg1/downloads/report/IPCC_AR6_WGI_Full_Report.pdf.
9. https://www.sparc-climate.org/publications/sparc-reports/sparc-report-no-8/.

climatology derived from the data of three limb viewing instruments (MLS is not measuring CH_4).

Instrument	Technique, spectral region	Vertical range, resolution	Availability	Number of species for climatology
HALOE	Solar occultation, MIR	Cloud top to 80 km, 3.5 km	10/1991 to 11/2005	More than 8 species
ACE-FTS	Solar occultation, MIR	5–62 km, 3–4 km	04/2004 to present	More than 18 species
MIPAS	Limb emission, TIR	Cloud top to 70 km, ~4 km	04/2002 to 04/2012	More than 17 species
Aura-MLS	Limb emission, MW, sub-mm	12–75 km, ~4 km	08/2004 to present	10 species

Table 11.1. *Limb viewing instruments that provided interesting climatologies for stratospheric profiles of chemical species (source: SPARC (2017))*

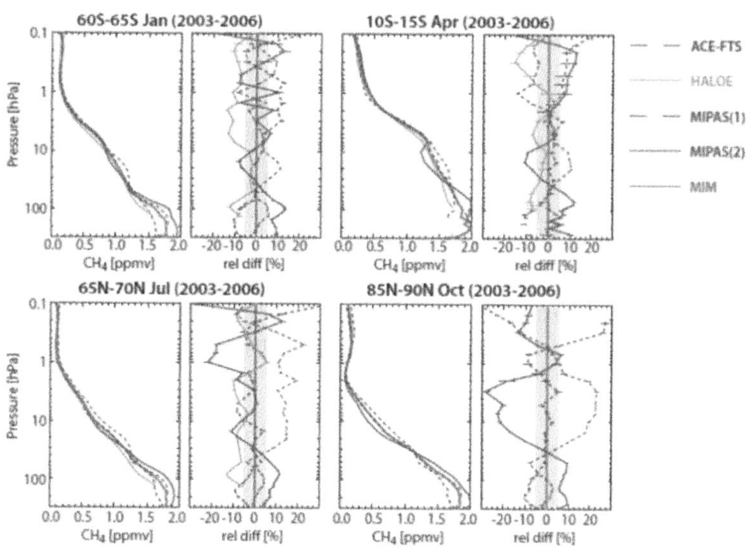

Figure 11.4. *Vertical profiles for monthly mean CH_4 for the period 2003–2006 and for four seasons. The monthly climatologies of the three instruments ACE-FTS, HALOE and MIPAS are compared to the multi-instrument mean (MIM), noting that MIPAS had two limb sampling modes depending on the year (source: SPARC (2017))*

11.5. Conclusion

It is impossible within this chapter dedicated to the stratosphere to cover all the aspects of stratospheric chemistry and we have chosen to present results focusing on two important species, that is, ozone and methane. We have presented selected observation results as well as provided some information on models and on instruments making the link between methods, models, instruments and observations, with results illustrating various aspects of stratospheric chemistry.

11.6. Acknowledgments

The authors thank Anne Boynard (LATMOS) for Figures 11.1 and 11.2. They also acknowledge figures extracted from IPCC and SPARC reports.

11.7. References

Books and articles

Bey, I., Jacob, D.J., Yantosca, R.M., Logan, J.A., Field, B., Fiore, A.M., Li, Q., Liu, H., Mickley, L.J., Schultz, M. (2001). Global modeling of tropospheric chemistry with assimilated meteorology: Model description and evaluation. *J. Geophys. Res.*, 106, 23,073–23,096.

Brasseur, G.P. and Solomon, S. (2005). *Aeronomy of the Middle Atmosphere: Chemistry and Physics of the Stratosphere and Mesosphere*. Springer, Dordrecht.

Hauglustaine, D.A., Hourdin, F., Jourdain, L., Filiberti, M.-A., Walters, S., Lamarque, J.-F., Holland, E.A. (2004). Interactive chemistry in the Laboratoire de Météorologie Dynamique general circulation model: Description and background tropospheric chemistry evaluation. *J. Geophys. Res.*, 109, D04314, doi: 10.1029/2003JD003957.

Lefèvre, F., Figarol, F., Carslaw, K.S., Peter, T. (1998). The 1997 Arctic ozone depletion quantified from three-dimensional model simulations. *Geophys. Res. Lett.*, 25(13), 2425–2428. doi:10.1029/98GL51812.

Websites

GEOS-Chem, https://acmg.seas.harvard.edu/geos_chem.

GMAO, https://gmao.gsfc.nasa.gov.

INCA, https://inca.lsce.ipsl.fr/index_UK.html.

IPCC (2013), https://www.ipcc.ch/site/assets/uploads/2018/02/WG1AR5_all_final.pdf.

IPCC (2021), https://www.ipcc.ch/report/ar6/wg1/downloads/report/IPCC_AR6_WGI_Full_Report.pdf.

IUPAC, http://publications.iupac.org/ci/2012/3405/pp2_2011-049-1-100.html.

JPL, https://jpldataeval.jpl.nasa.gov/pdf/NASA-JPL%20Evaluation%2019-5.pdf.

REPROBUS, http://espri.aeris-data.fr/reprobus.

SPARC, https://www.sparc-climate.org/publications/sparc-reports/sparc-report-no-8/.

WMO, https://library.wmo.int/doc_num.php?explnum_id=5705.

PART 3

Atmosphere and Climate

Introduction to Part 3 Atmosphere and Climate and the Contribution of Space

Paul POLI

European Organisation for the Exploitation of Meteorological Satellites (EUMETSAT), Darmstadt, Germany

With their high altitude standpoints, satellites have brought to scientists the new possibility of getting a glimpse of organized weather systems. These structures, and others of the atmospheric or ocean circulations, are difficult to fathom by one (or even many) observer(s) on the surface of the Earth. This difficulty stems from the two complementary views that always arise in the face of the observational problem of a fluid. Should observers follow the flow or should they observe the evolution of the overall patterns? These two points of view are, respectively, referred to as Lagrangian and Eulerian. In fact, satellites have contributed to developing both observational approaches.

Large-scale cloud formations in our atmosphere are of many types. We may think of tropical cyclones, with their characteristic eyes (cloud-free), and also polar lows, or seemingly less dramatic weather systems such as mid-latitude depressions, with their comma-shaped cloud organizations. When it comes to observing all of these, satellites have made tremendous contributions.

Figure I3.1 shows, for example, the Inter-Tropical Convergence Zone (ITCZ), all around the Earth, on December 3, 1978. This concept of observing from above was first tested using cameras flown by rockets. For example, Hubert and Berg (1955)

Satellites for Atmospheric Sciences 2,
coordinated by Thierry PHULPIN, Didier RENAUT, Hervé ROQUET and Claude CAMY-PEYRET.

analyzed pictures taken by a rocket on October 5, 1954 from 160 km altitude, observing the full extent of a tropical storm moving from Western Cuba and over the Gulf of Mexico. These early pictures held great promise for the satellite observations that followed.

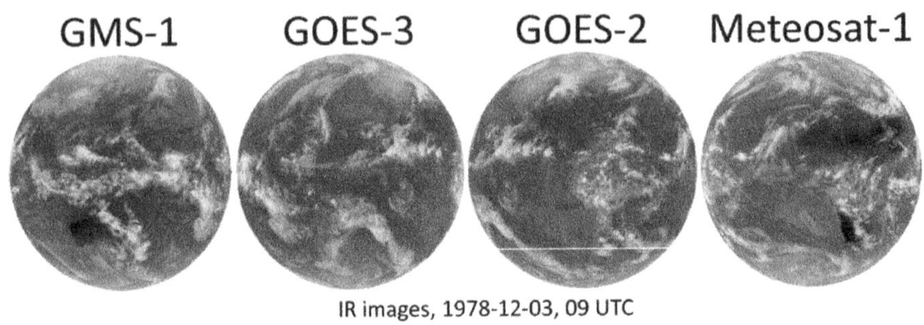

Figure 13.1. *Infrared images (1978) illustrating the early days of operational observations by geostationary satellites, operating from 36,000 km altitude. At the time, four satellites observed (respectively, from left to right) the Eastern Indian Ocean and Western Pacific (GMS-1), North America and Eastern Pacific (GOES-3), North and South America and Western Atlantic (GOES-2) and the Eastern Atlantic, Europe, Africa and Western Indian Ocean (Meteosat-1). Areas jointly observed by subsequent satellites are visible in the ITCZ around the equator. Forty-five years later, the international community operates a larger and more complete fleet of geostationary satellites, fully covering the equatorial ring (sources: NOAA, ESA, EUMETSAT and JMA)*

In addition, Earth observation satellites have been equipped with dedicated radio equipment to serve as relays of observing systems transmitting their observations from the field, thereby making it possible to collect in situ reports from previously unchartered areas. Along with tracking the location of the transmission apparatus, this has opened the way to mapping global ocean currents and surface winds. This greatly advanced the quest of mariners longing to understand the world's winds and currents, a pursuit that started the moment our ancestors were first able to set sail. For example, the World Ocean Circulation Experiment (WOCE) was specifically devised in the 1990s to acquire additional observations and develop ocean-predicting models. This involved in situ platforms tracked by satellites as well as new satellite missions such as European altimeters.

Another great discovery allowed by satellites concerns the vertical structure of the atmosphere. High-altitude weather balloons followed by rocket-sondes were first to collect in situ measurements, opening the way for investigators to assemble the data into vertical profiles. However, the geographical locations of these experiments were always limited, leaving aside great swathes of deserts, oceans and ice-covered areas. In the 1970s, several Nimbus satellites from NASA were equipped with additional instruments designed with scientists from the UK (in particular from the University of Oxford and the University of Reading). These instruments collected unprecedented measurements of radiation emanating from high altitudes, beyond those traditionally monitored by aircraft or even weather balloons. From these new measurements, scientists were able to map the global circulation of the middle atmosphere (stratosphere and lower mesosphere).

With later measurements by more recent satellites such as the Upper Atmosphere Research Satellite (Mote et al. 1996), a slow but steady pattern of upper-air circulation known as the "atmospheric tape recorder" was discovered, with water vapor entering the tropical stratosphere and slowly propagating upwards, at a pace that varies annually. A new understanding of the slow but steady movement of air parcels on our planet was unveiled, adding to the theory of the Brewer–Dobson circulation first formulated much earlier (Butchart 2014), whereby tropospheric air entering the stratosphere in the tropics slowly migrates up and towards the poles, before descending at higher latitudes. Today, with changes in the atmospheric composition including for some key trace gases of potent radiative effect, in situ measurements are still very much needed, continuing the tradition of long-range altitude balloons that started with the Eole experiment in 1971 (Morel 2002), albeit with much improved technology and addressing new contemporary challenges.

All of these "steady" features, taken as a whole, may be seen as "constant" characteristics, even though they are subject to cycles and their recurrences vary with seasons and locations. For that reason, they are often dubbed "climate" characteristics. Chapter 12 discusses the blurring of boundaries between weather, conceptualized as chaotic, and climate, thought to be stable, because of human-induced changes that upset the climate equilibrium.

Chapter 13 thereafter introduces the role of greenhouse gases in the climate system and how these gases are monitored from space. For this task challenges abound, requiring measurements of unprecedented accuracy to detect sources and sinks.

Finally, the last two chapters of this book expose the main characteristics of clouds and water vapor (Chapter 14) and precipitation (Chapter 15), and how they are monitored from space. While potentially appearing as purely academic topics, the availability of water has profound implications on where humans establish communities and overall world population development.

The reader will note that some atmospheric variables important from a climatic point of view and accessible to satellite measurement, such as temperature or aerosols, do not give rise to specific chapters in this Part 3. For these topics, the reader is referred to Parts 1 and 2 of this volume.

13.1. References

Butchart, N. (2014). The Brewer-Dobson circulation. *Rev. Geophys.*, 52. https://doi.org/10.1002/2013RG000448.

Hubert, L.F. and Berg, O. (1955). A rocket portrait of a tropical storm. *Mon. Wea. Rev.*, 83, 119–124. https://doi.org/10.1175/1520-0493(1955)083<0119:ARPOAT>2.0.CO;2.

Morel, P. (2002). L'expérience spatiale Éole et sa préparation (1962–1972). *La Météorol.*, 36, 42–53. https://doi.org/10.4267/2042/36205.

Mote, P.W., Rosenlof, K.H., McIntyre, M.E., Carr, E.S., Gille, J.C., Holton, J.R., Kinnersley, J.S., Pumphrey, H.C., Russell III, J.M., Waters, J.W. (1996). An atmospheric tape recorder: The imprint of tropical tropopause temperatures on stratospheric water vapor. *J. Geophys. Res.*, 101, 3989–4006. https://doi.org/10.1029/95JD03422.

12

Climate Monitoring

Paul POLI and Jörg SCHULZ
European Organisation for the Exploitation of Meteorological Satellites (EUMETSAT), Darmstadt, Germany

12.1. General concepts about the climate

12.1.1. *What is climate?*

Climate is what we expect, weather is what we get.

Mark Twain's quote above conveys the perception that weather phenomena fall within a range of expectations called climate. Interpreted differently, life-long series of weather events do not necessarily lead us to redefine what we hold as climate. Put simply, climate would be stable, whereas the weather is chaotic. This maxim, however, limits climate to its atmospheric manifestations. This omits the key roles of land, ocean and ice, now understood to all be integral parts of the climate system.

Notwithstanding human- or naturally induced changes of the present times, paleontologists have collected evidence that the Earth's climate was never completely stable. In fact, as far back as we are able estimate, our climate has exhibited large variations in temperature (Figure 12.1). These changes have accompanied displacements of human settlements throughout the centuries. Societal collapse or revolutions prompted by climate issues have also fascinated historians.

For a color version of all the figures in this chapter, see www.iste.co.uk/phulpin/satellites2.zip.

Satellites for Atmospheric Sciences 2,
coordinated by Thierry PHULPIN, Didier RENAUT, Hervé ROQUET and Claude CAMY-PEYRET.

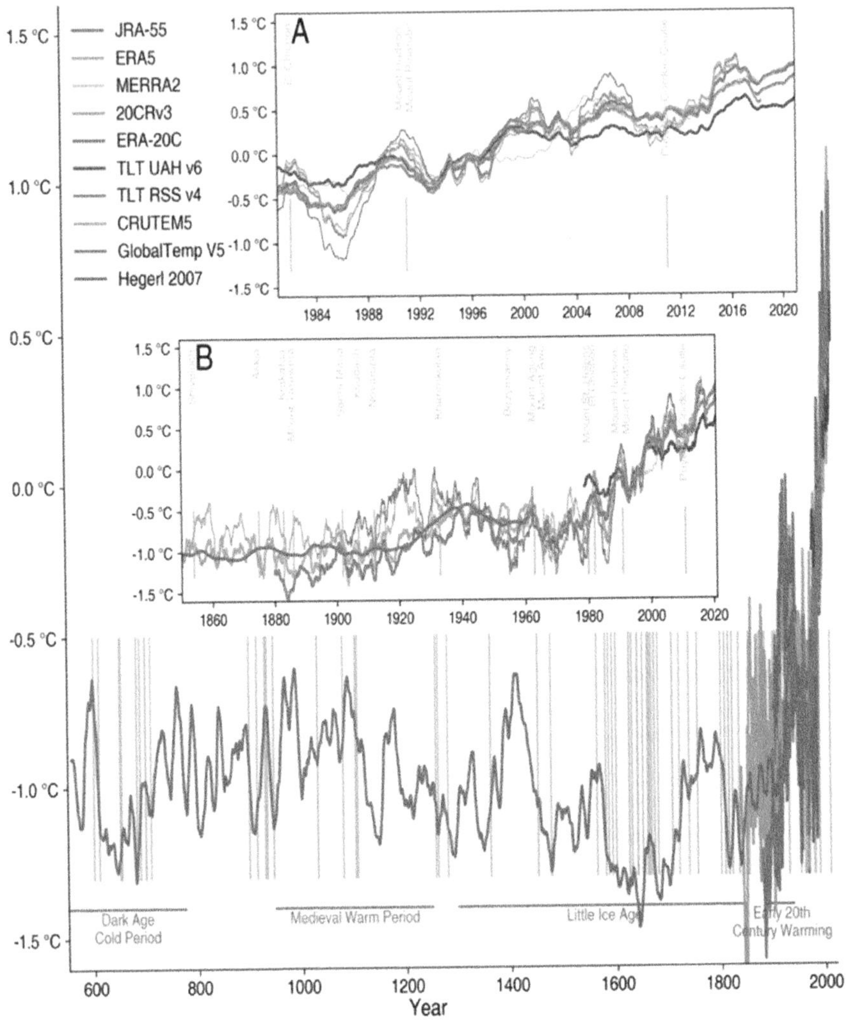

Figure 12.1. *Variations in air temperature may be estimated from various data sources. Timeseries from model-assisted climate reanalyses (JRA-55, ERA5, MERRA2, 20CRv3, ERA-20C), satellite data records (TLT UAH V6, TLT RSS V4), in situ data records (CRUTEM5, GlobalTempV5) and paleoclimate proxy records (Hegerl 2007) show three-year moving averages of air temperature anomalies with respect to 1981–2010 over extra-tropical Northern hemisphere land areas. Grey vertical lines indicate when major volcanic eruptions occurred. Inlay A (B) shows data since 1981 (1850, respectively). Data sources are presented in the References section. The climate reconstruction by Hegerl et al. (2007) is offset to match the 1881–1960 CRUTEM5 mean*

Today, the seemingly different themes of migration, environment and climate change have started to appear together. Whether prompted by sudden- or by slow-onset hazards, human migration has increased tremendously, usually in relation to conflicts. Because of incomplete knowledge of the climatic conditions on a local scale, even for historical times, our understanding of these events often falls short of explaining all nuances. Instruments have indeed collected measurements but only for about 200 years and only for some weather elements.

12.1.2. *Is climate limited to atmospheric phenomena?*

The atmosphere is only one of the components of the Earth system. Recognizing the multi-faceted nature of climate, the World Climate Research Programme (WCRP) was established in 1980 by three international coordination bodies: the World Meteorological Organisation (WMO), the International Council for Science and the Intergovernmental Oceanographic Commission (IOC) of UNESCO. The reasons for this multi-partite stakeholdership reside in the multiple connections that climate has with the various components of the Earth system. Figure 12.2 presents a schematic of the Earth system, separated into broad yet essential components relevant to climate.

The sun, whose activity is modulated by multi-year cycles, heats day-lit areas on Earth, while the other areas cool through radiation to outer space (at a low temperature). Understanding both requires physics, and advances in understanding are significant for both effects, thanks to astronomy and cosmology instruments in space. For example, the Solar and Heliospheric Observatory, stationed 1.5 million kilometers from the Earth, has investigated solar wind and the sun's outer layers (chromosphere, transition layer and corona), and enabled a deeper understanding of the interior structure of the sun. Launched earlier, the Cosmic Background Explorer had refined the temperature of the cosmic background to 2.725 Kelvin, with small fluctuations observed and believed to be due to density of matter in the early universe, confirming the Big Bang theory.

This hot–cold dipole of permanently shifting areas would make our planet inhabitable if it was not for its atmosphere, in the first place. It serves both to cool and to warm these respective areas, as the Earth rotates on itself and as it spirals around the sun each year, with the polar regions alternating between regimes of complete day or night for months. The gravitational pull of the sun, joined with that of the Moon, also generates tides, most visible near the coastlines. All of the components of the climate effectively control the diurnal and seasonal cycles' temperature amplitudes to ranges of values to which humans and other forms of life in the biosphere have adapted themselves, including in coastal regions when it comes to the effects of tides.

The atmospheric composition is responsible for the "insulation" effect, understood as early as 1824 by Joseph Fourier. Molecules of ozone in the middle atmosphere are broken down by solar radiation, thereby capturing radiation and greatly reducing the fraction of otherwise harmful ultra-violet radiation that reaches the Earth's surface. The appearance of an "ozone hole" over polar regions as witnessed in the late 1970s explains the international community's rapid response to the threat posed to society by increased ultraviolet radiation (see Chapter 11). With signatures of the Montreal protocol and later amendments, commitments have been made to ban the industrial production of chlorofluorocarbons (CFC) in 1987, hydrochlorofluorocarbons (HCFC) in 2007 and hydrofluorocarbons (HFC) in 2016.

At the same time, so-called greenhouse gases such as carbon dioxide, methane and water vapor reduce the transparency of the atmosphere. This results in an opaque medium (only at some wavelengths), which can absorb radiation (as heat). The atmosphere can then continue to serve as a transfer medium between strongly heated areas and strongly cooled areas. However, changes in composition affect this picture, as studied for example by Manabe and Wetherald (1967).

In 2021, the mean carbon dioxide concentration, largely driven by fossil fuel emissions, passed the 50% mark above pre-industrial levels. Methane concentration has also more than doubled, partly because of human activities (see Chapter 13). These human-induced evolutions, so-called anthropogenic, superpose on those induced by the biosphere (e.g. carbon uptake by forests over land and algae in the oceans), while the latter also reacts to these changes (e.g. coral bleaching).

Increased greenhouse gas concentrations have direct warming effects. Furthermore, this warming effectively enables the global atmosphere to hold more water vapor. An increase in the concentrations of these gases contributes to the lower atmosphere becoming more impermeable to infrared radiation and warming accordingly, although there are also climate feedback mechanisms that may act in the opposite direction (Held and Soden 2006). For example, aircraft-induced condensation trails and higher cloud fractions mean more solar radiation is reflected back to outer space. Another example concerns irrigation in (otherwise) dry areas, as it allows for a reduction in maximum temperatures during the day.

The atmospheric heat capacity is, however, small in comparison to that of the oceans. Owing to this difference in heat capacity, oceans transport heat on longer timescales. The global ocean circulation involves both upper currents that have been charted since the onset of global shipping, and much slower currents, which can run deep. However, understanding ocean currents is essential to find out where additional heat gets stored in the age of human-made climate change.

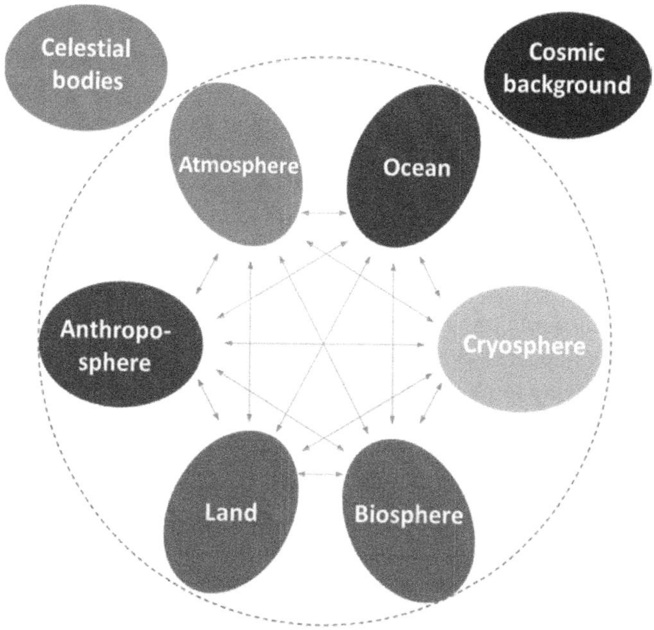

Figure 12.2. *Components of the Earth system relevant to climate timescales*

Three manifestations of oceanic warming are the melting of the Arctic sea ice, an increase in sea levels (owing to fresh water influx from sea ice and glaciers melting and also to thermal expansion) and ocean acidification. All three endanger vast amounts of species and upset the lives of coastal communities. As illustrated with an "anthroposphere" component in Figure 12.2, the "anthropocene" indicates that human actions are no longer separable from changes taking place in our environment (Waters et al. 2016).

12.1.3. *A question for Nobel Prize laureates: Is the climate stable?*

Some well-identified past climatic events brought catastrophic consequences. For example, the "year without a summer" over Europe and North America (1816) was likely caused by the Mount Tambora eruption (see other volcanic eruptions in Figure 12.1). The assumption of a rather oscillatory nature, with only slow changes, was anchored by meteorologists in the founding principles of the climate concept, in connection with a purely astronomic theory of climate and neglecting other factors. This concept called for climate normals in the 19th century, to describe the various elements of the climate averaged over long time periods, such as temperature minima and maxima expected in a given place and day of the year.

These practices are still used today in traditional climatology work, with regular publications of atlases, showing statistical summaries. These serve diverse applications, such as construction, for buildings to withstand strongest-observed winds or precipitation, for choosing renewable energy production sites (e.g. hydroelectric power, solar power, wind power) or agriculture, for example, to determine best-suited crops.

Climate measurements have continued to be collected solely for this purpose for a long time. In 1935, the International Meteorological Committee decided that Climatological standard normals were to be computed based on a fixed period, from 1901 to 1930, and updated every 30 years afterwards. In 2015, the WMO changed this definition, moving to more frequently updated means (every 10 years). This not only led to a fundamental change in perspective, requiring us to review the concept of climate observations. Indeed, one newcomer in the new definitions included a standard reference climate period (1961–1990), against which to assess long-term climate changes. This effectively recognized that oscillations about a position of equilibrium were not the only forces at play in the climate system. Methods have been developed to fingerprint the impact of human influence, as proposed, for example, by Hasselmann (1997), who was co-recipient of the 2021 Nobel Prize in Physics.

The proper international framework for tackling climate issues is recent. In 1992, the Rio de Janeiro Earth Summit produced three of the most important international scientific conventions. As we advance in time, these conventions occupy greater space in the public discourse. These are the United Nations Framework Convention on Climate Change (UNFCCC), the Convention on Biological Diversity and the United Nations Convention to Combat Desertification. While biodiversity loss, soil erosion and climate change have grown to be perceived as major threats to our societies, with inter-related links, their topics of concern have not abated at all, but all the contrary (Pörtner et al. 2021). Regarding climate, the UNFCCC relies on reports prepared by the Intergovernmental Panel on Climate Change (IPCC), who was co-recipient of the 2007 Nobel Peace Prize. The 6th Assessment Report (IPCC 2021) concludes that "human influence has warmed the atmosphere, ocean and land" and "the scale of recent changes across the climate system as a whole […] are unprecedented over many centuries to many thousands of years".

To support such work, a shifting paradigm in the understanding of the Earth's climate, observations are coordinated by the Global Climate Observing System (GCOS), facilitating the UNFCCC needs for systematic observations to assess and understand climate variability and change. These observations are now presented.

12.2. From space-based measurements to climate products

12.2.1. *Sensing the environment*

Collecting observations generally considers two targets: a geophysical state and natural processes. However, a natural process is merely an indication of changes in one or several geophysical states. For this reason, continuous (also called operational) monitoring of the climate is generally limited to physical states (such as temperature) or well-defined processes (such as precipitation). Observational campaigns allow scientists, mobilizing additional instruments, to study processes by looking at the detailed evolution of different geophysical variables.

These instruments deployed in situ are limited in their coverage, because every deployment and instrumented site requires logistics and generates maintenance costs (respectively). To maintain good-quality measurements, meteorological agencies regularly bring the instruments into their laboratories, for calibration* (or recalibration). So far, these aspects have prevented in situ systems from spreading too much outside populated areas, except for observations from ships, aircraft or a few automated observing systems such as ocean drifting buoys and profiling floats, shown in Figure 12.3. These capabilities may expand in the years to come with long-range unmanned aerial, underwater or surface drones as shown in this same figure. Because these surface-based platforms carry sensors that can be calibrated accurately in a controlled environment, they contribute to the GCOS.

To allow integration of all these data from various sources, and thus monitor climate changes across time and space on long timescales, it is essential that observations of physical states adopt international standards of measurements. These standards keep being refined as global knowledge improves. A case in point is the temperature scale. Among the several definitions proposed in previous centuries, Fahrenheit and Celsius are still used today in different parts of the world. The international temperature scale of 1990 codified the calibration standards for measuring thermodynamic temperature, with a number of fixed points, such as the triple point of hydrogen (at –259.35 degrees Celsius), the melting point of gallium (at 29.76 degrees Celsius), or the triple point of water (at 0.01 degree Celsius). This last temperature was conveniently used to define the Kelvin unit: 273.16 Kelvin.

This scale was in fact more a measurement standard, defining only some points but leaving all other points in between to interpolation. In 2019, the *Bureau International des Poids et Mesures* (BIPM) adopted a new definition of the unit of reference in the International System of Units (SI) to measure thermodynamic

*See Glossary.

temperature. In this definition, Kelvin is no longer based on the triple point of water but on the Boltzmann constant, which relates energy to temperature.

Figure 12.3. *Left: drifting buoy with high-precision calibrated platinum temperature sensor deployed along the coast of Brittany (source: NKE). Middle: ocean profiling float equipped with biogeochemical sensors (source: D. Luquet, IMEV). Right: Saildrone operating in the Pacific Ocean, measuring more than 20 different geophysical parameters at the air–sea interface (source: Saildrone)*

For collecting radiation measurements from space, efforts to consolidate standards started with the onset of multi-agency satellite systems prompted by the need to develop common practices for quantities, symbols, units and uncertainty* analysis. A pioneer was the Space-Based Observation Systems Committee on Standards in the United States, in 1988. Today, guidelines are managed internationally under the umbrella of the International System of Quantities, also known under ISO:80000, developed and promulgated jointly by the International Organization for Standardization (ISO) and the International Electrotechnical Commission (IEC).

Spatio-temporal references of the measurement are also important to enable later data exploitation. As the underlying standards evolve, so must the tagging associated with the corresponding data records. With the advent of the Global Positioning System, a global referential was adopted de facto, based on an ellipsoid representation of the Earth, the World Geodetic System 1984. Referencing data in the vertical has also progressed, referenced to a geoid, thanks to acquired knowledge of the Earth's gravity field, opening the way to accurate sea level monitoring. Similarly, Coordinated Universal Time has also evolved. Initially defined in 1960, close to the International Geophysical Year 1957–1958, this timescale has seen the insertion of 27 leap seconds since 1972, making it a discontinuous scale.

Operators of meteorological satellites regularly review and update their operating standards within the Coordination Group for Meteorological Satellites (CGMS). For

*See Glossary.

specific climate issues, a joint Working Group on Climate with the Committee on Earth Observation Satellites (CEOS) produces recommendations that affect how climate data are collected, validated and exploited to maximize benefits for climate research and services.

12.2.2. *The role of space-based observations*

Space-based observations have opened a new era in climate monitoring (see Chapter 1 of Volume 1). While surface-based observations continue the essential historical data records and maintain ties to laboratory references, they are naturally limited to their areas of deployment. Even though they may drift due to the ocean circulation, for example, drifting buoys or profiling floats, or with the upper-air circulation, for example, long-range air balloons, or when carried by ships or aircraft, the movement speeds of all these platforms are not comparable to those of low-Earth polar-orbiting satellites that circle the Earth at a speed of a few kilometers per second. Such speeds allow for great variability in sampling within a few hours of operations. Depending on the orbit choice and the instrument configuration, for example swath scanning (see Chapter 8 of Volume 1), global pictures may be compiled within hours.

Similarly, fixed observatories, like land stations or moored buoys, monitor their immediate environment. Ground-based remote sensing (such as radar) may extend the monitoring range to a cone of influence that surrounds the location by a few hundred kilometers at most (the horizon representing the limit of visibility). By comparison, a geostationary satellite, maintaining a (nearly) fixed position with respect to the Earth, can offer a more "continental" picture, albeit at high refresh rates, down to minutes, observing nearly one half of the planet. In practice, this ratio between the observed surface and the total surface of the Earth tends to be 40% or less. This limit is imposed by the viewing angle of the sensor in relation to the horizontal at the surface (the horizon also represents the limit). At angles greater than 70°, the atmosphere can no longer be assumed to be sensed from nadir, and data interpretation becomes problematic because the sensor design may not be adapted to such viewing conditions.

Other types of sensors have been developed to specifically observe the atmosphere from the side. This view, called limb sounding, enables the collection of data with high vertical resolution, at the expense of a lower horizontal resolution compared to the nadir view. Changes in the fine vertical structures can then be exploited to observe large-scale effects of climate change.

Another distinction among the various space-based sensing apparatus concerns passive versus active remote sensing systems (see Chapter 9 of Volume 1). Passive systems measure the emission or scattering of natural light or radiation (such as absorption by water molecules) or monitor the gravity field. Active systems exploit human-made signals to infer the difference imposed by the medium. Examples of such systems are the radio occultation of Global Navigation Satellite System (GNSS) signals, backscattering by ocean waves (scatterometer), Earth's surface (altimeter) and particles (lidar, radar).

The vast advantage of coverage offered by satellite-based systems is best understood when considering the observation of extensive geophysical variables. For such variables, the quantity increases with the size of the system. Prime examples are sea level rise (oceans volume expansion) or Arctic sea ice melting. Both may be appreciated in situ with repeated measurements at some fixed locations but only satellites may convey a more complete picture.

12.2.3. *The concept of essential climate variables*

Prompted by the globally accepted perception that the Earth's climate should no longer be assumed as necessarily stable, the GCOS has had to revisit how to measure the climate, that is, how to best observe a changing climate. The result of these discussions was to recommend monitoring a set of Essential Climate Variables (ECVs; see Table 12.1) (GCOS 2003). Each of these is "a physical, chemical or biological variable or a group of linked variables that critically contributes to the characterization of Earth's climate" (WMO 2016).

These variables cover several domains of the Earth's system. These distinct but interconnected domains are the atmosphere, the land and the ocean, along with ice- or snow-covered surfaces. Each of these physical components can be studied and understood in isolation but only until we reach an interface with the other domains. Such exchanges at the interfaces include mass (including carbon and water) and energy, through elementary processes such as radiation, evaporation and precipitation or friction. For example, the wind blowing over a sea surface generates waves; in this process, some energy is transferred from the atmosphere to the ocean.

When considering timescales, the atmosphere is generally the domain with the shortest memory. Its heat capacity is much lower than that of land surfaces, which in turn is much lower than that of the oceans. This means that air temperatures can change very rapidly, while land surfaces feature a longer latency, which can still vary faster than ocean temperatures. In turn, the timescales involved to form glaciers are on the order of centuries. This cascade of timescales explains why the different essential climate variables require different approaches for their monitoring.

Accessing all the Essential Climate Variables in Table 12.1 requires a combination of approaches: in-the-field measurements, whether measuring in situ or using local remote sensing, and also measurements from space. In fact, over half of the ECVs may only be fully observed from space, such as the Earth radiation budget.

The organization of these variables into domains represents a solid basis for optimizing international collaboration. A top-level ECV inventory[1] regularly updates existing and planned products.

Guided by internationally agreed requirements by GCOS, the aforementioned CEOS-CGMS Working Group on Climate coordinates their implementation with responsibilities shared across the world's space agencies. The decomposition of the climate system into a set of elementary, required products then serves the orientation of future developments in Earth observations.

Atmosphere	**Surface**	Precipitation, pressure, radiation budget, temperature, water vapor, wind speed and direction
	Upper air	Earth radiation budget, lightning, temperature, water vapor, wind speed and direction
	Atmospheric Composition	Aerosols, carbon dioxide, methane and other greenhouse gases, clouds, ozone, precursors for aerosols and ozone
Land	**Hydrosphere**	Groundwater, lakes, river discharge
	Cryosphere	Glaciers, ice sheets and ice shelves, permafrost, snow
	Biosphere	Above-ground biomass, albedo, evaporation from land, fire, fraction of absorbed photosynthetically active radiation (FAPAR), land cover, land surface temperature, leaf area index, soil carbon, soil moisture
	Anthroposphere	Anthropogenic greenhouse gas fluxes, anthropogenic water use
Ocean	**Physical**	Ocean surface heat flux, sea ice, sea level, sea state, sea surface currents, sea surface salinity, sea surface stress, sea surface temperature, subsurface currents, subsurface salinity, subsurface temperature
	Biogeochemical	Inorganic carbon, nitrous oxide, nutrients, ocean color, oxygen, transient tracers
	Biological/ Ecosystem	Marine habitats, plankton

Table 12.1. *Over 50 Essential Climate Variables are necessary to describe the Earth's climate elements (source: WMO GCOS). Monitoring these variables, with observations, is necessary to enable understanding and further to predict evolutions of the overall climate system, along with models*

1. https://climatemonitoring.info/.

12.2.4. *Observation-based products*

For each ECV, scientists wish to determine fully the spatio-temporal variations, at different scales. However, no single instrument or sensor, whether space- or surface-based, can suffice to provide spatial- and temporally-complete information. Consequently, several lines of products have been developed. Products that are spatio-temporally complete, that is, have no gaps in their definition range, the so-called level-3* or level-4* products by the space agencies, are sourced from two main streams.

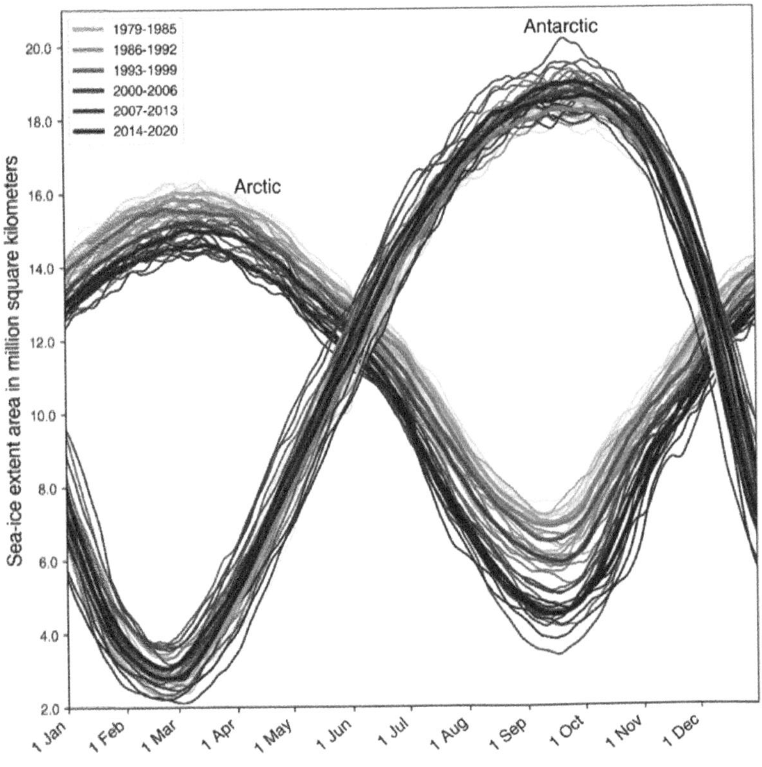

Figure 12.4. *Area extent of polar sea ice from 1979 to 2020. Thin blue lines show individual years (from light blue in 1979 to dark blue in 2020), and thick red lines show seven-year averages (see legend). The months of minimum extent are September and February/March for Arctic and Antarctic (respectively), with years of record minima in 2012 and 2017 (respectively). This highlights a fast pace of sea ice retreat in the Northern hemisphere (source: U.S. National Snow and Ice Data Center)*

*See Glossary.

The first stream of products uses observations. Such observations employ satellite data, in order for the products to cover large target domains, without gaps, and because many variables are impractical to measure in situ. However, the production may also employ data from instruments deployed in the field, which provide measurements in physical space (e.g. temperature, humidity). For practical exploitation, the satellite-based observations are processed into similar physical spaces, such as a layer-mean temperature or an atmospheric wind or cloud cover estimate (see Chapters 1–3). These products, still attached to the original date, time and location viewed by the satellite, are called level-2* products by the space agencies.

The satellite products that describe climate evolutions in physical space are called Climate Data Records* (CDR). They result from consistent processing of satellite sensor data into time series of uncertainty-quantified retrieved values of a geophysical variable or indicator (see Chapter 10 of Volume 1). Spatio-temporal references are required. The record length and quality must be appropriate for climate applications such as the analysis of climate variability at various scales and long-term change.

From level-2 CDR products, gap-filled (level-3) or temporal summaries (level-4) can be generated. An example is shown in Figure 12.4, showing the evolution of sea ice extent area over the polar regions. These particular data are obtained from long time-series of several generations of polar-orbiting sensors, by analysis of imagery.

12.2.5. *Model-assisted climate products*

In the 2000s, another stream of climate products emerged: climate reanalyses, such as featured in Figure 12.1. They assimilate not only in situ observations but also satellite observations, with the same techniques as employed for Numerical Weather Prediction, into models of the various components of the Earth system. In doing so, reanalysis systems combine a variety of observations and state-of-the-art models and bring the treatment of uncertainties into a single framework. This approach enables optimal retrieval of information content from the observations during the assimilation process (see Chapter 4).

The satellite products eligible for assimilation in climate reanalyses are limited to those for which developments have been made in the data assimilation. Other than this practical limitation, there is no a priori limit to the complexity and diversity of data that can be ingested in such systems. This leads to a highly diverse ingest of data as shown in Figure 12.5 for the latest reanalysis from ECMWF and the European Union Copernicus Climate Change Service (C3S), ERA5.

*See Glossary.

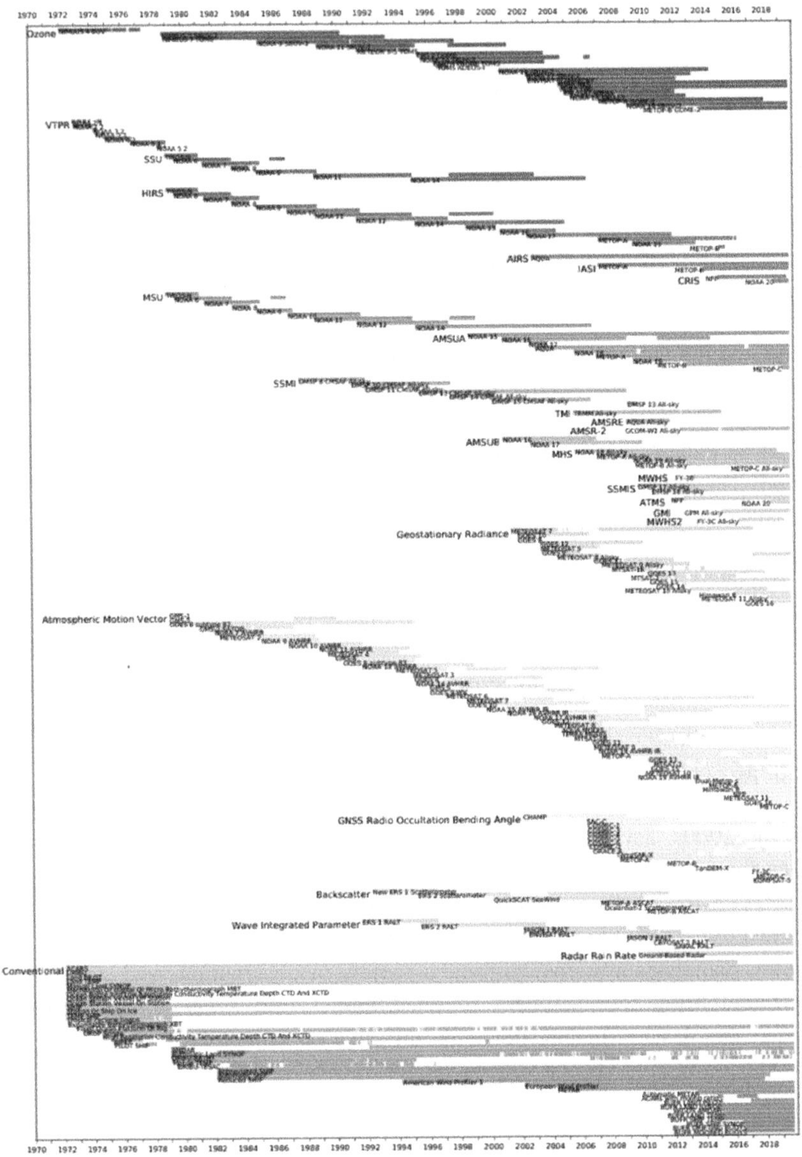

Figure 12.5. *Timeline of observations, as assimilated in ERA5, from top to bottom: satellite ozone sensors (purple), infra-red to microwave sounders on polar-orbiting satellites (dark blue to light blue), radiometers on geostationary satellites (cyan), atmospheric motion vectors (green), radio occultation (yellow), ocean scatterometers and altimeters (light orange) and surface-based or conventional observations (red) (source: C3S/ECMWF)*

Reanalyses present the dual advantage of separating the contributions of uncertainties from modeling and from observations, and also separating the modeling by Earth-system component. Thence, the reanalysis technique has become a vector of progress towards the realization of fully representative Earth digital twins[2].

12.3. Climate data records and uncertainty estimates

12.3.1. *Why reprocessing?*

Several reasons explain why measurements collected at one point in time become no longer directly suitable for exploitation after some time. First, the measurement units and scales, as well as the underlying spatio-temporal references, continue to evolve so that updated copies of the data need to be issued to reflect the new references. Second, other considerations on the data must be applied to keep them compatible with contemporary data analysis approaches. These include quality control of various types. Obviously wrong data or gross errors, even though flagged originally, can be improved thanks to accrued knowledge about physical effects that may have affected the measurement. Times when the instrument performed below expectations may thus be better identified. In addition, other quality controls can help users, guided by the latest science, in the most common applications at the time of preparation. These include, for example, the addition of flags to identify scenes affected by clouds or underlying sea ice, for instruments observing wavelengths sensitive to these natural occurrences, using the information obtained from other data sources.

Furthermore, as technology progresses, so do the data formats employed by an increasingly diverse population of data users. While retaining original data in their original formats is important to avoid data loss (no single data model making it possible to capture all cases), users will generally request access in the data formats contemporary at the time of their request (which can be quite different from the original data format or data model adopted at the time of data creation).

Finally, methods to derive geophysical variables (parameter-specific inversion schemes or reanalysis) are improving over time due to improved knowledge on processes in the Earth system domains from dedicated measurements and modeling activities. This represents another reason for reprocessing observations into ECV data records employing the enhanced quality control above.

2. Digital twin: numerical or digital representation of the Earth or one of its components.

In order to allow for consistent data usage, all of these points call for regular reprocessing into consistent, contemporary referentials and data practices, to maintain these data records in a state that can be summed as "ready for analysis".

12.3.2. *Calibration*

To enable data analysis in a comparable manner, regardless of source, area and date, observations must be calibrated a priori to SI physical units. The BIPM defines calibration as a two-step operation. First, calibration establishes, under specified conditions, a relation between the measurement quantity values and corresponding indications, both with associated uncertainties. Second, this relation is applied to obtain a measurement result from an indication, where the indication can be a raw sensor reading, for example, sensor count.

Therefore, like reanalysis, calibration can be improved to benefit from advances in our fundamental understanding of physics and the measurements. This leads to reprocessing the original data and/or improving the way the data are to be compared with models in order to benefit from the latest scientific knowledge, even after a "best reprocessed record" should have been achieved at a certain point in time.

Each reprocessing generates a Fundamental Data Record (FDR), a consistently processed time series of sensor observations calibrated to physical units, located in time and space, possibly with suitable corrections and uncertainty characterization.

Even a short FDR can prove essential in expanding our knowledge of climate, for example, by filling gaps in the data records for some particular Essential Climate Variable. This may also ensure a consistent representation of variability in climate products including reanalyses. Another impetus may be to provide a reference to the reanalysis data assimilation, in the case of variables/times/regions that are not observed by other sensors. In all cases, retention of original data, instrument design and operations metadata* is an important prerequisite. "Forgotten" satellite data can also sometimes be rescued from decaying media.

For each of the various frequency bands observed from space, several missions have usually been flown. For atmospheric sounding of thermodynamic variables (see Chapter 1), several of the most-often observed wavelengths are in the infrared. The trend for new instruments has been towards increased spatio-temporal and spectral resolutions, especially with hyperspectral sounders that have become more widespread since the 2000s. Owing to various limitations such as sensitivity to

*See Glossary.

clouds, these data alone are not sufficient to construct a comprehensive picture of the climate, but they are used alongside microwave sounders, as in Figure 12.1. For temperature and water vapor sounding, key frequency bands have been monitored in the microwave around 50–60 GHz spectral range, covering absorption and emission from the oxygen rotational band and around 22 GHz (both since the 1970s) and 183 GHz (since the 1990s) spectral ranges, respectively. International collaboration remains essential to preserve these frequencies for environmental monitoring, free of unintentional interference by third-party radio systems (see Chapter 7 of Volume 1).

Exploiting the breadth of data collected by Earth-observation satellites since the 1960s, for example, as introduced in Chapter 1 of Volume 1 and inventoried by WMO's Observing Systems Capability Analysis and Review Tool (OSCAR[3]), necessitates knowledge of the frequencies monitored by each instrument and the instrument's performance. Indeed, one desirable property for any climate record is performance stability. This is more likely with self-referenced stable sensing apparatus, as we would be able to maintain in a laboratory. However, no such system exists fully at present in space. Approaching this is GNSS radio occultation because GNSS signal frequencies are calibrated by atomic clocks. Also proposed are concepts of better-calibrated hyperspectral reference radiometers in the infrared spectral range.

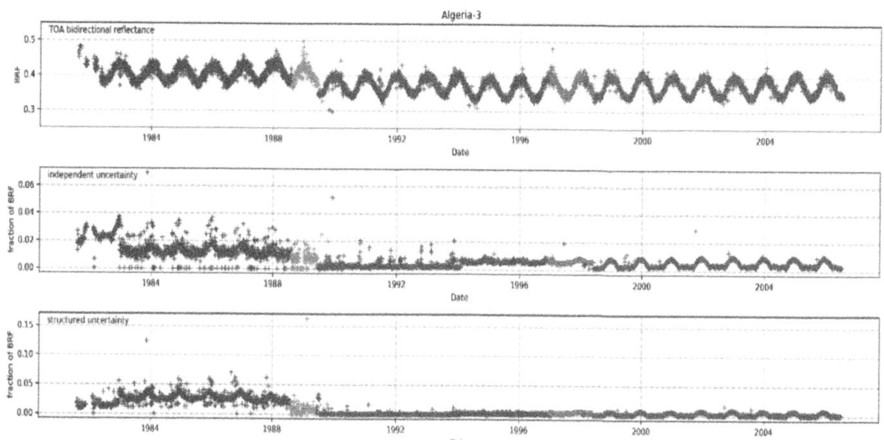

Figure 12.6. *Time-series of (top) recalibrated top-of-atmosphere clear-sky reflectance from Meteosat First Generation satellites (from 1981 to 2017, Meteosat-2 to Meteosat-7, shown with various colors) as well as (middle) independent uncertainties and (bottom) structured uncertainties, for the so-called Algeria-3 calibration site. Results obtained by Rüthrich et al. (2019) (image reproduced thanks to the Creative Commons Attribution License, initially published by Remote Sensing)*

3. https://space.oscar.wmo.int/.

However, for past records, we must rely on ground and/or comparison of space records to assert data record stability. For this, calibration transfer techniques have progressed significantly, for example, within the Global Space-based Inter-Calibration System (see Chapter 7 of Volume 1). When reprocessing past sensors, in the absence of in-space onboard calibration, one solution is to revert to more advanced instruments. An example is shown in Figure 12.6. The temporal degradation of the Meteosat First Generation Visible and Infrared Imager was inferred by comparison with modern instruments on later-generation satellites. This reprocessing estimated uncertainty at the individual observation level, separating between random errors or noise (so-called independent uncertainty) and spatially- and temporally-correlated errors (so-called structured uncertainty). This topic is discussed in more detail in the next section.

12.3.3. *Uncertainty*

Producing an FDR is the first step towards the construction of a Fundamental Climate Data Record (FCDR). A record such as this builds on multiple-satellite or multiple-sensor FDR series and refines the underlying data to reach a sufficient length and quality to be directly useful for climate applications. This may involve inclusion of reversible corrections, for example, obtained by harmonization (removing non-climatic changes) or homogenization (removing differences between various sensors).

After initial/operational data processing has been carried out a first time (possibly years or decades ago), several improvements may be obtained to revisit the original process. Effects that can be assessed and corrected a posteriori include, for example, instrument detector ageing effects (as illustrated in the previous section), improved spectroscopy and modeling of effects of lunar intrusions into cold-sky calibration views. Consequently, even after applying recalibration, time series from multiple satellites or sensors will feature remaining differences between each other, because some errors in the measurements always remain uncharacterized and undetermined to some extent. Figure 12.7 presents a simplified and idealized view of a series of steps that do not always follow sequentially from one another, and that may be re-applied depending on program or project resource opportunities.

In this pursuit of excellence, understanding uncertainties takes on a particular meaning. In data assimilation, uncertainty-characterized data are ideally required to allow for optimal usage of the data. While the BIPM has a clear set of definitions on how to estimate uncertainties for measurements, in a controlled environment and with traceable calibration references, uncertainties are more difficult to estimate for measurements from space.

The issues are multiple. First, the operating environment is not fully known once deployed as a spacecraft loaded with electronics, reflectors and antennas poses challenges of electromagnetic interactions in the harsh radiation environment of space. Second, satellite sensors are not brought back to Earth, preventing rigorous in-laboratory evaluation after the mission, for example to measure any potential drift in characteristics of essential components, such as optical element transmissivity, antenna gain or blackbody emissivity. Nonetheless, attempts are made to take final calibration measurements before deorbiting a satellite, for example by controlled heating and cooling of instruments taking calibration readings or a satellite maneuver. This allows microwave radiometers to observe, under various angles, the deep space isotropic radiation at 2.725 K from the cosmic background.

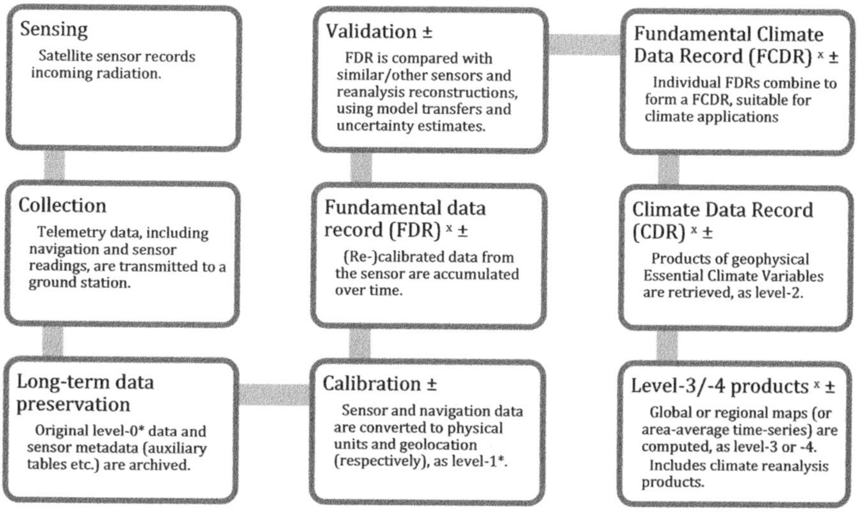

Figure 12.7. *Flowchart from satellite remote sensing to directly-usable climate products. Each step involves quality controls. Crosses (x) indicate products that can be ingested into reanalyses. The ± symbols indicate steps when uncertainty analysis may be applied. Not shown for clarity are the numerous feedback improvement loops, that is, reproducing a previous step with an improved procedure, whenever new findings justify the effort, provided essential mission data and metadata were safeguarded from decay*

The process of identifying and quantifying sources of uncertainties in measurements and downstream data products is rather slow because instruments and processing chains are initially designed to the state-of-the-art capabilities and understanding of the physics. Consequently, improvements beyond this require

*See Glossary.

feedback from field campaign and field data, as well as analysis of long time series. In this regard, new missions allow for new breakthroughs. For example, the reprocessing campaigns of the infrared and microwave sounder into climate data records now benefit from improved usage of all metadata, where the temperature of the instrument or the emissivity of the reflector antenna are included in the calibration process, which was not the case during the lifetime of the instruments.

All of the space agencies involved in Earth observations and mentioned in this book have set-up activities so their past data records contribute to climate applications. In addition, we may note efforts by agencies or institutions to exploit data beyond their own satellites or instruments. Examples in Europe include the ESA Climate Change Initiative (Plummer et al. 2017), the European Union C3S (Thépaut et al. 2018), as well as EUMETSAT and its Climate Monitoring Satellite Applications Facility (Schulz et al. 2009).

12.4. The usage of climate data records in science and services

The needs for climate data records are multiple. Institutional needs are expressed by and to national and international agencies whose missions are cross-cutting the environmental spheres. The GCOS ECV and OSCAR inventories mentioned earlier assist in capturing the evolutions of climate products, on the one hand, and of satellite programs, on the other hand. This mapping enables the identification of potential measurement gaps in the future that would threaten the continuation of climate data records for a given ECV. In addition, they support optimization of the space-based part of the global observing system through the coordinated activities of space agencies.

The understanding and monitoring of climate variability and its evolution, at different scales, are a general pursuit of climate services in collaboration with the academic community. The resulting publications feed the IPCC Assessment Reports, which guide changes in public policies at the highest levels of government. For example, the 6th Assessment Report (IPCC 2021) used 238 data products, with about half of them generated from satellite data. Another scientific use of satellite climate data is to verify and improve climate models. In the end, the discrepancies found from the comparisons represent another source of information for climate data producers to identify product deficiencies or understand new needs from data users.

These applications of Climate Data Records are not exclusive. Other users include industries for which weather exposure is a critical factor, such as the agricultural or the renewable energy sectors. For example, the availability of wind or solar radiation are largely predictable within a few hours, but knowing the production capability of a given site for the next decades is an important element to take into account during the financial modeling step, before deciding to invest in site installations. Such data users generally welcome the packing of satellite monitoring information into model-assisted climate products such as climate reanalysis, as it enables them to directly exploit the multi-variate representations of the "weather and climate cube".

With the generalization of coding in education curriculum and the growing availability of computing resources, the uptake of climate data is growing to unprecedented levels. This, however, calls for increased education and training in the advanced use of satellite climate data, so that users understand the potential, and also the limitations, by applying the associated quality and uncertainty information.

Figure 12.8. *Cover page of the WMO State of the Climate for the year 2020 (source: WMO)*

Other use cases of climate data records are the regular updates at national or international levels, such as the WMO report on the State of the Climate (Figure 12.8). By regularly summarizing events of the past year into indicators and maps, with excerpts generally distilled via press, television or social network relays, the public at large can better apprehend the scale of the issues posed by a changing climate.

12.5. Looking ahead

The pioneer space observers of the Earth's atmosphere have delivered more than they probably expected. As promised, their initial missions helped to explain the basic features of the atmospheric general circulation, and to even expose the organization of large-scale extreme events such as tropical cyclones (see Chapter 6). In addition, their legacy enabled the developments of digital Earth representations with a time dimension (such as reanalyses). These data and possibly new digital representations will form the backbone of a 21st century global society, where scarcity of resources such as groundwater will demand more careful monitoring.

In this future, maintaining the traceability of space instruments to standards will become easier. Indeed, several missions have been proposed to provide benchmark measurements for some of the main frequency bands monitored by Earth observation satellites: at visible to far-infrared wavelengths (Climate Absolute Radiance and Refractivity Observatory), proposed by Wielicki et al. (2013) and in the solar reflective range (Traceable Radiometry Underpinning Terrestrial and Helio Studies), proposed by Fox and Green (2000). However, for passive microwave observations, traceability of sufficient accuracy will not be achievable within the next decade as the radiometric uncertainties reached using current in-laboratory standards from National Metrology Institutes are at the same level as those required from satellite sensors in orbit. Further activities are needed to develop the necessary infrastructure, tailored to climate needs in readiness for its use in climate observing systems.

With the adoption of ambitious targets to reduce greenhouse gas emissions under the Paris Climate Agreement of 2015, several signatory countries have now started to explore ways of routinely monitoring carbon emissions and more accurately pinpointing regional climate change patterns since the pre-industrial era. To achieve this task will require minutiae in planning and in data analysis. The European Union Copernicus program plans to take an active role in this arena with the Copernicus Carbon Dioxide Monitoring CO2M mission (see Chapter 4 of Volume 1). There is no doubt that all of the lessons learnt from climate reprocessing approaches will contribute to this new challenge. The data collected will further continue to feed the developments in our understanding of the climate system. With climate information having a role to play in all 17 Sustainable Development Goals (WMO 2017), understanding the climate is a prerequisite for appropriate allocation of resources for the world's population.

12.6. References

Fox, N. and Green, P. (2020). Traceable radiometry underpinning terrestrial- and helio-studies (TRUTHS): An element of a space-based climate and calibration observatory. *Rem. Sens.*, 12, 2400. https://doi.org/10.3390/rs12152400.

GCOS (2003). The second report on the adequacy of the global observing system for climate in support of the UNFCCC. Report, WMO/TD-No. 1143 and GCOS-82, WMO Secretariat, Geneva.

Hasselmann, K. (1997). Multi-pattern fingerprint method for detection and attribution of climate change. *Clim. Dyn.*, 13, 601–611. https://doi.org/10.1007/s003820050185.

Held, I.M. and Soden, B.J. (2006). Robust responses of the hydrological cycle to global warming. *J. Clim.*, 19, 5686–5699. https://doi.org/10.1175/JCLI3990.1.

IPCC (2021). Climate change 2021: The physical science basis. In *Contribution of Working Group I to the Sixth Assessment Report of the Intergovernmental Panel on Climate Change*, Masson-Delmotte, V., Zhai, P., Pirani, A. (eds). Cambridge University Press.

Manabe, S. and Wetherald, R.T. (1967). Thermal equilibrium of the atmosphere with a given distribution of relative humidity. *J. Atmos. Sci.*, 24, 241–259. https://doi.org/10.1175/1520-0469(1967)024<0241:TEOTAW>2.0.CO;2.

Plummer, S., Lecomte, P., Doherty, M. (2017). The ESA climate change initiative (CCI): A European contribution to the generation of the global climate observing system. *Rem. Sens. Environ.*, 203, 2–8. https://doi.org/10.1016/j.rse.2017.07.014.

Pörtner, H.O., Scholes, R.J., Agard, J., Archer, E., Bai, X., Barnes, D., Burrows, M., Chan, L., Cheung, W.L., Diamond, S. et al. (2021). IPBES-IPCC co-sponsored workshop report on biodiversity and climate change [Online]. Available at: https://doi.org/10.5281/zenodo.4782538.

Rüthrich, F., John, V.O., Roebeling, R., Quast, R., Govaerts, Y., Woolliams, E.R., Schulz, J. (2019). Climate data records from Meteosat First Generation part III: Recalibration and uncertainty tracing of the visible channel on Meteosat-2-7 using reconstructed, spectrally changing response functions. *Remote Sens.*, 11(10), 1165. https://doi.org/10.3390/rs11101165.

Schulz, J., Albert, P., Behr, H.-D., Deneke, H., Dewitte, S., Dürr, B., Fuchs, P., Gratzki, A., Hechler, P., Hollmann, R. et al. (2009). Operational climate monitoring from space: The EUMETSAT satellite application facility on climate monitoring (CM-SAF). *Atmos. Chem. Phys.*, 9, 1687–1709. https://doi.org/10.5194/acp-9-1687-2009.

Thépaut, J.-N., Pinty, B., Dee, D., Engelen, R. (2018). The copernicus programme and its climate change service. *IEEE Intl. Geosci. Rem. Sens. Symp.* https://doi.org/10.1109/IGARSS.2018.8518067.

Waters, C.N., Zalasiewicz, J., Summerhayes, C., Barnosky, A.D., Poirier, C., Galuszka, A., Cearreta, A., Edgeworth, M., Ellis, E.C., Ellis, M. et al. (2016). The Anthropocene is functionally and stratigraphically distinct from the Holocene. *Science*, 351(6269). https://doi.org/10.1126/science.aad2622.

Wielicki, B., Young, D.F., Mlynczak, M.G., Thome, K.J., Leroy, S., Corliss, J., Anderson, J.G., Ao, C.O., Bantges, R., Best, F. et al. (2013). Climate absolute radiance and refractivity observatory (CLARREO): Achieving climate change absolute accuracy in orbit. *Bull. Amer. Meteorol. Soc.*, 1519–1539. https://doi.org/10.1175/BAMS-D-12-00149.1.

WMO (2017). White paper on the contribution of the global framework for climate services to transforming our world: The 2030 agenda for sustainable development [Online]. Available at: https://library.wmo.int/doc_num.php?explnum_id=3586.

12.7. References of the data sources cited in Figure 12.1

20CRv3: Slivinski, L.C., Compo, G.P., Whitaker, J.S., Sardeshmukh, P.D., Giese, B.S., McColl, C., Allan, R., Yin, X., Vose, R., Titchner, H. et al. (2019). Towards a more reliable historical reanalysis: Improvements for version 3 of the twentieth century reanalysis system. *Quart. J. Roy. Meteorol. Soc.*, 145, 2876–2908. https://doi.org/10.1002/qj.3598.

CRUTEM5: Osborn, T.J., Jones, P.D., Lister, D.H., Morice, C.P., Simpson, I.R., Winn, J.P., Hogan, E., Harris, I.C. (2021). Land surface air temperature variations across the globe updated to 2019: The CRUTEM5 dataset. *J. Geophys. Res: Atmos.*, 126, e2019JD032352. https://doi.org/10.1029/2019JD032352.

ERA-20C: Poli, P., Hersbach, H., Dee, D.P., Berrisford, P., Simmons, A.J., Vitart, F., Laloyaux, P., Tan, D.G.H., Peubey, C., Thépaut, J.-N. et al. (2016). ERA-20C: An atmospheric reanalysis of the twentieth century. *J. Clim.*, 29, 4083–4097. https://doi.org/10.1175/JCLI-D-15-0556.1.

ERA5: Hersbach, H., Bell, B., Berrisford, P., Hirahara, S., Horányi, A., Muñoz-Sabater, J., Nicolas, J., Peubey, C., Radu, R., Schepers, D. et al. (2020). The ERA5 global reanalysis. *Quart. J. Roy. Meteorol. Soc.*, 146, 1999–1949. https://doi.org/10.1002/qj.3803.

GlobalTempV5: Zhang, H.-M., Lawrimore, J.H., Huang, B., Menne, M.J., Yin, X., Sánchez-Lugo, A., Gleason, B.E., Vose, R., Arndt, D., Rennie, J.J., Williams, C.N. (2019). Updated temperature data give a sharper view of climate trends. *Eos Trans. Amer. Geophys. Union*, 100. https://doi.org/10.1029/2019EO128229.

Hegerl, G.C., Crowley, T.J., Allen, M., Hyde, W.T., Pollack, H.N., Smerdon, J., Zorita, E. (2007). Detection of human influence on a new, validated 1500-year temperature reconstruction. *J. Clim.*, 20, 650–666. https://doi.org/10.1175/JCLI4011.1.

JRA-55: Kobayashi, S., Ota, Y., Harada, Y., Ebita, A., Moriya, M., Onoda, H., Onogi, K., Kamahori, H., Kobakashi, C., Endo, H. et al. (2015). The JRA-55 reanalysis: General specifications and basic characteristics. *J. Meteorol. Soc. Japan.*, 93, 5–48. https://doi.org/10.2151/jmsj.2015-001.

MERRA2: Gelaro, R., McCarty, W., Suárez, M.J., Todling, R., Molod, A., Takacs, L., Randles, C.A., Darmenov, A., Bosilovich, M.G., Reichle, R. et al. (2017). The modern-era retrospective analysis for research and applications, Version 2. *J. Clim.*, 30, 5419–5454. https://doi.org/10.1175/JCLI-D-16-0758.1.

TLT RSS V4: Mears, C.A. and Wentz, F.J. (2017). A satellite-derived lower-tropospheric atmospheric temperature dataset using an optimized adjustment for diurnal effects. *J. Clim.*, 7695–7718. https://doi.org/10.1175/JCLI-D-16-0768.1.

TLT UAH V6: Spencer, R.W., Christy, J.R., Braswell, W.D. (2017). UAH version 6 global satellite temperature products: Methodology and results. *Asia-Pacific J. Atmos. Sci.* https://doi.org/10.1007/s13143-017-0010-y.

13

Anthropogenic Greenhouse Gases: CO_2 and CH_4

Cyril CREVOISIER
Laboratoire de Météorologie Dynamique, IPSL, Palaiseau, France

Recent IPCC reports have clearly established the causal link between global warming and an increase in the atmospheric concentration of greenhouse gases (GHGs) influenced by human activities. The main contributors to this warming are carbon dioxide (CO_2), methane (CH_4) and nitrous oxide (N_2O), which contribute 56%, 24% and 8%, respectively, to the additional greenhouse effect[1] and whose anthropogenic emissions have been constantly increasing since the beginning of the industrial era. While the average concentrations of GHGs in the atmosphere are now well measured by ground-based observatories, a full understanding of the associated biogeochemical cycles remains uncertain. This chapter presents the contribution of spatial observation to the monitoring of changes in the atmospheric content of anthropogenic greenhouse gases and their emissions, focusing mainly on CO_2 and CH_4, with N_2O remaining a poorly observed gas with no real observation perspective at present.

For a color version of all the figures in this chapter, see www.iste.co.uk/phulpin/satellites2.zip.

1. The additional greenhouse effect represents the share of the atmospheric greenhouse effect, a predominantly natural phenomenon, caused by human activities.

Satellites for Atmospheric Sciences 2,
coordinated by Thierry PHULPIN, Didier RENAUT, Hervé ROQUET and Claude CAMY-PEYRET.

13.1. Monitoring anthropogenic greenhouse gases

13.1.1. *Biogeochemical cycles*

Generally speaking, the atmospheric abundance of a gas is governed by the balance between its sources (which emit it) and its sinks (which absorb it). Sources and sinks can be located on the Earth's surface or in the atmosphere. Interactions between the different reservoirs (atmosphere, continent and ocean) of these gases form the biogeochemical cycles. These cycles are composed of a natural part that is disturbed by human activities, mainly in the form of additional emissions.

CO_2 is exchanged cyclically between three reservoirs: the atmosphere, oceans and the continental biosphere. Natural fluxes equalize over about a year. The main anthropogenic CO_2 emissions are those generated by the combustion of fossil resources (~91%) and cement production (~9%), for a total of 36.4 $GtCO_2.year^{-1}$ for the years 2000–2010 (Le Quéré et al. 2015). They are estimated on the basis of statistical data on fuel production. Fossil emissions are constantly increasing at the global scale. Human activities also lead to changes in land use and thus disrupt exchanges with the continental biosphere. Slightly less than half (~44%) of the CO_2 emitted remains in the atmosphere, where the growth rate is between 2 and 3 $ppm.year^{-1}$ over the last few decades; the other half is absorbed by the terrestrial biosphere and oceans, 30% and 26%, respectively, thus decreasing the rate of climate change. But there are great uncertainties about the fate of these sinks.

On a global scale, methane emissions are of the order of 500–600 $TgCH_4.year^{-1}$ with considerable uncertainty about individual processes and the regional scale. These emissions are of biogenic origin (wetlands, inland waters, livestock, rice cultivation, landfills), thermogenic (exploitation and transport of fossil fuels, natural degassing of surfaces) or pyrogenic (combustion of biomass and biofuels). Between 50% and 65% of emissions are anthropogenic. In the coming decades, global warming will lead to climate feedbacks (gradual melting of permafrost, changes in precipitation patterns and impacts on flooded areas) that will affect methane emissions. The atmospheric methane sink is mainly due to the OH hydroxyl radical in the troposphere (85–90%) supplemented by the oxidation of methane in the stratosphere and in the marine boundary layer.

13.1.2. *Determination of gas sources and sinks*

Two complementary approaches can be used to estimate surface GHG fluxes (sources and sinks) on a regional or global scale: the bottom-up or top-down approaches. The bottom-up approach consists of identifying all natural phenomena controlling sources and sinks at the local level by direct flux measurements and then extrapolating them at the regional to global level. It is based on the idea that once fundamental processes governing the operation of sources and sinks are known and understood, they can be taken into account in numerical models that will be used to predict the spatial and temporal variations of gas fluxes over large regions. This approach is generally limited by poor understanding of certain processes and the difficulty of extrapolating local measures.

The top-down approach, often referred to as "atmospheric inversion", consists of estimating gas fluxes at the surface by studying the spatial and temporal variations in the concentration that they generate. This is a classic inverse modeling problem since it consists, from the observed effects (gas variations), of inferring the causes that give rise to them (surface sources and sinks), as well as their uncertainties. The associated direct problem is to estimate, using an atmospheric transport model, the concentrations of gases in the atmosphere for which the sources and sinks at the surface are known. Such a model numerically solves the continuity equation for an atmospheric constituent (e.g. a trace gas such as CO_2) assuming that the meteorological fields are known. The top-down approach therefore makes use of measurements of the concentration of the gas studied, and possibly of other associated gases, and these are measurements characterized by a certain spatial and temporal sampling.

Atmospheric inversion has been widely used in recent decades to estimate surface CO_2 or CH_4 fluxes at various spatio-temporal scales (LeQuéré et al. 2015; Saunois et al. 2016). These studies were mainly based on the use of in situ data (surface stations, flux tower, aircraft and boats). However, this approach is limited by the low spatial coverage and high heterogeneity of surface observation networks, as well as by the errors associated with atmospheric transport and chemistry modeling.

13.1.3. *The global observation network*

Atmospheric concentrations of GHGs have been measured directly by surface stations since 1958 for CO_2, since 1972 for N_2O and since 1979 for CH_4 by a set of networks grouped within the Global Atmosphere Watch (GAW) of the World Meteorological Organization (WMO), which now includes more than 140 stations.

These surface observations are complemented by airborne measurements giving access to vertical variations in GHGs. Regular scientific campaigns are organized on dedicated sites, mainly in North America, Siberia and over the Amazon. Commercial aircraft are also equipped with in situ measuring devices (the CONTRAIL program or IAGOS European Research Infrastructure). The majority of airborne measurements are carried out at the cruising altitudes of the aircraft, that is, between 9 and 12 km, the remainder corresponding to vertical profiles obtained during the take-off or landing phases of an aircraft at airports. To reach higher altitudes, in situ measuring instruments have been developed to be released under meteorological balloons, such as the AirCore, which now perform regular measurements of profiles between the surface and 35 km of altitude at a few points of the globe.

The current global observation network is characterized by a high degree of spatial heterogeneity. Thus, while Western Europe, North America and Japan are well covered by measurement stations, the tropics and the southern hemisphere are particularly under-sampled, despite their significance in understanding carbon and methane cycles. The intensification of measures in the tropical, Asian and oceanic regions is essential for the implementation of carbon flux monitoring aimed at informing policy makers (Ciais et al. 2014). This explains the strong interest in space-based GHG measurements carried out by various instruments on board satellites that could ensure global and continuous coverage of the Earth.

13.2. Contribution of spatial observation of greenhouse gases

13.2.1. *Specificities of greenhouse gas observation*

Observing GHG from space is based on instruments measuring radiation emitted by a natural source (the sun or the terrestrial system) or a manufactured source (lidar remote sensing*) which, after having passed through the different layers of the atmosphere, contains information on the thermal structure and composition of the atmosphere. The measurements of atmospheric radiation, essentially carried out in the infrared range, are sensitive to the concentration of CO_2 over all or part of the atmospheric column. However, they do not contain enough information to resolve this concentration over the entire vertical profile and rather give access to integrated content* over the entire thickness of the atmosphere, denoted as Xgas, with a certain vertical sensitivity to the variations of the gas in each pressure layer. Figure 13.1 shows the typical weighting functions[2] of instruments measuring at nadir: measuring the solar radiation reflected by the Earth's surface in the shortwave infrared gives access to a total column* of gas and measuring the thermal infrared radiation

*See Glossary.

2. Weighting functions are used to characterize the transmission properties of the atmosphere changing with altitude, which allows for better identification of the atmospheric layers that have contributed the most to the signal measured by the space instrument (see Chapter 9 of Volume 1 for further explanation).

emitted by the terrestrial system gives access to a partial column* of gas, representative of the mid-to-upper troposphere.

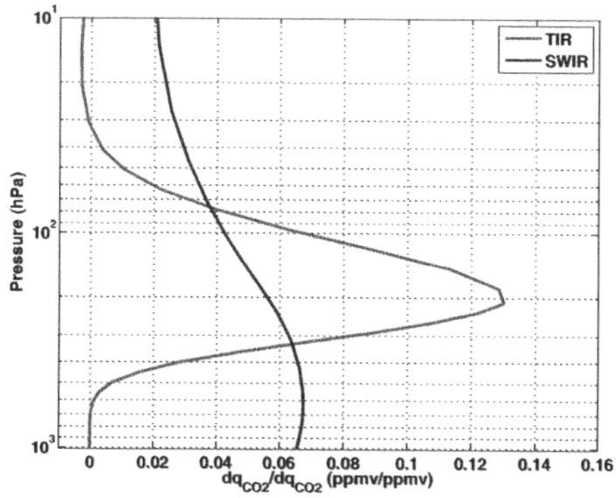

Figure 13.1. *Example of weighting functions for CO_2 in the case of nadir measurement performed in the shortwave infrared (blue), which gives access to a column sensitive to the whole atmospheric column, or thermal infrared (red), which gives access to a column sensitive to the mid-to-upper troposphere*

In a pioneering study, Rayner and O'Brien (2001) showed interest in estimating CO_2 fluxes by using integrated columns obtained from spatial measurements in addition to surface concentration measurements. Indeed, their study highlighted the major advantage of the global coverage of satellite measurements compared to the sparse coverage of ground station networks, provided that accurate and unbiased measurements are available.

Indeed, because of the long lifetime of GHGs in the atmosphere, the background level (the proportion of well-mixed gas) is very high compared to the spatio-temporal gradients of the gas. The measurement of the concentration must therefore be very precise and above all unbiased. Thus, it is necessary to measure CO_2 with an accuracy of a few ‰, better than 1 part per million (ppm) for a total gas column of about 410 ppm.

Moreover, with the exception of regions of high convection, gas fluxes at the surface essentially have an impact on concentrations in the atmospheric boundary layer. The concentration signal is then transported, within a few days, to the upper layers of the atmosphere. This dilution complicates the interpretation of gas variations observed along the atmospheric column. As shown in Figure 13.2, for the same July day, large horizontal variations in CO_2 visible in surface concentration

*See Glossary.

measurements, which result from the combination of surface CO_2 fluxes and surface transport, are very much reduced (by more than a factor of 10!) when looking at the variations detected on a total or partial column of CO_2. Thus, any measurement of the gas column from space is useful for constraining GHG emission inventories only if it achieves the precision and accuracy necessary to detect and quantify the small (< 0.25%) variations in XCO_2 and XCH_4 associated with natural sources and sinks.

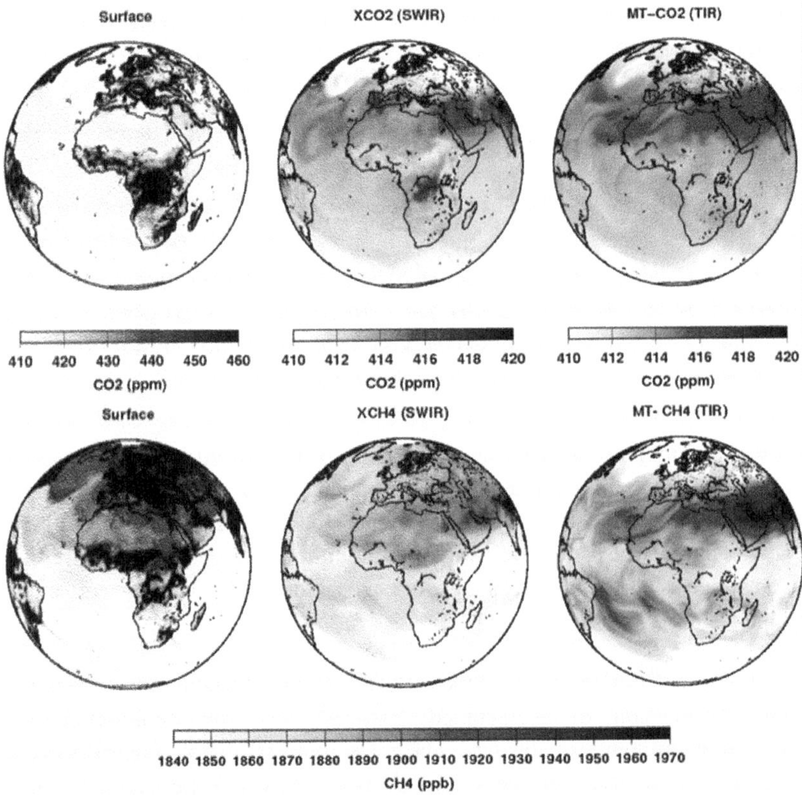

Figure 13.2. *For the month of July 2019, CO_2 field (top) and CH_4 field (bottom) at the surface (left), typical total column of shortwave infrared measurements (middle) and typical integrated column of thermal infrared measurements (right). Based on forecasts from the Copernicus Atmospheric Monitoring Service*

In situ and spatial observations are thus highly complementary and have their own advantages and disadvantages. In situ observations provide a highly accurate and unbiased measurement of the atmospheric mixing ratio* of GHGs, but remain

*See Glossary.

unevenly distributed over time and space. Satellite observations greatly improve spatial and temporal coverage but at the cost of less precision of individual measurements than for in situ measurements (10–100 times less) and of more difficult biases to establish.

13.2.2. *Particularly rich spatial programming*

Over the past decade, the monitoring of the carbon cycle and, more generally, of greenhouse gases has been recognized as a major challenge by space agencies. Figure 13.3 shows the temporal coverage of past, present and short-term planned space missions that provide information on CO_2, CH_4 and N_2O. The ultimate objective is to quantify and locate the sources and sinks of these gases through accurate mapping that can be derived from assimilation models coupled with spatial observations. One of the underlying political and societal goals is to better manage and anticipate actions to limit emissions or deforestation.

Since the early 2000s, the methane observation system from space has benefited from global coverage through the launch of SCIAMACHY**, followed by GOSAT in 2009, to which IASI** has been added in three successive missions since 2007, and finally TROPOMI**/Sentinel-5P in 2018. The observation of CO_2 began with the exploitation of the first generations of TOVS infrared sounders, followed by the second-generation AIRS** and IASI sounders. The launch of GOSAT in 2009 and then OCO-2** in 2013 marked the beginning of the launch of missions dedicated to the first greenhouse gas influenced by human activities.

Various space missions are planned in the short and medium term to provide essential observations for monitoring and understanding CO_2 and CH_4 trends. Near-infrared missions will provide total column measurements, by day, typically over a period of 5–10 years. Of particular note are the MicroCarb** mission of CNES (see Chapter 6 of Volume 1) and CO2M on Sentinel-7 (see Chapter 4 of Volume 1) of the European Union and ESA, the latter having the objective of participating in the monitoring of global inventories of anthropogenic CO_2 emissions set up within the framework of the Paris Agreement, with two to three satellites. The UVNS/Sentinel-5 mission will take over from TROPOMI/Sentinel-5P for a period of 20 years. The thermal infrared missions IASI and then IASI-NG** will provide partial column measurements over a total period of more than 40 years, day and night. It should be noted that the IASI-NG and UVNS/Sentinel-5 missions will be carried on the same MetOp-SG-A-1/2/3 platforms, making it possible to envisage a combination of these measurement techniques. Finally, the geoCARB** mission,

**See Appendix 3.

selected in December 2016 by NASA should have been the first geostationary mission to measure total CO_2 and CH_4 columns but it was cancelled in November 2021. Beyond these various state missions, missions have recently been launched by private operators (GHGSat) and are the subject of many projects under development.

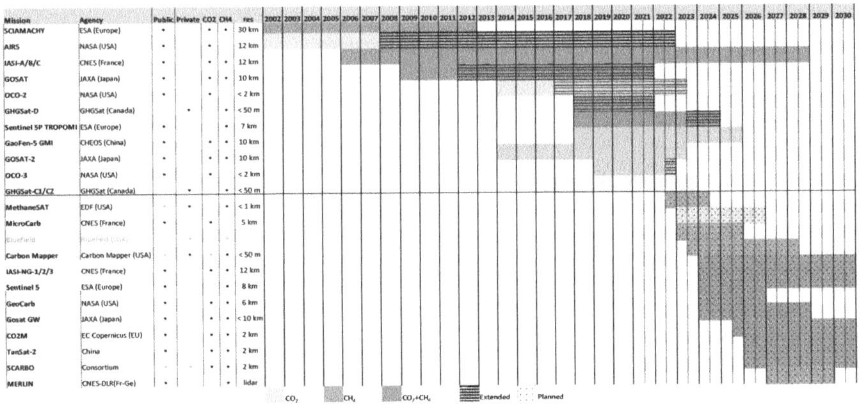

Figure 13.3. *Past, present and future missions observing CO_2 (yellow), methane (green) or both (red)*

13.3. Measurement techniques

The main techniques used to study greenhouse gases from space are divided into four types that depend on the spectral domain studied, the source of radiation (natural or not) and the aiming technique (nadir or limb, see Chapter 9 of Volume 1). Depending on these choices, each type of instrument gives access to different information on the atmospheric abundance of gases: total integrated column, partial integrated column and vertical profile.

The spectral bands covered by instruments are chosen as a function of spectral regions in which the atmospheric constituents to be studied have absorption lines. The two main spectral regions used are shortwave infrared radiation (SWIR) between 0.7 and 3.0 μm and thermal infrared radiation (TIR) between 3.0 and 15.5 μm. The measurement principles for the three main types of space missions for GHG observation at nadir are shown in Figure 13.4.

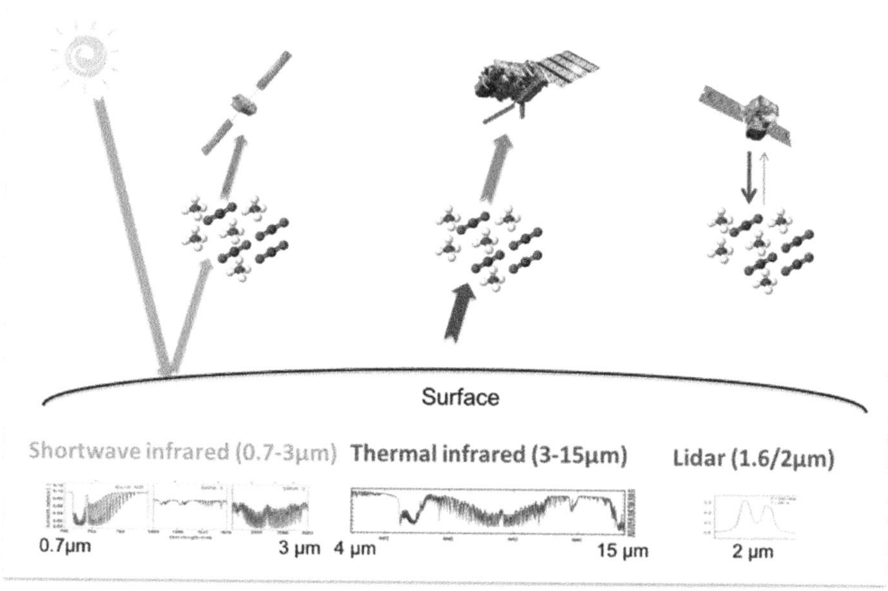

Figure 13.4. *The three measurement principles aiming at the nadir for greenhouse gas missions*

13.3.1. *Passive observations in the infrared range*

Observations in the thermal infrared are mainly sensitive to the concentration of CO_2 in the mid-to-upper troposphere with a maximum sensitivity towards ~12 km or ~200 hPa. They therefore make it possible to access integrated columns on the mid-troposphere, denoted, for example, as MT-CO_2 (Figure 13.1). Being based on the measurement of the infrared radiation emitted by the terrestrial system, these instruments, called infrared sounders, carry out observations day and night, on land and at sea.

In a pioneering study, Chédin et al. (2003) conducted the first estimates of CO_2 columns from space using first generation HIRS** meteorological sounders on board NOAA polar satellites. Beyond trends and the seasonal cycle, this work has revealed a particularly strong signature of CO_2 emissions from tropical biomass fires. The precision reached was 1%–2%. This work was then extended to new generation instruments that allow for measuring spectra with high spectral resolution and cover the entire thermal infrared spectrum (Crevoisier et al. 2018). AIRS instruments

**See Appendix 3.

on board Aqua; IASI on board MetOp-A, -B and -C satellites and TES** are examples. For IASI, random errors in CO_2 and CH_4 estimates are now estimated at 1 ppm and 8 ppb, respectively, at the observation point scale. Figures 13.5 and 13.6 illustrate the time series of contents integrated into the tropospheric average of CO_2 and CH_4 estimated from IASI observations since 2007. The trends and seasonal cycles of the two gases are clearly visible and show, in particular, a 5% increase over 10 years in CO_2 and CH_4 in the tropospheric average.

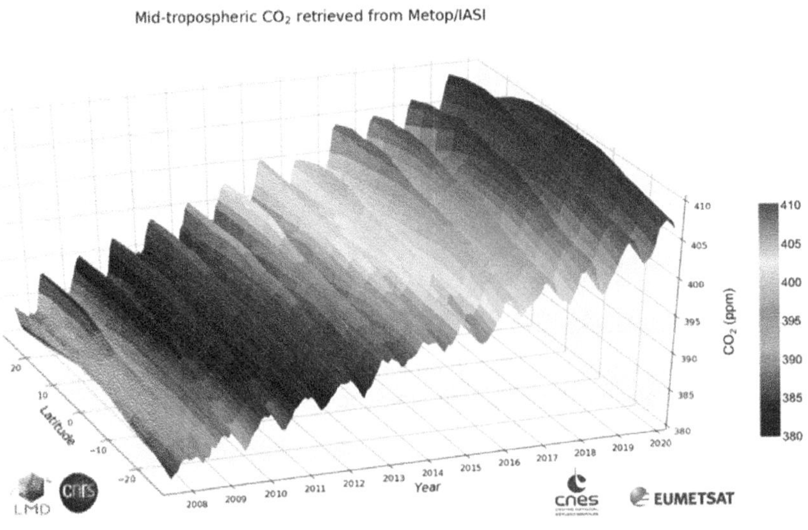

Figure 13.5. *Monthly variation of integrated MT-CO_2 measured by IASI from 2007 to 2020 as a function of latitude*

The maps obtained clearly show signals related to gas sources. In addition to emissions from biomass fires, there are methane emissions from Asian rice fields that are rapidly transported through the troposphere by the monsoon during the summer months. However, as illustrated in Figure 13.2, TIR measurements are not very sensitive to GHG concentrations in the lower troposphere and the planetary boundary layer. Their use in estimating surface fluxes is therefore strongly impacted by the transport errors of the atmospheric model used in the "top-down" approach. As a result, these measurements have rarely been used to estimate monthly fluxes at regional to hemispheric scales (Cressot et al. 2014). CO_2 and CH_4 estimates from IASI are now assimilated daily to provide CO_2 and CH_4 forecasts as part of the Copernicus Atmospheric Monitoring Service (Massart et al. 2014).

**See Appendix 3.

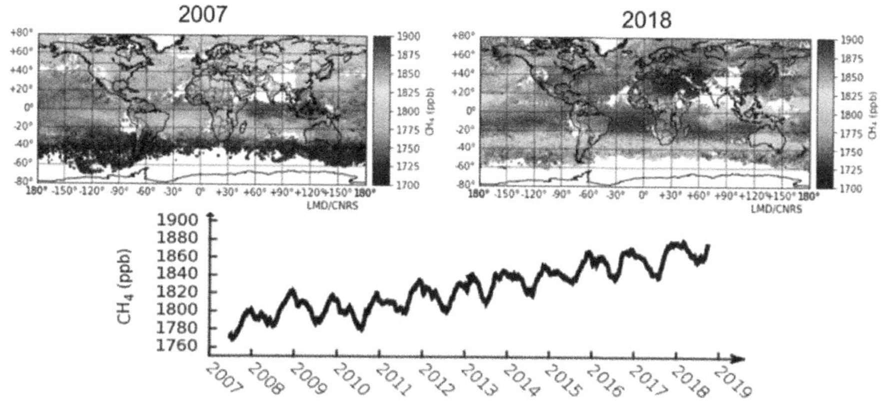

Figure 13.6. *Monthly variation of the integrated MT-CH_4 columns measured by IASI from 2007 to 2018 and annual map in 2007 and 2018*

13.3.2. *Passive observations by solar reflection*

Passive sounders operating in shortwave infrared (SWIR) make observations with high spectral resolution of the solar radiation reflected by the surface of the Earth, following a bidirectional path, by the so-called differential absorption approach. Absorption lines of the different atmospheric gases are identified on the recorded spectra, and their depth is related to the gas charge along the entire atmospheric column. This technique makes it possible to restore integrated CO_2 and CH_4 contents sensitive to the concentration of the gas over the entire thickness of the atmosphere. Thus, the XCO_2 estimated from shortwave infrared radiation are called (improperly) "total columns of CO_2".

In this spectral band, measurements are possible only by day and mainly above continental surfaces. In addition to the use of bands around 1.6 and 2.0 μm sensitive to CO_2 and CH_4 variations, the sensors also measure spectra in the A band of molecular oxygen (O_2). Indeed, the latter provides information necessary to calculate the total abundance of the column of dry air (making it possible to calculate Xgas) and the surface pressure, and to provide information on the presence, vertical distribution and total optical depths of the clouds and aerosols that disturb the radiation.

The ESA SCIAMACHY instrument on board Envisat provided information on CO_2 and CH_4, with uneven quality, from 2003 to 2012 (Frankenberg et al. 2008).

The TANSO-FTS** instrument of the Japanese GOSAT and GOSAT-2 missions has been providing CO_2 and CH_4 observations since 2009 (Yokota et al. 2009). Finally, the American missions OCO-2 and OCO-3**, the latter on board the international space station, are entirely dedicated to the study of CO_2 (Crisp et al. 2004). The time series of XCO_2 and XCH_4 estimated from these missions are plotted in Figure 13.7. Following these, the CNES MicroCarb mission is expected to be launched in 2024, with two innovations: coverage of the oxygen absorption band at 1.27 μm to improve the consideration of aerosols in the retrieval of XCO_2 and a mode of observation aiming specifically at areas of high anthropogenic emissions. Finally, in 2027, ESA should launch the CO2M mission of the Copernicus program of the European Union, the objective of which is to monitor anthropogenic emissions using a spectroimager.

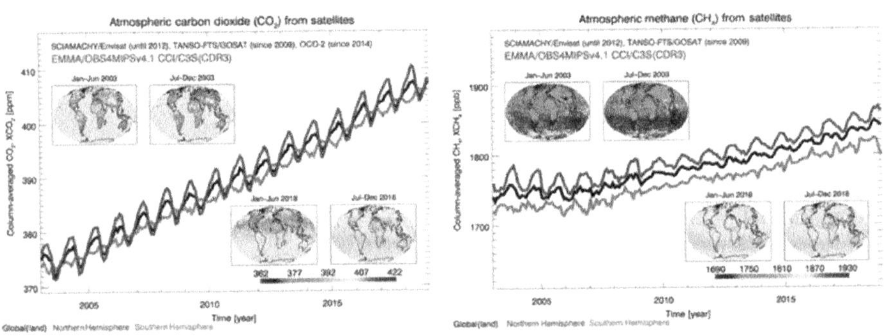

Figure 13.7. *Monthly variation of XCO_2 (left) and XCH_4 (right) in three latitude bands obtained from observations of SCIAMACHY, GOSAT and for CO_2, OCO-2 (adapted from Reuter et al. 2020)*

The performances achieved on XCO_2 are typically random errors for a sampling point between 0.4 and 1.2 ppm (0.1% and 0.3%) and systematic biases between 1 and 2 ppm (0.25% and 0.5%) over most of the globe. This performance approaches the levels needed to quantify natural sources and sinks of CO_2 on a regional scale, to detect CO_2 gradients in large urban areas, and in some cases, to quantify emissions from large coal-fired power plants. For XCH_4, random errors for a sampling point are close to 13 ppb and systematic biases are between 4 and 7 ppb. These observations are used to estimate some local anthropogenic emissions. The TROPOMI/Sentinel-5P instrument provided for the first time CH_4 and CO over a wide swath (~2,600 km) at moderate spatial resolution (~7 km x 7 km at nadir), producing two-dimensional "images" of XCH_4 and XCO.

**See Appendix 3.

13.3.3. *Passive observations by solar occultation*

Also based on measurements in the infrared domain, instruments carrying out limb measurements by solar occultation can give access to a vertical profile of gas concentration covering mainly the stratosphere and sometimes the top of the troposphere. Thus, the ACE-FTS** instrument on board the SciSat platform gives access to CO_2 profiles between 5 and 25 km with a vertical resolution of the order of 2 km and an accuracy of about 1 ppm (Foucher et al. 2011), as well as CH_4 profiles with an uncertainty in the upper troposphere–lower stratosphere of 10%. The SCIAMACHY measurements on board Envisat were also reanalyzed to provide profiles of the two gases at altitudes between 17 and 45 km (Noël et al. 2016). CO_2 estimates have an accuracy of 8–12 ppm (2–3%) and CH_4 measurements have an accuracy of 50–100 ppb (5–10%).

Since these solar occultation measurements do not reach altitudes below 5 km, they cannot be directly used to study the spatial and temporal distributions of CO_2 and CH_4 associated with surface fluxes. Nevertheless, they provide valuable information for the study of atmospheric transport, trends and processes of greenhouse gases in the upper troposphere or above.

13.3.4. *Active observations using lidar*

Atmospheric sounding with an active instrument is based on the same principle as that used by passive instruments studying solar radiation reflected from the Earth's surface. The main difference is that the radiation source is artificial, a laser, and not natural. Different techniques are possible depending on the laser source used: laser spectrometer (continuous laser source), PRN lidar (amplitude-modulated continuous laser source) and pulsed lidar (pulsed laser source). It is this latter approach that seems the most promising in terms of observing greenhouse gases from space. Using the laser signal backscattered by particles and molecules of the atmosphere, a lidar DIAL measurement can theoretically give access to a vertical profile of the mixing ratio of CO_2 or CH_4 (see Chapter 9 of Volume 1). However, for a spatial system, this backscattered signal is too weak; it is therefore the ground echo (or from the top of homogeneous and optically thick clouds) that is used. In order to achieve the accuracy necessary for observing greenhouse gases, it is necessary to reduce the vertical resolution of the measurement. Therefore, this active measurement gives access to a "weighted total column" of gas, similar to that obtained by exploiting the solar radiation reflected at the surface.

**See Appendix 3.

Lidar can measure by day and night, including in thin clouds (cirrus) or in partly cloudy skies (cumulus). Not limited by the angle of incidence of solar radiation at the Earth's surface, a lidar can also provide measurements at high and medium latitudes throughout the year. In addition, a lidar has the advantage of providing additional information on vertical structure and atmospheric composition (aerosols and clouds).

After the pioneering work carried out under the A-SCOPE proposal for an ESA Earth Explorer mission that was not selected due to low technological maturity, the first space mission based on an active instrument should be the Merlin mission (Ehret et al. 2017) designed by the CNES and the DLR and planned to be launched in 2027. Any further development of active missions, in particular for CO_2, requires the development and maturation of efficient laser sources and detectors.

13.4. From radiation measurement to gas flux at the surface

When monitoring emissions or more generally surface gas fluxes, a two-step approach is needed. In a first step, radiances measured by the instrument are interpreted in terms of the average mixing ratio over all or part of the atmospheric column, as well as the associated uncertainty and vertical sensitivity. This step is based on an iterative loop between the direct and reverse radiative transfer models. In a second step, the weighted columns of gases are used to derive surface sources and sinks and their associated uncertainties through atmospheric inversion, which is generally based on an atmospheric transport model.

13.4.1. *From radiation measurement to gas concentrations*

13.4.1.1. *Inverse radiative transfer*

Space instruments measure the radiative spectra at a given spectral resolution and on a given field of view. These spectra contain different information on the thermodynamics and composition of the atmosphere. The mixing ratio of the gas is obtained by interpreting the radiation absorption observed in measurements. Inference algorithms, based on the inversion of the radiative transfer, are used to estimate atmospheric and surface parameters (e.g. concentration of trace gases, temperature profile, albedo or optical depth of scattering particles), which make it possible to best explain the hyperspectral measurements made from space. For each observation point, not only the quantity of gas (total or partial column, profile) but also the vertical sensitivity of the estimate (Figure 13.1) and the associated uncertainty are estimated.

Estimation algorithms use different statistical inference methods that address the fundamentally poorly-posed problem of estimating gas columns. The most commonly used for GHGs are the optimal estimate based on Bayesian inference and neural approaches. These algorithms are based on direct radiative transfer modeling, which is fundamentally dependent on spectroscopic databases containing parameters for calculating atmospheric gas absorption.

In the case of GHGs, estimation is carried out in the absence of clouds and large layers of aerosols, which greatly disrupt measurements. In the case of measurements of radiation reflected from the surface of the Earth in SWIR, measurements are also quite sensitive to the presence of low-diffusing particles (thin clouds of the cirrus type, aerosols) in the optical path, which can then considerably disturb gas column estimates if they are not taken into account. Since accurate multiple scattering calculations are too time-consuming for operational estimates, all inference algorithms are based on radiative transfer codes that implement various approximations to accelerate scattering modeling.

The design of a gas column estimation algorithm has a critical influence on the overall performance of the observation system. Systematic and random errors in estimated concentrations relative to the actual (but unknown) state of the atmosphere have a particular impact on the reduction of uncertainty and bias in greenhouse gas flux reversal studies (Chevallier et al. 2007).

13.4.1.2. *Validation*

In order to identify, characterize and mitigate the impact of instrumental, geophysical and methodological biases in space-based estimates of greenhouse gas columns, validation protocols must be put in place. In situ measurements of CO_2 and CH_4 on ground or by aircraft play an essential role in this activity, including the validation of growth rates and certain spatial-temporal gradients; however, they cannot be used directly to validate integrated columns, since they describe gas concentrations at a single point or along the flight path, where gas columns estimated from space are deduced from an optical path in the atmosphere that takes into account the entire atmospheric column.

The main source of validation of missions based on the observation of solar radiation reflected at the surface of the Earth in the SWIR is the TCCON network (Total Carbon Column Observing Network; Wunch et al. (2011)), which groups together about twenty stations distributed on the surface of the globe, each equipped with a Fourier transform spectrometer, which provides a spectrum with very high spectral resolution of the incident solar radiation by aiming directly at the sun. By inverse radiative transfer, it is possible to extract from these spectra total columns of greenhouse gases. Because they measure direct sunlight rather than sunlight scattered from the surface and atmosphere, TCCON spectrometers have much greater sensitivity to CO_2 and CH_4 than space-based instruments and much lower uncertainties over the length of the optical path. In a now systematic manner, XCO_2 and XCH_4 column estimates from space are compared with columns measured on TCCON sites. The overall agreement obtained is generally better than ~1 ppm for XCO_2 and ~6 ppb for XCH_4. In the thermal infrared field, a similar effort is exerted on the basis of measurements made by Fourier transform spectrometers operating in the infrared of the Network for the Detection of Atmospheric Composition Change (NDACC).

In addition to these fixed stations, compact and mobile Fourier transform interferometers have recently been developed. We can mention the EM27/SUN instruments, combined within the international COCCON (COllaborative Carbon Column Observing Network) or CHRIS (Compact High-Spectral-Resolution Infrared Spectrometer) consortium, which makes it possible to carry out measurements in both TIR and SWIR and they can therefore be used for validating missions covering these two spectral ranges.

Whether measured from the ground or from space, weighted columns of gas must be validated regularly by accurate and unbiased in situ measurements that can be linked to international standards. With this in mind, measurements of vertical gas concentration profiles by airplane or, even better, by balloon are organized near the TCCON stations or as close as possible to the passage of satellites as in the case of the AirCore-Fr program under which monthly releases of AirCore, an atmospheric sampler under a weather balloon, are carried out on three stations in France during the passage of satellites of interest. Finally, field campaigns bringing together all the instruments developed by the international community to carry out measurements on ground, on aircraft, or under a balloon are organized to validate space missions in different parts of the world. This is particularly the case with CoMet initiatives led by the DLR and MAGIC led by the CNRS and the CNES.

13.4.2. *From concentration to fluxes*

Once gas columns are estimated and validated, this data is combined with atmospheric wind estimates to estimate surface fluxes of gases in a top-down approach (see section 13.1.2). Three approaches are possible, depending on the nature of the fluxes studied, the spatial and temporal resolution available and knowledge of the atmospheric environment.

When looking at a well-defined source where the wind profile is well known and rather constant over the space–time domain in question, mass balance methods are used. They are typically based on the difference between the gas fluxes entering and leaving the domain of interest. In the case where atmospheric transport is more complex and it is necessary to take into account, for example, convection or advection phenomena, it is possible to use a method modeling the dispersion of plumes; it amounts to measuring the shape of the plume and evaluating the increase in the atmospheric concentration of CO_2 observed in the plume in order to calculate, thanks to knowledge of the wind speed, the emission of the source point associated with the plume (Bovensmann et al. 2010). Finally, in a regional to global approach, the preferred method will be that of atmospheric inversion, which is based on four-dimensional modeling (latitude, longitude, altitude and time) of wind and gas distribution (Bousquet et al. 2000). The first two approaches are mainly used to derive local emissions (emissions from coal-fired power plants or large urban areas) from total columns obtained from observations of solar radiation reflected at the surface. Global atmospheric inversion is used to estimate all gas sources and sinks on a continental scale, and is based on observations of total or partial columns of gas, sometimes combined with in situ measurements.

13.4.3. *Main limitations*

Various processes can introduce random or systematic errors in the gas column estimates made from spatial observation: measurement errors (instrument noise or errors in the radiometric, spectroscopic or geometric calibration of the instrument), error in the radiative transfer inversion algorithms often resulting from excessive simplifications (uncertainties of the absorption cross section (see Chapter 9 of Volume 1) of gases or aerosols, atmospheric reflectance or sphericity, a priori, etc.), sampling or representativeness bias (limitations of spatial or temporal coverage imposed by the performance of the instrument, the geometry of the orbit, the presence of clouds or the lack of signal).

With regard to direct modeling of radiative transfer, the two main limitations are as follows: (i) poor representation of scattering particles on the optical path – any error on the optical properties or on the vertical distribution of optically-thin clouds and aerosols will introduce uncertainty on gas column estimates; and (ii) poor spectroscopy. For example, since the number of CO_2 or CH_4 molecules along the optical path between the top of the atmosphere, the surface and the instrument is estimated from the amount of sunlight absorbed by these molecules, small uncertainties (~0.1%–1%) in the absorption cross-sections of these molecules can introduce systematic biases in the estimated number of molecules.

In terms of spatial measurement resolution and coverage, three factors may introduce a representativeness bias: (i) spatial sampling strategies adopted by these first generation sensors have low spatial resolution (SCIAMACHY), collect soundings at discrete points separated by large distances (GOSAT) or along narrow ground tracks separated by large distances (OCO-2); (ii) optically-thick clouds and aerosols prevent total (for more than 90% of soundings) or partial (for more than 70% of soundings) column measurements; and (iii) passive remote sensing systems that measure solar radiation reflected from the surface of the Earth in SWIR can only provide data on the sunny hemisphere and scarcely cover high latitudes during the winter. In addition, spatial measurements are generally carried out at a fixed time (twice a day in the thermal infrared, once a day around noon local time in near infrared), which makes it impossible to follow the diurnal variations of gases.

All of these biases can then in turn cause significant systematic errors in gas flux estimates. Fortunately, only biases that are spatially and temporally correlated on the space and time scales of interest, and not corrected in the analysis process, can produce persistent biases in the estimated fluxes. Joint improvement of flux estimation methods, atmospheric transport models and space mission performance is needed to significantly reduce uncertainties in national emission inventories or to monitor carbon cycle trends associated with climate change.

13.5. Challenges for the future

Spatial observation of anthropogenic greenhouse gases has provided essential global data for monitoring changes in atmospheric concentrations of gases, identifying selected emission mechanisms and evaluating model results and assumptions. Building on these building blocks, the future development of a global architecture for monitoring atmospheric concentrations of CO_2, CH_4 and N_2O to improve our knowledge of surface gas fluxes will need to meet two objectives: (i) reducing the uncertainty of national emission inventories and identifying additional opportunities to reduce emissions; and (ii) monitoring changes in the natural carbon cycle caused by human activities (deforestation, ecosystem degradation, etc.) and climate change (drought, permafrost melting, etc.).

The coming decade thus promises to be extremely rich in concepts for observing the atmospheric concentration of greenhouse gases. Whether they are already programmed or under study, these concepts explore many parameters, the main ones of which are detailed below.

13.5.1. *Towards the observation of anthropogenic emissions by spatial imagery*

Most of the satellite missions planned or under study for the coming decade have the objective of participating in monitoring global inventories of anthropogenic CO_2 and CH_4 emissions provided for every five years by the Paris Agreement. The main difficulty is to isolate, in measurements, the signature of anthropogenic emissions, which is intertwined with the broader signature of natural gas variations. Any monitoring of anthropogenic fluxes therefore requires the ability to separate the contributions of these two types of fluxes in atmospheric concentrations.

The preferred way to achieve this is the imaging of gas columns carried out using instruments with a wide swath and fine spatial resolution. These observations make it possible to distinguish plumes of XCO_2 or XCH_4 coming from anthropogenic point sources (coal-fired power plants) or very localized sources (megacities) and to deduce their emissions. In particular, the CO2M mission, a space component of the future European system for monitoring anthropogenic emissions of greenhouse gases, is based on the imaging of the total CO_2 column with a spatial resolution of about 2 km and a swath of 250 km. This spatial resolution will be obtained at the cost of a compromise on the spectral resolution with respect to the OCO-2 or MicroCarb missions. In addition to measurements in the three usual near-infrared spectral bands (O_2 at 0.76 μm and CO_2 at 1.6 and 2.05 μm), CO2M will have a visible imager to estimate the total NO_2 column, whose anthropogenic emission plumes are more easily detectable than those of CO_2. Finally, a multi-angle polarimeter, dedicated to aerosols, will also be present to constrain these parameters and reduce their deleterious impacts on XCO_2 estimates. The CO2M mission will have two satellites to increase the frequency of revisiting. Other concepts under consideration include the Space Carbon Observatory (SCARBO) project, which pushes the compromise between spatial and spectral resolution even further by proposing to measure only CO_2- or CH_4-sensitive truncated interferograms, in order to obtain an extremely miniaturized instrument.

13.5.2. *Reducing spatio-temporal sampling biases*

The coverage and frequency of revisiting current space missions are limited by the presence of optically thick clouds and the need for solar radiation for SWIR missions. The first of these two problems can be mitigated to some extent in partially cloudy regions if data from all available sensors can be combined, as these sensors will have different passage times and some clouds will move between their observation hours. To go further, three approaches currently under consideration are:

observation from a platform in geostationary orbit, launch of constellations of satellites and active methods.

Satellite constellations are also envisaged to increase the revisiting frequency. This approach is the basis of several concepts (CO2M, GHGSat**, SCARBO). It is often linked to the need to miniaturize instruments and platforms in order to make these constellations economically viable.

Active sensors (lidars), such as the MERLIN mission, should make it possible to carry out soundings that can be exploited in partially cloudy areas, based on the ability of lidars to discriminate very well between clouds and aerosols. In addition, lidars will provide two valuable assets in the study of greenhouse gases: (i) biases on gas columns will be much lower than those of passive spectrometers operating in SWIR; and (ii) better coverage of emissions during the night and at high latitudes during the winter when sunshine is limited.

13.5.3. *Towards an operational greenhouse gas monitoring service*

An operational system for monitoring the fluxes of greenhouse gases at the surface, based on spatial and in situ observation, is under development (Ciais et al. 2015). The European Sentinel-5P and Sentinel-5 missions for CH_4 and CO2M for CO_2 will form the backbone of the future GHG measurement constellation in terms of total columns. For partial column requirements in the mid-troposphere, the European IASI-NG mission will ensure continuity of data over more than 20 years. These missions will be complemented by various contributions combining geostationary satellites and constellations of small satellites.

To feed a fully operational greenhouse gas emission monitoring service, such as the Copernicus CO_2 service under development, observations will need to be processed routinely, have clearly defined performance indicators, well-defined users and interfaces and a robust and sustainable architecture in terms of funding.

At the same time, many private actors have emerged in the field of GHG satellite observations by developing their own missions. Their main driver is the commercial potential of GHG satellite data. These efforts are generally aimed at the development of satellite constellations and focus almost exclusively on local-scale methane monitoring with a very high resolution approach (< 50 m), to the detriment of a relaxed accuracy compared to reference missions. CH_4 monitoring has a significant monetary value for the oil and gas industry. The current objective is to

**See Appendix 3.

detect accidental CH_4 leakage with very high resolution imaging suitable for single point source observations. CO_2 tracking has a monetary value through carbon taxation and emissions trading. The market for GHG satellite products is still in its infancy, with only GHGSat having the first satellites in orbit. The overall market is expected to grow over the next few decades as new climate change legislation is adopted and GHG emission reporting requirements increase.

A combination of different objectives and different actors is therefore emerging for future space missions. On a global to regional scale, the needs will mainly come from the scientific community working on biogeochemical cycles. At the national level, these needs will be met by the need to monitor gas emissions within the framework of the Paris Agreement and to reduce uncertainties on fluxes, in particular for CH_4 and N_2O. In addition to reducing uncertainties, the impact of the various mitigation strategies put in place will also need to be observed. Finally, at the local level, detection and monitoring of high emission points will also be needed.

It is the combination of different types of spatial observations, combining missions in low (polar) or high (geostationary) orbit, and providing additional information on the distribution of greenhouse gases, which will make it possible to complete the ground observation network effectively. Together with emission inventories, atmospheric transport models and processing, archiving and trading infrastructure, the space segment thus has the capacity to become a key component of the global monitoring system for anthropogenic greenhouse gas emissions.

13.6. References

Bousquet, P., Peylin, P., Ciais, P., Le Quéré, C., Friedlingstein, P., Tans, P.P. (2000). Regional changes in carbon dioxide fluxes of land and oceans since 1980. *Science*, 290, 1342–1346.

Bovensmann, H., Buchwitz, M., Burrows, J.P., Reuter, M., Krings, T., Gerilowski, K., Erzinger, J.A. (2010). Remote sensing technique for global monitoring of power plant CO_2 emissions from space and related applications. *Atm. Meas. Tech.*, 3, 781–811. doi:10.5194/amt-3-781-2010.

Chédin, A., Serrar, S., Scott, N.A., Crevoisier, C., Armante, R. (2003). First global measurement of midtropospheric CO_2 from NOAA polar satellites: Tropical zone. *J. Geophys. Res.*, 108(D18), 458.

Chevallier, F., Bréon, F.-M., Rayner, P.J. (2007). The contribution of the orbiting carbon observatory to the estimation of CO_2 sources and sinks: Theoretical study in a variational data assimilation framework. *J. Geophys. Res.*, 112, D09307. doi:10.1029/2006JD007375.

Ciais, P., Dolman, A.J., Bombelli, A., Duren, R., Peregon, A., Rayner, P., Miller, C., Gobron, N., Kinderman, G., Marland, G. et al. (2014). Current systematic carbon-cycle observations and the need for implementing a policy-relevant carbon observing system. *Biogeosciences*, 11, 3547–3602. doi :10.5194/bg-11-3547-2014.

Cressot, C., Chevallier, F., Bousquet, P., Crevoisier, C., Dlugokencky, E.J., Fortems-Cheiney, A., Frankenberg, C., Parker, R., Pison, I., Scheepmaker, R.A. et al. (2014). On the consistency between global and regional methane emissions inferred from SCIAMACHY, TANSO-FTS, IASI and surface measurements. *Atmos. Chem. Phys.*, 14, 577–592. doi:10.5194/acp-14-577-2014.

Crevoisier, C. (2018). Use of hyperspectral infrared radiances to infer atmospheric trace gases. *Comprehen. Rem. Sens.*, 345–387. doi: 10.1016/B978-0-12-409548-9.10396-3.

Crisp, D., Atlas, R., Breon, F.-M., Brown, L., Burrows, J., Ciais, P., Connor, B.J., Doney, S.C., Fung, I.Y., Jacob, D.J. et al. (2004). The orbiting carbon observatory (OCO) mission. *Advances in Space Research*, 34, 700–709.

Ehret, G., Bousquet, P., Pierangelo, C., Alpers, M., Millet, B., Abshire, J., Bovensmann, H., Burrows, J., Chevallier, F., Ciais, P. et al. (2017). MERLIN: A French-German space lidar mission dedicated to atmospheric methane. *Rem. Sens.*, 9(10), 1052. https://doi.org/10.3390/rs9101052.

Foucher, P.-Y., Chédin, A., Armante, R., Boone, C.D., Crevoisier, C., Bernath, P. (2011). Carbon dioxide atmospheric vertical profiles retrieved from space observation using ACE-FTS solar occultation instrument. *Atmos. Chem. Phys.*, 11, 2455–2470. doi:10.5194/acp-11-2455-2011.

Frankenberg, C., Bergamaschi, P., Butz, A., Houweling, S., Meirink, J.F., Notholt, J., Petersen, A.K., Schrijver, H., Warneke, T., Aben, I. (2008). Tropical methane emissions: A revised view from SCIAMACHY onboard ENVISAT. *Geophys. Res. Lett.*, 35, L15811. doi:10.1029/2008GL034300.

Le Quéré, C., Moriarty, R., Andrew, R.M., Canadell, J.G., Sitch, S., Korsbakken, J.I., Friedlingstein, P., Peters, G.P., Andres, R.J., Boden, T.A. et al. (2015). Global carbon budget. *Earth Syst. Sci. Data*, 7(2), 349–396. doi: 10.5194/essd-7-349-2015.

Massart, S., Agusti-Panareda, A., Aben, I., Butz, A., Chevallier, F., Crevoisier, C., Engelen, R., Frankenberg, C., Hasekamp, O. (2014). Assimilation of atmospheric methane products in the MACC-II system: From SCIAMACHY to TANSO and IASI. *Atmos. Chem. Phys.*, 14, 6139–6158 doi:10.5194/acp-14-6139-2014.

Noël, S., Bramstedt, K., Hilker, M., Liebing, P., Plieninger, J., Reuter, M., Rozanov, A., Sioris, C.E., Bovensmann, H., Burrows, J. (2016). Stratospheric CH_4 and CO_2 profiles derived from SCIAMACHY solar occultation measurements. *Atmos. Meas. Tech.*, 9, 1485–1503. doi:10.5194/amt-9-1485-2016, 2016.

Rayner, P.J. and O'Brien, D.M. (2001). The utility of remotely sensed CO_2 concentration data in surface source inversions. *Geophys. Res. Lett.*, 28, 175–178.

Reuter, M., Buchwitz, M., Schneising, O., Noel, S., Bovensmann, H., Burrows, J., Boesch, H., Di Noia, A., Anand, J., Parker, R.J. et al. (2020). Ensemble-based satellite-derived carbon dioxide and methane column-averaged dry-air mole fraction data sets (2003–2018) for carbon and climate applications. *Atm. Meas. Tech.*, 13(2), 789–819.

Saunois, M., Bousquet, P., Poulter, B., Peregon, A., Ciais, P., Canadell, J.G., Dlugokencky, E.J., Etiope, G., Bastviken, D., Houweling, S. et al. (2016). The global methane budget 2000–2012. *Earth Syst. Sci. Data*, 8, 697–751. doi:10.5194/essd-8-697-2016.

Wunch, D., Toon, G.C., Blavier, J.-F.L., Washenfelder, R.A., Notholt, J., Connor, B.J., Griffith, D.W., Sherlock, V., Wennberg, P.O. (2011). The total carbon column observing network. *Philos. T. R. Soc. A*, 369, 2087–2112. doi:10.1098/rsta.2010.0240.

Yokota, T., Yoshida, Y., Eguchi, N., Ota, Y., Tanaka, T., Watanabe, H., Maksyutov, S. (2009). Global concentrations of CO2 and CH4 retrieved from GOSAT: First preliminary results. *SOLA*, 5, 160–163. doi:10.2151/sola.2009-041.

14

Clouds and Water Vapor

Hélène BROGNIEZ[1], Laurence PICON[2] and Dominique BOUNIOL[3]
[1] *LATMOS, Université de Versailles St Quentin-en-Yvelines, Guyancourt, France*
[2] *LMD, Sorbonne Université, Paris, France*
[3] *CNRM, Université de Toulouse, Météo-France, CNRS, France*

14.1. Atmospheric water cycle and climate

Water, in any of its states: vapor, liquid or ice, is a key component of our climate system. It is the coexistence of these three phases of water on Earth that makes it such a special planet. In fact, the amount of water in the atmosphere represents less than 0.01% of the total water on Earth (Gleick 1996). However, this small amount is fundamental for the redistribution of heat and energy in the climate system and in the global water cycle: from the evaporation of liquid water at the surface to its return to the ground in the form of precipitation after passing through complex cloud processes that strongly influence the radiative fluxes through the atmosphere.

Measuring the moisture content of the atmosphere in order to follow its distribution and understand its modes of variability is very complicated: tiny amounts of water vapor are difficult to measure despite technological innovations; cloud structures are the result of scale interactions from microphysical droplets and ice crystals to mesoscale* cloud cell aggregates, themselves integrated into large-scale atmospheric circulation. The precipitation associated with these cloud structures is very heterogeneous and extremely difficult to monitor and predict.

For a color version of all the figures in this chapter, see www.iste.co.uk/phulpin/satellites2.zip.
*See Glossary.

Satellites for Atmospheric Sciences 2,
coordinated by Thierry PHULPIN, Didier RENAUT, Hervé ROQUET and Claude CAMY-PEYRET.

Today, it remains unclear as to how the global atmospheric water cycle will adapt to global warming. But what is more important for our daily lives, and more difficult for scientists, is how the atmospheric cycle of water will evolve on a regional and local scale under the influence of human-made global warming.

The water cycle is clearly visible from space, even though the naked eye perceives only the atmospheric structures associated with the condensed phase of the water. This is illustrated in Figure 14.1 which shows two images taken by the European geostationary satellite Meteosat Second Generation (MSG) in two spectral bands. The visible band (0.8 μm) makes it possible to clearly see the cloud windings at mid-latitudes as well as the mesoscale convective systems* that are developing in Africa; clouds characterize in particular a part of the atmospheric dynamics of greater or lesser scale which produces convective instabilities. Hurricane Larry was also clearly visible in the middle of the North Atlantic Ocean. In addition, the infrared (IR) band at 6.3 μm, also called the "water vapor" band, shows the meanders of atmospheric circulation associated or not with clouds.

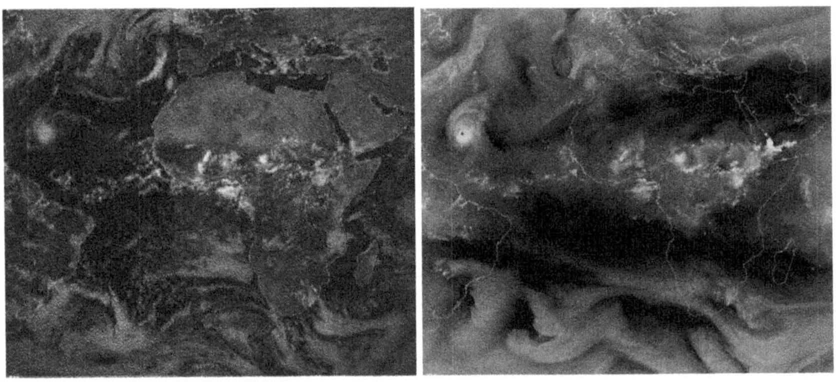

Figure 14.1. *MSG images obtained on September 5, 2021 at 12:00 UTC. Left: 0.8 μm. Right: 6.3 μm (CC BY-SA 3.0 IGO, © EUMETSAT, 2021)*

14.2. Observations of water vapor

The water molecule is a triatomic, nonlinear molecule with a strong dipole nature, which gives it three fundamental modes of vibration associated with three modes of rotation about the axes of the molecule. The combination of multiple rotational and rotational-vibrational transitions produces intense and particularly numerous absorption lines that appear from the near-IR range to microwaves, as illustrated in Figure 14.2.

*See Glossary.

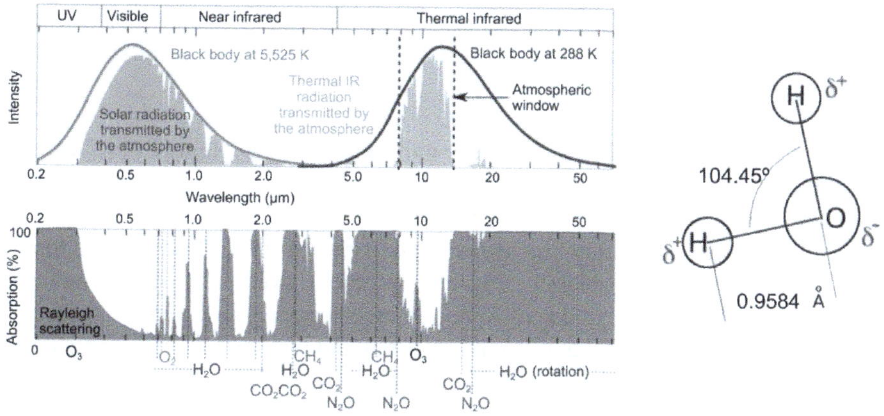

Figure 14.2. *Left: Spectra of emission (top) and absorption (bottom) by the atmosphere of solar and terrestrial radiation (adapted from Poitou (2013)). The absorption spectrum is calculated in the absence of clouds and aerosols. Right: diagram of the water molecule showing the characteristics of the dipole*

In IR, the strongest absorption band is situated at 6.3 μm and is associated with a deformation of the molecule with wings which extend towards 5 μm and 9 μm. Two other less extensive modes of rotation-vibration are located on either side of 2.7 μm, and are associated with symmetrical and asymmetrical stretches. As shown in Figure 14.2, other molecules in the atmosphere have capacities to absorb IR radiation. Some are particularly well located, such as that at 15 μm which corresponds to absorption by CO_2 (linear molecule, not having a mode of rotation) located close to the maximum IR radiation emission from Earth.

Water vapor is the most important natural greenhouse gas; it contributes up to 60% of the natural greenhouse effect! Thus, water vapor is the second heat source after the sun for heating Earth's surface. As a result, water vapor is considered an Essential Climate Variable (ECV) by GCOS, and various actions aim to provide long time series to enable local and global trend studies of atmospheric water vapor content.

Climate and weather conditions are highly dependent on the distribution of water vapor in the atmosphere. In terms of climate change, water vapor provides the greatest feedback, as it more than doubles surface warming due to atmospheric CO_2 (Held and Soden 2000). In the tropics, water vapor from the lower troposphere is the best indicator of vertical motion and precipitation, making it a central element for understanding tropical weather, tropical climate, tropical climate change and cloud feedback (Sherwood et al. 2010). In extratropical regions, the transport of water vapor to the poles via storm paths, for example, in the form of "atmospheric rivers" which often accompany the rapid intensification of storms, is frequently associated with severe and extreme weather events.

Whether it is for weather prediction, to understand the evolution of cloud cells or for climate change studies, observing the distribution of atmospheric water vapor at any point in the atmosphere is thus a central issue. There are almost five orders of magnitude on the concentration of water vapor between the surface and the top of the meteorological atmosphere. Evapotranspiration over continental surfaces, atmospheric dynamics or cloud formation reinforce horizontal and vertical gradients. Unfortunately, the accuracy required at all spatial and temporal scales is difficult to achieve, leading to the use of different types of measures, each with its own strengths and weaknesses, according to the objectives targeted (Table 14.1).

	Techniques	First measurements
Satellite	IR sensors on geostationary or moving platforms	1977: MVIRI** on METEOSAT-1
	MW sensors on moving platforms	1987: SSM/I** on DMSP
	VIS/NIR sensors on moving platforms	1996: Polder** on Adeos
	Limb sounding sensors (radio occultation) on moving platforms	1995: GPS-MET**
In situ	Radiosondes on tropospheric balloons for sensor measurements of the thermodynamic vertical profile	First releases in the 1930s
	Commercial aircraft equipped with sensors	1993: European MOZAIC program
Ground	Meteorological station hydrometers and GPS ground stations for continuous monitoring at a given point	18th century for hygrometers 1992 for GPS in France

Table 14.1. *Methods for measuring atmospheric water vapor (see also Kämpfer (2012) for more details or the WMO OSCAR portal*[1]

Regular measurements at high altitude are carried out by the meteorological services with radiosondes, which offer direct and vertically-resolved thermodynamic measurements of an atmospheric column or by airliners. Radiosounding and airplane measurements are supplemented by indirect measurements provided by satellites. Given the absorption spectrum of water vapor, many spectral ranges are exploited to provide global measurements: from near infrared to microwaves (hereafter MW) and

**See Appendix 3.

1. https://space.oscar.wmo.int/.

thermal infrared (hereafter IR). These direct and indirect measurements are assimilated with other meteorological fields in numerical models (see Chapter 4) and thus contribute to the improvement of weather predictions.

Current sensors make it possible to observe the spatial distribution of water vapor thanks to three parameters: the vertical specific humidity profile (q, in kg/kg), the relative humidity of the upper troposphere (UTH for Upper Tropospheric Humidity, in %) and the total water vapor content (TCWV for Total Column Water Vapor, in kg/m^2). These sensors are deployed either in polar orbits, or in geostationary orbits, or else in inclined orbits making it possible to reinforce the temporal sampling of a latitude band (e.g. the SAPHIR** sounder on the Franco-Indian satellite Megha-Tropiques, which has an inclination of 20° on the equator; Roca et al. (2019); see also Chapter 8 of Volume 1).

14.2.1. *Passive sensors*

Passive sensors are by far the majority in the panorama, and the estimation of water vapor from their measurements is based on the use of the absorption band (in particular in the 6.3-µm band) and the emission band (in particular at 183.31 GHz). The observations are thus measurements of radiation (radiances or brightness temperatures, see Chapter 9 of Volume 1), the intensity of which is associated with the vertical distribution of water vapor in the atmospheric column. This approach thus requires the development of inversion methods, which are based on: (i) a good knowledge of the radiative transfer in these high absorption/emission bands; (ii) on the use of line-by-line or broadband radiative transfer models whose uncertainties are documented; and (iii) on a representation of the radiance/humidity relationship (see Chapter 10 of Volume 1). This last point is crucial, and must be the most representative of atmospheric conditions because it is on this point that the quality of the inversion method at the global scale is based. Measurements (radiances or brightness temperatures) are also assimilated into numerical weather prediction models (see Chapter 4 of Volume 1).

In the IR domain, the 6.3 µm band is by far the most used; IR sensors thus provide, since the 1980s, the longest time series on atmospheric humidity. Broadband sensors such as HIRS** instruments and GOES or MSG satellite imagers can be distinguished from hyperspectral sensors such as AIRS**, IASI** and CrIS** (used as sounders). The latter, which began in 2002 with the AIRS instrument, offer higher vertical resolution and lower uncertainty than broadband sensors. However, for inversion, spatial sampling is limited to non-cloudy

**See Appendix 3.

atmospheric conditions, and methods for detecting cloudy scenes must be developed for each sensor. The impact for climate studies is discussed later (section 14.2.3).

The visible (0.885 μm) and near IR (0.9 μm) domains are also used and offer excellent horizontal resolution (~1 km, for example for the MODIS** instrument launched in 1999). It is only possible to use them when the sun illuminates the scene, since the principle is based on the reflection of solar radiation in these absorption bands, and thus (as well as for the IR) the estimation of the water vapor content can only be done for situations without clouds and with little aerosol content (the droplets or crystals of clouds and aerosols scatter the solar radiation; other measurement channels are used to filter these situations). Figure 14.3 illustrates the questions associated with spatial sampling, for a 1-day and 1-month accumulation of orbits of the Aqua satellite carrying the MODIS instrument. Analysis of such time series makes it possible to study both the distribution of water vapor as well as the representation of clouds in numerical models (climate model outputs or re-analyses).

In the MW domain, the main lines used are located at 22.7 GHz and 183.31 GHz. The advantage of the MW domain over the IR domain is that humidity estimates are also allowed in the case of situations involving clouds, as long as these are not highly precipitating because raindrops or large ice crystals that can be found at the top of storms scatter the MW radiation (Ulaby et al. 2014). Thus, we find imagers that use the water vapor absorbing line at 22 GHz for TCWV measurements (such as SSMI/I, AMSR-E**, GMI** or TMI** instruments). However, this measurement is limited to ice-free oceans because the very high variability in the emissivity of the soil, affected by vegetation, soil moisture or even by its electromagnetic properties, has long been a barrier to the exploitation of these measurements over continents (see also Chapter 9 of Volume 1). Many recent studies have demonstrated their contribution in an assimilation system for weather forecasting. Sounders, for their part, sample the line at 183 GHz for vertical humidity profile estimates with a lower vertical resolution than infrared hyperspectral sounders, which leads to missing the strong vertical gradients associated with the cloudiness and atmospheric dynamics (Stevens et al. 2017). Uncertainties persist on the absorption line at 183 GHz and on its modeling (Brogniez et al. 2016) and work on the spectroscopy of the water molecule in the MW domain, such as that of Tretyakov (2016), underlines the essential role of fundamental physics in the treatment of passive measurements.

**See Appendix 3.

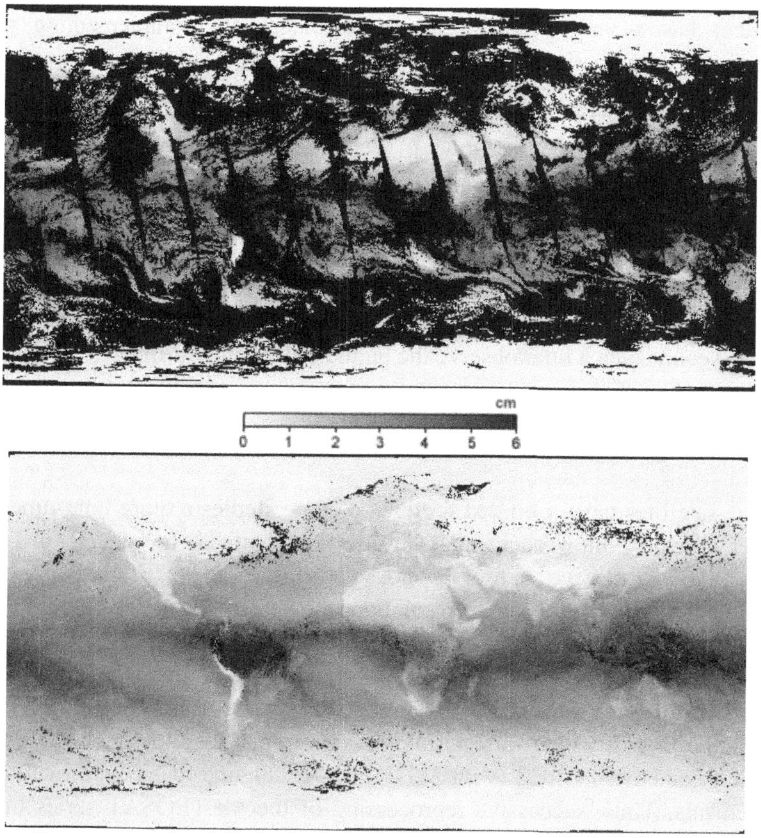

Figure 14.3. *Total water vapor content (TCWV in cm, expressed as the height of water column) estimated from the measurements of the MODIS instrument of the Aqua satellite. The top map is for February 1, 2021, the bottom map is for February 2021. The fields in black are missing data fields. (© NASA Earth Observatory)*

14.2.2. *Active sensors*

The only current active space-based sensors for water vapor are the Global Navigation Satellite System (GNSS) sensors by radio occultation (see Chapter 4 of Volume 1). As the refractivity of the atmosphere is affected by the concentration of water vapor, extraction of this "wet" delay allows for a very precise measurement of the water vapor with a high vertical resolution (~200 m), and very good stability over time. The main drawback remains its horizontal resolution (~200 km) since the measurement is carried out on an oblique path. The assimilation of water vapor measurements by radio occultation for weather forecasting (Météo-France,

MetOffice) has a very positive effect, advantageously complementing passive measurements (Cardinali and Healy 2014).

Like space lidars for clouds and aerosols (in-flight in 2022) or greenhouse gases (under development), there is a wish to develop space lidars for water vapor profiling combining vertical (~200 m) and horizontal (between 5 and 50 km depending on the altitude) resolutions that have never been seen before. Mention may be made of the QSAT and ATLAS projects proposed to ESA by consortia of European scientists. However, the cost of developing the various elements associated with technologies with limited know-how remains the main barrier at present to seeing such a lidar observe the humidity of our atmosphere.

14.2.3. *Homogenization and intercomparison*

While satellites have a limited lifetime, climate studies require long time series. However, sensors follow each other but are not identical. In general, differences between two successive sensors whose objective is to offer continuity of measurements concern the spatial resolution, the sensitivity of the sensors (i.e. the measurement uncertainty) or the shape of the spectral response of the instrument in the measurement band. This involves processing on measurements in order to offer a homogeneity of time series. For example, the series of European METEOSAT satellites, the first of which was put into operation in 1977 and which saw the third generation launched in 2022 (with MTG), has undergone many modifications and improvements. Thus, successive reprocessing of the METEOSAT series makes it possible to exploit the measurements of its "water vapor" channel at 6.3 μm and thus carry out climatic analyses of the variability of atmospheric humidity (mainly for the tropical Atlantic and Africa, for example, Brogniez et al. 2006). Space agencies are actively involved in this work on ECVs and water vapor in particular; mention may be made of the roles of the EUMETSAT CM-SAF and ESA (Climate Change Initiative Water Vapor Project), which propose long series of water vapor based on a combination of intercalibrated instruments with homogeneous scene processing (see also Chapter 12).

Intercomparison exercises between different sensors are particularly useful to take advantage of complementarities between different datasets. For example, the analyses carried out by the "Water Vapor Assessment" group of the GEWEX (G-VAP) project of the World Climate Research Programme (WCRP) focused on the stability, sampling, sensitivity and homogeneity of several atmospheric water vapor databases covering at least 10 years of data records. The G-VAP group[2]

2. http://gewex-vap.org/.

quantified the strengths and weaknesses of the different atmospheric water vapor archives, according to different complementary axes in order to propose a coherent framework for climate variability studies of the water cycle. Figure 14.4 illustrates, for example, seasonal differences in upper tropospheric humidity (UTH) estimated from measurements in the IR or in the MW. The main differences are due to the measurement itself, and more specifically to the difference between the IR and MW domains for the interpretation in "moisture" content, which leads to artificially drying (or humidifying according to the reference) the atmosphere. Indeed, these two spectral domains are not sensitive in the same way to the presence of droplets forming clouds; while clouds are opaque to IR radiation, thus masking the moisture signal, they can be transparent in the MW domain (this "transparency" depends on the size of the droplets, see also section 14.2.1 and Chapter 10 of Volume 1), thus allowing enhanced sampling of the atmosphere. This makes knowledge of measurement crucial in climatology and climate trend studies.

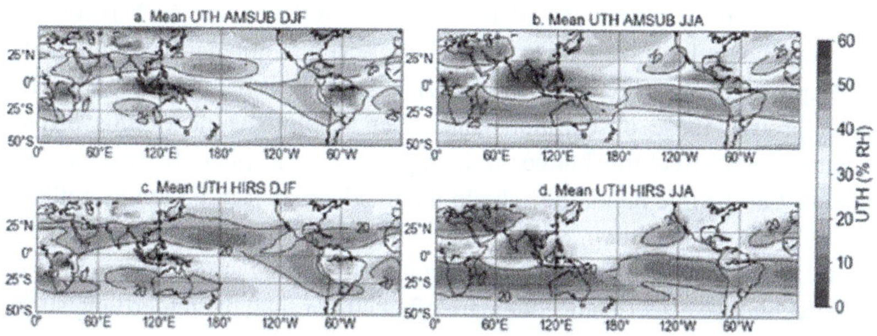

Figure 14.4. *Average upper tropospheric relative humidity (UTH, in %), for the winters (a and c: December–January–February, DJF) and summers (b and d: June–July–August) of the period 1999–2018, as measured in the MW domain (AMSU-B series, top) and in the IR domain (HIRS series, bottom) (source: Tivig et al. (2020))*

14.3. Observation of cloud properties

A cloud consists of droplets of liquid water and/or ice crystals suspended in saturated air aggregated around particles of various origins (spray, sand, soot, etc.) that promote condensation and are called "cloud condenstation nuclei". These cloud particles are often referred to as hydrometeors* and are moving within the cloud. Water in clouds can remain liquid at negative temperatures (0 to –40°C, this is called supercooled water) and cloud droplets can be represented as spheres whose radius is variable (from a few microns to 50 microns) as well as their density (70–500 droplets per cm^3) depending on the cloud type. For a cloud to form, it is necessary that the moist air (transparent in the visible wavelengths to which the eye is sensitive) saturates. This saturation is obtained through atmospheric motions generally causing cooling, for example, the passage of a mountain barrier, the superposition of two air masses of different temperatures in a front (the cold air mass being under the warmer air mass) or large-scale ascents related to the distribution of large air masses. The phenomena related to water phase changes involve considerable amounts of energy; clouds are therefore very efficient actors in the transport of water and energy. On the contrary, they interact with radiation by strongly influencing the Earth's radiation budget.

The Earth radiation budget is a key parameter in the study of climate and its evolution. It aims at quantifying at the top of the atmosphere the difference between the density of energy flux received from the sun and that lost by reflection and thermal emission. On average over the globe and over a long period, the radiation balance must cancel out if the planet is in a stable climate (energy conservation). This amount has been measured by space-based broadband passive remote sensing instruments (such as ERBES**, CERES**, ScaRaB**, GERB**) for several decades. However, the most recent data show a slight imbalance in the radiation budget (Stephens et al. 2012; Loeb et al. 2018), without it being possible to directly determine through these observations the exact absolute value of this difference. Clouds particularly affect this budget at the top of the atmosphere by two antagonistic mechanisms. On the one hand, they shield the incident solar radiation by reflecting it (parasol effect) because their albedo (see Chapter 9 of Volume 1) is generally important. But, at the same time, they moderate the thermal emission from the planet to space (greenhouse effect), since they are often colder than the underlying surfaces.

PARASOL EFFECT/GREENHOUSE EFFECT.– The parasol effect and the greenhouse effect are two inseparable effects of the atmosphere. The parasol effect is a return to space of incident solar radiation, and it is produced by the reflection or the scattering by clouds and some aerosols. Its consequence is a loss of absorbed solar energy leading globally to a cooling of the surface. The greenhouse effect comes from a decrease in the upwelling IR emission from the surface to space caused by clouds, certain aerosols and gases, which results in IR radiation being stored by the atmosphere. Its global consequence is a warming of the surface. While the parasol effect prevents overheating during the day, the greenhouse effect limits cooling at night. For clouds, the amplitude of these two effects depends on the altitude of the clouds, their geometric thickness, their droplet concentration, the phase of the water (liquid, ice, mixing) and some clouds, such as some low-altitude oceanic clouds, will have a parasol effect (therefore cooling) that dominates while other clouds, such as certain types of cirrus clouds, will have a greenhouse effect (therefore warming) that dominates.

**See Appendix 3.

Depending on the nature of the clouds, these two effects do not have the same magnitude. In order to quantify these processes, the cloud radiative effect is calculated (Figure 14.5).

RADIATIVE EFFECT OF CLOUDS.– The Cloud Radiative Effect (CRE) represents the change in Earth's radiation balance if all clouds disappeared from the atmosphere. This takes into account the effects of clouds in the solar (shortwave, $\lambda < 4$ μm) and thermal (longwave, $\lambda > 4$ μm) domains. In practice, it is estimated by calculating the difference between the radiation balance of a cloudless Earth and the radiation balance measured at the top of the atmosphere, which includes all scenes over a long period.

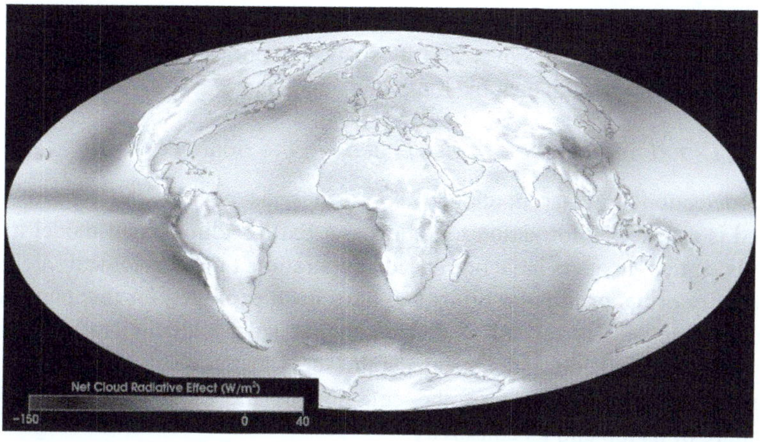

Figure 14.5. *Net radiative effect of clouds at the top of the atmosphere from CERES instrument products on the Aqua and Terra missions and averaged over the period 2000–2015. (© NASA[3])*

If clouds suddenly disappeared from the Earth's atmosphere without further modification, recent measurements show that the planet would gain around 20 W/m^2 on average, because the parasol effect leading to a loss of energy far outweighs the greenhouse effect leading to an energy gain (Loeb et al. 2018). In the context of climate change, clouds participate in mechanisms that can amplify or reduce global warming; these are cloud feedbacks, which are very complex because of the diversity of cloud characteristics. The main characteristics of clouds and the most commonly used methods for measuring them are described in the context of nowcasting (see Chapter 5). For the study of climate, the use of these same methods differs in two aspects: first, the notion of real-time measurements is not essential and

3. https://svs.gsfc.nasa.gov/30603.

allows for less operational developments, but above all, algorithms must be able to be applied to data series over large regions and long periods, sufficiently stable to determine their statistical parameters (see also Chapter 12). These data are then analyzed to determine a significant evolution of the climate system. They also serve as validation and calibration for climate modeling, which are the only tools available for estimating future climate change.

14.3.1. *Observations using passive instruments*

Observations from passive imagers provide access to 2D cloud cover* “from above”. They can approximately resolve the vertical dimension and are generally not sensitive to overlapping cloud scenes. On the contrary, they are very useful tools for characterizing global cloud cover and its spatio-temporal variability, as well as its effect on the radiation budget at the top of the atmosphere.

14.3.1.1. *Cloud detection and classification in the optical domain*

The detection of clear or cloudy sky scenes, as well as the type of clouds observed is essential information to understand Earth’s climate and model it. Since the 1980s, optical imagers have been used in operational meteorological satellites, in geostationary orbits or in polar orbits and have been exploited for this purpose. The visible measurement is related to the amount of solar radiation reflected by the cloud, thus to its albedo and optical depth (see Chapter 9 of Volume 1). At IR wavelengths, it is a function of the emission of the cloud, and therefore, the temperature of its top and its emissivity.

For optical sensors, clouds appear brighter than the surface, but only during the day; their presence can be assimilated to a maximum in a short time series. Day and night, they are detected in infrared because they are generally colder than the surface of the globe; their presence can be assimilated to a minimum in a short time series. Their very rapid evolution also results in greater temporal and spatial variances of optical signals than in cloudfree scenes. These properties make it possible to develop algorithms for detecting clear (cloudfree) skies, which is important for many applications, or on the contrary, the presence of clouds (see Chapter 9 of Volume 1). However, detection difficulties persist when the contrast, within the pixel, between the cloud and the underlying surface is too low; this is the case of clouds over snow surfaces or ice, low clouds in very humid regions, edges of clouds or clouds not covering the entire pixel, thin clouds present above a surface known as heterogeneous, or coastal pixels including a mixture of oceanic and continental surfaces.

*See Glossary.

In regions with high cloudiness (e.g. tropical regions), long-wave radiation going out of the atmosphere (OLR*) is often used as an indicator of cloud activity. To go further in the cloud documentation, we can try to identify the types of clouds automatically, in a way close to that used by the meteorologist from the ground. These methods use statistical algorithms, mixing radiances and their spatio-temporal variances at several wavelengths, to identify a dominant cloud type in the scene under study. In a very simplified way, it is a question of exploiting three-dimensional histograms of radiances obtained at two different wavelengths. The two wavelengths are associated with the axes of the diagram while the isolines of the distribution are associated with a "cloud class". The algorithm then assigns the values of a pixel in both wavelengths to a class by a proximity criterion. The actual calculation is more complex, since the histogram used is generally more than two dimensions due to the introduction of spatial and temporal variances of multiple wavelengths. This type of classification, sometimes also using neural networks, is, however, computationally costly and, with the multiplication of spectral channels of recent imagers, operational processes use threshold methods instead. For example, EUMETSAT's NowCasting SAF uses cloud classifications (Chapter 5) to establish a cloud detection threshold for each measurement channel. These thresholds will depend on the viewing angle, the surface type (land/ocean), the surface temperature and water content. They are delivered in a look-up table (LUT).

14.3.1.2. *Determination of cloud parameters*

When a cloud scene is detected, different algorithms determine physical parameters characteristic of clouds: cloud cover (percentage of pixels declared cloudy for a given type of cloud), altitude (or pressure) and temperature of their top, visible optical depth, IR emissivity, phase of hydrometeors, amount of liquid or solid water and approximate size of cloud particles. These techniques are the basis of cloud climatologies for analyzing long time series for climate studies. The methods are identical to those described in Chapter 5. Some of their principles and limitations are recalled here.

IR radiance is related to the temperature of the top of the cloud and its emissivity. Auxiliary data describing the vertical distribution of atmospheric temperature and humidity are used in a radiative transfer code adapted to the instrument to calculate the altitude that will best correspond to the measured radiance. Again, for an opaque cloud, these methods are effective, but multi-spectral information is necessary to take into account the cases of semi-transparent or subpixel clouds.

---

*See Glossary.

The phase (liquid or ice) of cloud hydrometeors is assigned based on the temperature of cloud tops and liquid water and ice content found using a particle model and a radiative transfer model. Using the differences at two IR wavelengths also gives information about the phase. The near infrared gives additional information for the search for the effective average radius of cloud particles, which is then carried out through pre-calculated empirical relationships. Many instruments improve these determinations, as follows.

– Observation of the same scene from different angles of view, as allowed by the MISR** multi-angle spectro-imager, gives a stereoscopic vision of the scene and improves the retrieval of the altitude of the clouds.

– The measurement of polarization in the visible range, provided, for example, by the POLDER instrument, provides information on the phase of hydrometeors, due to the difference in light scattering by non-spherical ice crystals and spherical water droplets. This measurement is much less empirical than thresholds used for conventional imager data.

– IR sounders, used operationally to restore vertical temperature and humidity profiles such as those of the TOVS** family (see Chapter 10 of Volume 1), also provide information on atmospheric opacity and allow for better cloud detection through the joint use of cloud-insensitive microwave data. On the contrary, several methods based on spectral differences of channels in the CO_2 absorption band between 13 and 15 μm in the case of clear and cloudy skies have been developed to improve the identification of high and thin clouds (cirrus) including for sounders with high spectral resolution (HIRS, AIRS or IASI).

– Relatively low-frequency microwaves are transmitted more efficiently in clouds than IR, except during precipitation. Over the sea, the increase in brightness temperature is directly related to the amount of liquid water in clouds. Over continents, this contrast signal is more complex to detect and quantify. It is then possible to determine the water content of the cloud much more directly than in the optical domain, over oceans. Combined with an empirical approach to precipitation, this technique has been applied to MW imagers such as SSM/I, TMI, AMSR** or AMSR-E (Figure 14.6).

– Finally, solar occultation sounders in the visible, due to the wide optical path of their sight and their vertical resolution, are extremely sensitive to the presence of clouds and help to determine the occurrence of opaque clouds; the use of two simultaneous wavelengths in the visible and near IR makes it possible, on a statistical basis, to identify very thin cirrus clouds.

**See Appendix 3.

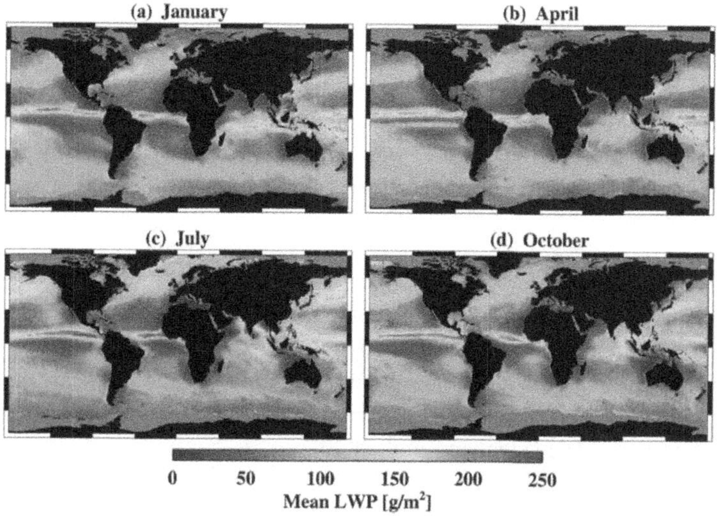

Figure 14.6. *Average value of liquid water cloud path (in g/m^2) from 1988 to 2005 over the oceans for January, April, July and October. Missing areas in gray (adapted from O'Dell et al. (2008))*

14.3.2. *Observations using active instruments*

Passive observations provide only limited information on the vertical distribution of hydrometeors. However, vertical distribution and cloud overlap[4] directly determine the intensity and vertical profile of radiative heating. Active observations in the visible (lidar) or microwave (radar) range have been deployed to compensate for this lack. The particularity of these wavelength ranges is to have the ability to penetrate clouds, depending on their optical depths, and precipitation thus providing detailed information on their vertical structure. A new era for observation in the atmospheric column opened in the late 1990s with the launch of the first spaceborne radar dedicated to measuring precipitation on board the Japanese–American TRMM satellite. Due to its frequency and sensitivity, this radar provides little information on cloud stratification. The next two decades were rich in launches of instruments capable of vertically probing clouds, whether lidars on board the LITE, ICESat, CALIPSO satellites (see Chapter 3 of Volume 1), or more recently CATS and Aeolus, or radar on board CloudSat. The latter, coupled with the lidar on board CALIPSO within the international convoy "A-Train", allowed for the unprecedented

4. Cloud overlap: several clouds can overlap in an atmospheric column and this superposition modifies the heating of the atmosphere. Cloud overlap is associated with a given area and provides the percentage of overlay between cloud types in the column.

documentation of cloudiness, from the finest clouds to thick precipitating clouds (Stephens et al. 2018). The succession of these pioneering experiments was partly guaranteed by the launch of the GPM-Core whose radar combines two frequencies to document both liquid and solid precipitation. In addition, ESA's EarthCARE mission (see Chapter 4 of Volume 1) will combine radar and lidar and is expected to expand the perspective of the radar–lidar combination with the first space-based Doppler radar.

These observations are generally made from relatively low orbits over limited periods of time of the order of a few years and, apart from radars dedicated to the study of precipitation, do not have swaths. The new generation of small satellites such as CubeSat, SmallSat and Nanosat and the miniaturization of sensors should in the near future propose new multi-instrument strategies for the observation of the atmospheric water cycle (Stephens et al. 2020).

14.3.2.1. *Space lidars*

A lidar measures the amount of energy of a light pulse backscattered by atmospheric targets (see Chapter 9 of Volume 1). Lidars are powerful tools for probing optically thin clouds and aerosols suspended in the atmosphere. The lidar signal (proportional to the second moment of the particle size distribution) is particularly sensitive to the concentration of targets and their size. The signal is therefore completely extinguished when the concentration is high, such as at the top of the clouds of the boundary layer or in thick clouds. Nevertheless, this property makes it a powerful tool to detect with great accuracy the altitude of the cloud tops. The measurement of depolarization, providing information on the shape of the target, makes it possible to identify the phase of the hydrometeors. Figure 14.7 (left) corresponds to a typical scene observed by the lidar of CALIPSO.

The first step in processing these observations is to separate the signal from the clouds from that of other components (aerosols, clear skies or surfaces). To this end, Vaughan et al. (2009) developed a cloud detection algorithm in lidar data. Other parameters such as optical depth and extinction can also be obtained. The ability to detect clouds, their altitude and their phase offered the possibility of building cloud climatologies on a global scale quite comparable to those obtained from passive observations (see previous sections). The ability of CALIPSO's lidar to detect cloud phases has shown that a large majority of clouds over the Southern Ocean contain liquid supercooled water and that it contributes significantly to radiative fluxes at the top of the atmosphere, particularly south of 60°S. This analysis helped to explain a persistent bias in cloud representation in general circulation models.

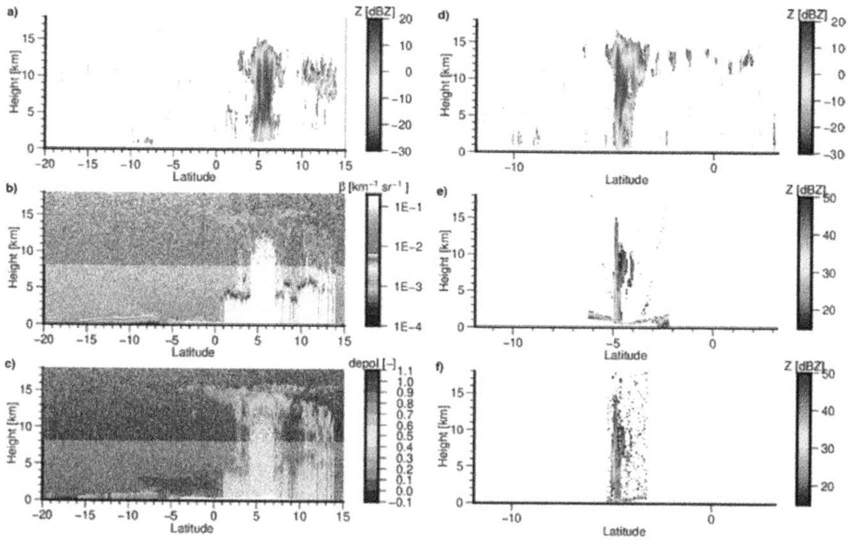

Figure 14.7. *Latitude/altitude transects observed by active instruments. On the left for June 21, 2008, with (a) the 94-GHz reflectivity measured by CloudSat's radar, (b) the 532-nm lidar backscatter coefficient and (c) the corresponding depolarization ratio measured by CALIPSO's lidar. On the right for June 2, 2014, with (d) 94 GHz reflectivity measured by CloudSat's radar, reflectivities (e) at 14 GHz and (f) 35 GHz measured simultaneously by the GPM-Core Observatory radar*

More specific uses for model evaluation have also been developed. In this context, mention may be made of the GOCCP product (Chepfer et al. 2010). In this case, CALIPSO observations are aggregated on a monthly scale on a 2° x 2° grid offering an alternative to the low revisiting and absence of CALIOP** lidar swath. Thus, in each grid point, statistics of cloud characteristics are computed, such as the distribution of signal intensities or cloud cover. In parallel, a simulator of the CALIOP instrument is fed with cloud parameters simulated by the model making it possible to derive in the latter quantities directly comparable to those deduced from the observations. The use of this type of simulator makes it possible to take into account instrumental limitations in comparisons and to simulate what the lidar would have observed if it had flown over the atmosphere simulated by the model. Figure 14.8 presents such a comparison which shows that the models simulate a cloud mode at high altitude, but with properties (optical depth, altitude, etc.) that may differ from one model to another. These comparisons also show the difficulty for models to simulate both optically thin clouds at high altitude and optically thick clouds near the surface.

**See Appendix 3.

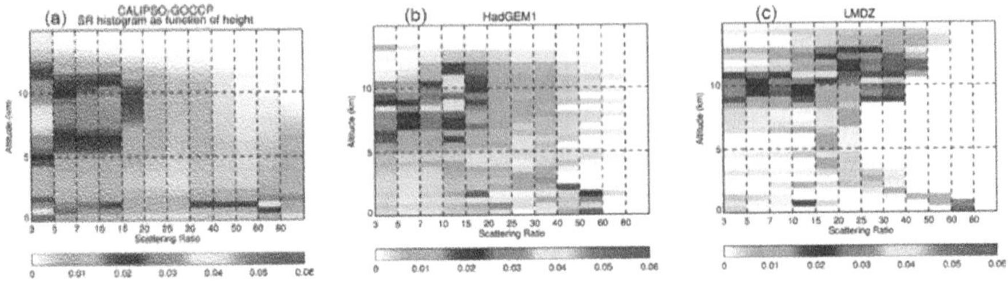

Figure 14.8. *Lidar scattering ratio histogram as a function of altitude observed by CALIPSO lidar (a) compared to that obtained from two general circulation models (b: English model HadGEM1; c: French model LMDZ) for the North Pacific region (0°–60°N,160°E–125°W) for the period of September–November 2006 (adapted from Bodas-Salcedo et al. (2011))*

14.3.2.2. *Spaceborne radars*

Radar measures signal backscattered by atmospheric targets in the MW domain (see Chapter 9 of Volume 1). When the targets are small in front of the wavelength, this power is proportional to the moment of order 6 of the particle size distribution. The size of hydrometeors therefore plays an important role in the backscattered signal, and it is the largest scatterers that have the highest contribution to the measured signal. When the size of the targets approaches the size of the wavelength, an oscillation of the backscatter cross section (Mie scattering) is observed, leading to a saturation effect in the measured powers at the highest frequencies. The measurement of the phase shift of the electromagnetic wave allows for measuring the velocity of the target in the direction of sampling that will be called Doppler velocity. At present (in 2023), there is no spaceborne radar with this capability.

A trade-off between antenna size, transmitted power and sensitivity is necessary to allow radars to be put into orbit. Thus, the lowest frequencies (14 GHz or 35 GHz) are used for precipitation sampling and the highest (94 GHz) for documenting cloud properties. This ability to sample different aspects of cloud systems can be seen in Figure 14.7 (right).

The limited ability of GPM-Core's radar to document ice clouds and low precipitation is shown on the bottom two panels of Figure 14.7. In the same way, the attenuation of CloudSat's radar signal when the cloud becomes too thick or precipitation too intense is visible on the top panel, in these situations, the radar signal can even be affected by multiple scattering. However, these high frequencies are powerful tools for documenting cloud processes and low-intensity precipitation (Ulaby and Long 2014). Many algorithms have been developed for the retrieval of cloud geophysical parameters from reflectivity measured by CloudSat especially for water and ice content, sometimes adding a thermodynamic constraint such as temperature profile for the estimation of precipitation and parameters such as optical depth. These parameters can feed radiative transfer algorithms allowing each observed cloud profile to obtain an estimate of radiative heating rates (L'Ecuyer et al. 2008).

RADIATIVE HEATING RATE.– The rate of radiative heating, or radiative cooling, is determined by the net difference $\Delta F(z)$ between the incoming and outgoing radiative fluxes ($\Delta F = F_{incoming} - F_{outgoing}$) which are defined for each layer of the atmosphere (altitude z, thickness Δz). A gain in radiative energy ($\Delta F(z) > 0$) leads to warming of the z-layer, while a loss of radiative energy ($\Delta F(z) < 0$) leads to cooling.

These inversions constitute the first observation on a global scale of the macrophysical, microphysical and radiative properties of clouds and have allowed a new view on a number of processes. Examples include the characterization of different cloud regimes affecting the Southern Ocean, the understanding of the physical properties of liquid water clouds (Stephens et al. 2019) or the relationship between convection intensity and the properties of clouds detrained[5] from deep convection (Bouniol et al. 2016) highlighting the differentiated behavior of deep convection between continental and oceanic surfaces.

Like lidar observations, space-based radar observations have been used to assess the representation of clouds in climate models, showing their difficulties in simulating clouds (Jiang et al. 2012). This data also made it possible to propose improvements in their representation of physical processes.

14.3.3. *Multi-instrument synergy for the establishment of cloud climatologies*

The properties of observed clouds are crucial to properly assess the water cycle and are necessary for modelers to evaluate the representation of clouds in different models. Hence, cloud climatologies have been established over relatively long periods and intercompared with each other.

5. Cloud detrainment corresponds to the horizontal spread of a cloud that has developed vertically following strong updrafts and is subject to a horizontal wind. The opposite phenomenon is the entrainment of air outside the cloud within the cloud.

14.3.3.1. *Radar–lidar synergy studies*

Both radar and lidar have the ability to document cloud properties but they also have weaknesses. For example, radar alone cannot accurately document the size of hydrometeors because it is less sensitive to small hydrometeors. On the contrary, lidar penetrates the thinnest clouds well but suffers from a strong attenuation in the other clouds (Figure 14.6, left). Their combined use, enabled by their tandem flight within the A-Train, makes it possible to advantageously exploit their complementarity providing the most complete view of the vertical distribution of cloudiness.

This joint documentation showed the predominance of low, thin clouds in the marine boundary layer from high latitudes to subtropical latitudes. In terms of stratification, situations are complex in almost half of the cases with clouds spread over different layers of the atmosphere. This more complete description of the cloud profile is now used to calibrate the inversion algorithms used for passive sensors.

Exploiting the complementarity of radar and lidar observations of the A-Train, Delanoë and Hogan (2008) developed an algorithm for the retrieval of the microphysical properties of ice clouds capable of operating on regions common to both instruments, and on parts seen by only one of them. Building on the first version established by L'Ecuyer et al. (2008), this more complete documentation of cloud stratification allowed for a new estimation of radiative heating rates taking into account thin clouds only detected by lidar (Henderson et al. 2013). The vertical distribution of radiative heating is the preferred way for clouds to feed back to atmospheric circulation. Figure 14.9 shows that depending on the cloud type, there is then a warming effect (multi-layer: M.L., cirrus: Ci or Altostratus: As) or cooling (stratocumulus: Sc or stratus: St). On an annual average and on a global scale, the strongest cloud impact is related to multi-layer clouds, which is particularly important in the tropics.

These complementary observations are therefore powerful tools for documenting cloud processes on a global scale. The limitation of these observations is due to the orbit of the non-geostationary satellites which they are on, limiting their sampling capacity of the diurnal cycle (especially in the tropics) or medium- to large-scale meteorological objects. Combination strategies have therefore been developed to document the cloudiness associated with the Madden–Julian oscillation (Madden and Julian 1971) or the evolution of microphysical properties during the life cycle of tropical convective systems (Bouniol et al. 2016).

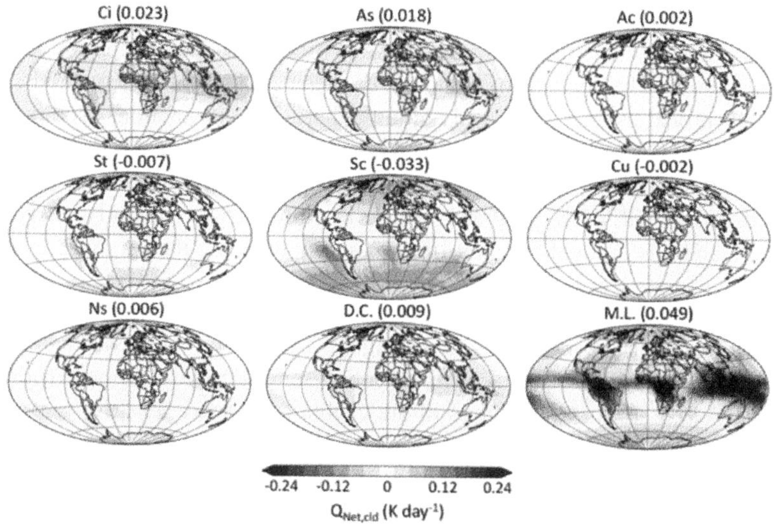

Figure 14.9. *Radiative heating rate in K/day related to different cloud types on an annual average. Global mean values are given in parentheses (source: Hang et al. (2019))*

14.3.3.2. *Studies in optical-lidar synergy*

Since the 1980s, the International Satellite Cloud Climatology Project (ISCCP; Rossow and Schiffer (1999)) has been producing data on geophysical variables, globally, from infrared and visible observations, from operational geostationary and polar satellites. This climatology has been updated and currently version H of the ISCCP series is available to the scientific community (Young et al. 2018). This series integrates optical domain data from Japanese, European, Indian and US geostationary satellites, which have the advantage of sampling the diurnal cycle, as well as NOAA polar satellites, which complement the measurements. The spatial sampling is about 10 km, and data is available every 3 hours for this latest update. This series has long served as a reference for modelers. Threshold methods, as described above, are used to estimate the cloud cover, the temperature and pressure of the top of the cloud, its optical depth and water content, its main phase and its nature. Clouds are classified as high, medium and low according to the pressure of their top and according to their visible optical depth (optically thick, thin or very thin). Another database of about the same dimension, but derived from AVHRR** radiometers on moving orbits (PATMOS-x), has also been developed (Heidiger et al. 2014). However, as mentioned above, the difficulties associated with optical

**See Appendix 3.

imaging have encouraged space teams to develop alternatives. Also other series of cloud parameters, sometimes shorter, were extracted by the different methods mentioned above. The analysis of cloud cover in the 2000s by active methods, more precise on vertical aspects, removed some of these indeterminations (see previous section). It has become important to intercompare information from these different methods and instruments, and this has been the subject of an international program, promoted by the GEWEX program, which has produced a monthly analysis of differences and has brought together under a single database the set of 12 available datasets (Stubenrauch et al. 2013).

All comparisons are summarized in Figure 14.10. The average annual cloud cover varies from 0.73, for instruments detecting semi-transparent clouds (lidar and IR sounders) to 0.56, for radiometers in the visible only, more sensitive to optically thick clouds. The overall annual average is 0.68 ± 0.03. The last three columns of the figure give the relative percentage of high, medium and low clouds in relation to total coverage. The difference between instruments is the greatest for high relative cloud cover according to their sensitivity to semi-transparent clouds.

Taking into account attribution errors, about 40–50% of clouds are high clouds, about 15% are medium clouds and 40% are low altitude clouds in a single layer. About 20% of high clouds are above lower clouds, according to active measurements.

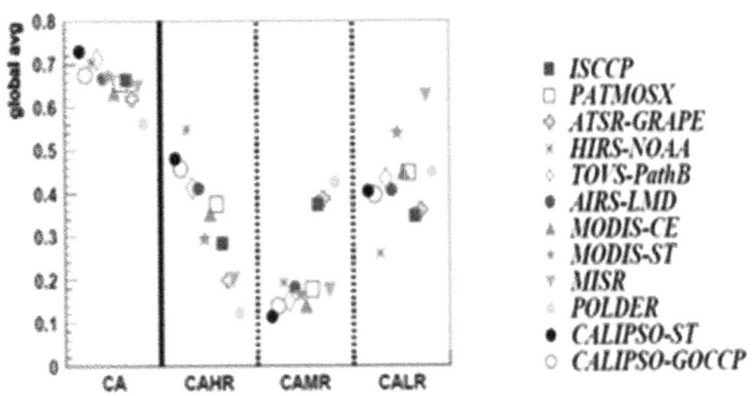

Figure 14.10. *Global average cloud cover (%) from left to right: for all clouds (CA), for high (CAHR), medium (CAMR) and low (CALR) altitude clouds. The different acronyms correspond to the results of 12 climatologies from passive (first 10) and active (two CALIPSO lines) instruments (adapted from Stubenrauch et al. (2013))*

In general, oceanic regions are 10–15% cloudier than continents. Moreover, there is high latitudinal variability. The typical value of interannual variability is 3%, which is 10 times lower than the daily variability. The mean cloud optical depth in the visible domain on global average varies between 4 and 10. The average IR emissivity value fluctuates between 0.6 and 0.8. The average effective radius of the particles is around 14 μm and 25 μm, respectively, at the top of the clouds of liquid water and solid ice. The average amount of water for liquid water clouds ranges from 30 to 60 g.m^{-2}, while clouds with icy tops contain 60–120 g.m^{-2}.

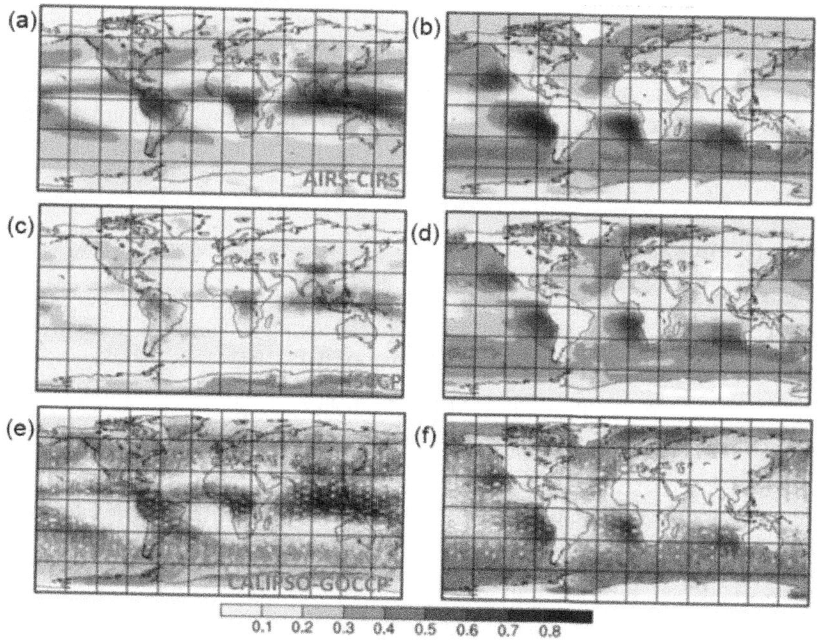

Figure 14.11. *Geographic distribution of annual cloud cover for high-altitude (left) and low-altitude (right) clouds from databases: AIRS-CrIS (2003–2015, top), ISCCP-D (1983–2009 middle) and CALIPSO-GOCCP (2007–2008, bottom) (source: Stubenrauch et al. (2017))*

Figure 14.11 shows the geographical distribution of cloud cover as an annual average for high and low altitude clouds, estimated by three climatologies from the AIRS sounder, ISCCP version D and CALIOP lidar on CALIPSO. This figure shows a very big difference on the climatologies of high clouds (left column). It should be remembered, however, that the comparison of cloud altitudes conducted with CALIOP lidar shows that in the case of very thin cirrus clouds, the altitude restored by passive methods is a "radiometric" altitude lower than the actual altitude of the cloud top. On the contrary, situations detecting high clouds correspond for passive instruments to all possible cases of overlapped layers, whereas this is not the case for the detection of low clouds. These geographical distributions clearly show the important coverage by high clouds in tropical regions and a predominance of low maritime clouds off the eastern edges of the oceans.

The joint use of active and optical passive measurements is therefore fundamental to determine cloud parameters. It should be noted that passive measurements often offer longer time series or finer spatial and temporal resolutions more suitable for climate studies, making them indispensable. This combination of passive and active measures was also used for the determination of cloud classifications (Sèze et al. 2014) and for the understanding of cloud feedbacks (Vaillant du Guélis 2018).

14.4. References

Bodas-Salcedo, A., Webb, M.J., Bony, S., Chepfer, H., Dufresne, J.-L., Klei, S., Zhang, Y., Marchand, R., Haynes, J., Pincus, R. et al. (2011). COSP satellite simulation software for model assessment. *Bull. Amer. Meteor. Soc.*, 92(8), 1023–1043.

Bouniol, D., Roca, R., Fiolleau, T. (2016). Macrophysical, microphysical, and radiative properties of tropical mesoscale convective systems over their life cycle. *J. Clim.*, 29, 3353–3371.

Brogniez, H., Roca, R., Picon, L. (2006). A clear sky radiances archive from Meteosat "water vapor" observations. *J. Geophys. Res.*, 111, D21109.

Brogniez, H., Roca, R., Picon, L. (2009). A study of the free tropospheric humidity interannual variability using Meteosat data and an advection-condensation transport model. *J. Clim.*, 22, 6773–6787.

Brogniez, H., English, S., Mahfouf, J.-F., Behrendt, A., Berg, W., Boukabara, S., Buehler, S.A., Chambon, P., Gambacorta, A., Geer, A. et al. (2016). A review of sources of systematic errors and uncertainties in water vapor information derived from observations at 183GHz. *Atmos. Meas. Tech.*, 9, 2207–2221.

Cardinali, C. and Healy, S. (2014). Impact of GPS radio occultation measurements in the ECMWF system using adjoint based diagnostics. *Quart. J. Roy. Meteor. Soc.*, 140(684), 2315–2320.

Chepfer, H., Bony, S., Winker, D., Cesana, G., Dufresne, J.-L., Minnis, P., Stubenrauch, C.J., Zeng, S. (2010). The GCM-oriented CALIPSO cloud product (CALIPSO-GOCCP). *J. Geophys. Res.*, 115, D00H16.

Delanoë, J. and Hogan, R.J. (2008). A variational scheme for retrieving ice cloud properties from combined radar, lidar, and infrared radiometer. *J. Geophys. Res.*, 113, D07204.

Elsaesser, G.S., Del Genio, A.D., Jiang, J.H., van Lier-Walqui, M. (2017). An improved convective ice parameterization for the NASA GISS global climate model and impacts on cloud ice simulation. *J. Clim.*, 30, 317–336.

Gleick, P. (1996). Water resources. *Encyclopedia of Climate and Weather*, 2, 817–823.

Hang, Y., L'Ecuyer, T.S., Henderson, D., Matus, A.V., Wang, Z. (2019). Reassessing the effect of cloud type on Earth's energy balance in the age of active spaceborne observations. Part II: Atmospheric heating. *J. Clim.*, 32, 6219–6236.

Heidiger, A., Foster, M., Walter, A., Zhao, X. (2014). The pathfinder atmospheres–extended AVHRR climate dataset. *Bull. Amer. Meteor. Soc.*, 95(6), 909–922.

Held, I.M. and Soden, B.J. (2000). Water vapor feedback and global warming. *Annu. Rev. Energy Environ.*, 25, 441–475.

Henderson, D.S., L'Ecuyer, T., Stephens, G., Partain, P., Sekiguchi, M. (2013). A multisensor perspective on the radiative impacts of clouds and aerosols. *J. Appl. Meteor. Climatol.*, 52, 853–871.

Jiang, J.H., Su, H., Zhai, C., Perun, V.S., Del Genio, A., Nazarenko, L.S., Donner, L.J., Horowitz, L., Charles, S., Cole, J. et al. (2012). Evaluation of cloud and water vapor simulations in CMIP5 climate models using NASA "A-Train" satellite observations. *J. Geophys. Res.*, 117, D14105.

Kämpfer, N. (2012). *Monitoring Atmospheric Water Vapour: Ground-Based Remote Sensing and In-situ Methods*. Springer, New York.

L'Ecuyer, T.S., Wood, N.B., Haladay, T., Stephens G.L., Stackhouse Jr., P.W. (2008). Impact of clouds on atmospheric heating based on the R04 CloudSat fluxes and heating rates data set. *J. Geophys. Res.*, 113, D00A15.

Loeb, N.G., Doelling, D.R., Wang, H., Su, W., Nguyen, C., Corbett, J.G., Liang, L., Mitrescu, C., Rose, F.G., Kato, S. et al. (2018). Clouds and the Earth's radiant energy system (CERES) energy balanced and filled (EBAF) top-of-atmosphere (TOA) edition 4.0 data product. *J. Clim.*, 31, 895–918.

Madden, R. and Julian, P. (1971). Detection of a 40–50 day oscillation in the zonal wind in the tropical Pacific. *J. Atmos. Sci.*, 28, 702–708.

O'Dell, C.W., Wentz, F.J., Benartz, R. (2008). Cloud liquid water path from satellite-based passive microwave observations: A new climatology over the global oceans. *J. Clim.*, 1721–1739.

Poitou, J. (2013). Composition atmosphérique et bilan radiatif. *Reflets de la physique*, 33, 28–33.

Roca, R., Chambon, P., Gosset, M., Bouniol, D., Brogniez, H., Cloché, S., Fiolleau, T., Raberanto, P., Dejus, M. (2019). Le bilan scientifique de la mission Megha-Tropiques après 8 ans dans l'espace. *La Météorologie*, 107, 36–44.

Rossow, W.B. and Schiffer, R.A. (1999). Advances in understanding clouds from ISCCP. *Bull. Amer. Meteor. Soc.*, 80, 2261–2287.

Schröder, M., Lockhoff, M., Shi, L., August, T., Bennartz, R., Borbas, E., Brogniez, H., Calbet, X., Crewell, S., Eikenberg, S. et al. (2017). GEWEX water vapor assessment (G-VAP). WCRP report 16/2017, World Climate Research Programme (WCRP), Geneva.

Sèze, G., Pelon, J., Derrien, M., Le Gléau, H., Six, B. (2014). Evaluation against CALIPSO lidar observations of the multi-geostationary cloud cover and type datasets assembled in the framework of the Megha-Tropiques mission. *Quart. J. Roy. Meteor. Soc.*, 141(688), 774–797.

Sherwood, S.C., Roca, R., Weckwerth, T.M., Andronova, N.G. (2010). Tropospheric water vapor, convection, and climate, *Rev. Geophys.*, 48, RG2001.

Stephens, G.L., Li, J., Wild, M., Clayson, C.A., Loeb, N., Kato, S., L'Ecuyer, T., Stackhouse Jr., P.W., Lebsock, M., Andrews, T. et al. (2012). An update on Earth's energy balance in light of the latest global observations. *Nat. Geosci.*, 5(10), 691–696.

Stephens, G.L., Winker, D., Pelon, J., Trepte, C., Vane, D., Yuhas, C., L'Ecuyer, T., Lesbock, M. (2018). CloudSat and CALIPSO within the A-Train: Ten years of actively observing the earth system. *Bull. Amer. Meteor. Soc.*, 99(3), 568–582.

Stephens, G.L., Christensen, M., Andrews, T., Haywood, J., Malavelle, F.F., Suzuki, K., Jing, X., Lebsock, M., Li, J.-L.F., Takahashi, H. et al. (2019). Cloud physics from space. *Quart. J. Roy. Meteor. Soc.*, 145(724), 2854–2875.

Stephens, G.L., Freeman, A., Richard, E., Pilewskie, P., Larkin, P., Chew, C., Tanelli, S., Brown, S., Posselt, D., Peral, E. (2020). The emerging technological revolution in Earth observations. *Bull. Amer. Meteor. Soc.*, 11(3), E274–E285.

Stevens, B., Brogniez, H., Kiemle, C., Lacour, J.-L., Crevoisier, C., Kiliani, J. (2017). Structure and dynamical influence of water vapor in the lower tropical troposphere. In *Shallow Clouds, Water Vapor, Circulation, and Climate Sensitivity*, Pincus, R., Winker, D., Bony, S., Stevens, B. (eds). Springer, Cham.

Stubenrauch, C.J., Rossow, W.B., Kinne, S., Ackerman, S., Cesana, G., Chepfer, H., Di Girolamo, L., Getzewich, B., Guignard, A., Heidinger, A. et al. (2013). Assessment of global cloud datasets from satellites: Project and database initiated by the GEWEX Radiation Panel. *Bull. Amer. Meteor. Soc.*, 94, 1031–1049.

Stubenrauch, C.J., Feofilov, A.G., Protopapadaki, S.E., Armante, R. (2017). Cloud climatologies from the infrared sounders AIRS and IASI: Strengths and applications. *Atmos. Chem. Phys.*, 17, 13625–13644.

Tivig, M., Grützun, V., John, V., Buehler, S. (2020). Trends in upper-tropospheric humidity: Expansion of the subtropical dry zones? *J. Clim.*, 33, 2149–2161.

Tretyakov, M.Y. (2016). Spectroscopy underlying microwave remote sensing of atmospheric water vapor. *J. Mol. Spectrosco.*, 328, 7–26.

Ulaby, F., Long, D., Blackwell, W., Elachi, C., Fung, A., Ruf, C., Sarabandi, K., Zebker, H., van Zyl, J. (2014). *Microwave Radar and Radiometric Remote Sensing*. University of Michigan Press, Ann Arbor.

Vaillant de Guélis, T., Chepfer, H., Noel, V., Guzman, R., Bonazzola, M., Winker, D.M. (2018). Space lidar observations constrain longwave cloud feedback. *Nat. Sci. Rep.*, 8, 16570.

Vaughan, M., Powell, K., Kuehn, R.E., Young, S.A., Winker, D.M., Hostetler, C.A., Hunt, W.H., Liu, Z., McGill, M.J., Getzewich, B.J. (2009). Fully automated detection of cloud and aerosol layers in the CALIPSO lidar measurements. *J. Atmos. Oceanic Technol.*, 26, 2034–2050.

Young, A.H., Knapp, K.R., Inamdar, A., Hankins, W., Rossow, W.B. (2018). The international satellite cloud climatology project h-series climate data record product. *Earth Syst. Sci. Data*, 10, 583–593.

15

Precipitation

Vincenzo LEVIZZANI[1] and Christopher KIDD[2,3]

[1] National Research Council of Italy, Institute of Atmospheric Sciences and Climate (CNRISAC), Bologna, Italy

[2] Earth System Science Interdisciplinary Center, University of Maryland, United States

[3] NASA Goddard Space Flight Center, Greenbelt, United States

15.1. Need for global precipitation measurements

Precipitation is a key component of the energy and water cycle through the evaporation of water, condensation and deposition as precipitation (Levizzani and Cattani 2019). Precipitation is extremely variable in space and time, the characteristics of which are just as vital as the amount so that quantities such the frequency and intensity are necessary to correctly describe this essential component of the water cycle (Trenberth et al. 2003). At the same time, a large number of applications require precipitation measurements to be as accurate as possible (Kucera et al. 2013; Kirschbaum et al. 2017) and over a range of spatial (local to global) and temporal (instantaneous to climate) scales (Michaelides et al. 2009). Climate-oriented applications, in particular, challenge our capability to measure precipitation across the globe, with spatial and temporal resolutions, latencies, spatial homogeneity and stability that address the user requirements. Many scientific studies rely upon information on global precipitation, often contending with a range of values.

Conventional observations of precipitation from surface-based measurements, including rain gauges and weather radars, form the basis of many global precipitation products. However, the distribution of gauges over the Earth's surface

---

For a color version of all the figures in this chapter, see www.iste.co.uk/phulpin/satellites2.zip.

Satellites for Atmospheric Sciences 2,
coordinated by Thierry PHULPIN, Didier RENAUT, Hervé ROQUET and Claude CAMY-PEYRET.

is generally confined to land surface areas (Kidd et al. 2017) and the timely availability of data from these gauges is often problematic. The most comprehensive gauge-based precipitation data set is the Global Precipitation Climatology Centre (GPCC) product available from 1901 to the present. The GPCC product incorporates gauge data from national networks as well as gauges that are included in other global or regional data sets, such as the Global Historical Climatology Network (GHCN) and the Asian-region APHRODITE data set. However, such global and regional gauge-based data sets must be generated with care to reduce errors and uncertainties in the final product.

The overall accuracy of the "standard" accumulation gauge is subject to discussion. Importantly, it should be noted that there are random errors in any gauge measurement since rain gauge errors are highly dependent upon the local rainfall characteristics. At the gauge scale, errors arise due to the deformation of the wind field around the gauge itself. Wind-induced losses in thunderstorms cause undercatch when using tipping bucket rain gauges in high intensity rain events and high winds. Some of these issues have been addressed through improving the design of gauges, although such design changes have generally yet to be implemented and therefore challenges remain in providing reliable point rainfall measurements, not least since different gauges lead to different responses. However, on a global scale, the distribution of gauges is largely confined to land areas with a dense population: consequently, much of the Earth's surface has poor or no gauge information (Kidd et al. 2017).

The development of the weather radar has provided a significant source of precipitation information, particularly for near real-time monitoring and short-term weather forecasting. Radars provide frequent and regular spatial information on precipitation at regional scales, although errors and uncertainties arise from variable reflectivity* – rain rate relationships. While such radar observations may be used to generate longer term products, careful processing is required to avoid artifacts in the final product. Despite the usefulness of radar-derived precipitation products, the distribution of weather radars typically matches that of the denser gauge networks, and consequently, large areas of the Earth's surface are not covered by either gauge or radar measurements.

Oceans are sparsely covered by scattered ship observations, buoys and radars on small islands which have been made available through the International Comprehensive Ocean–Atmosphere Data Set (ICOADS), the Global Summary of the Day (GSOD), the Pacific Rainfall Database (PACRAIN) and ship-based measurement campaigns, such as the Ocean Rain And Ice-phase precipitation measurement Network (OceanRAIN).

*See Glossary.

Development of Numerical Weather Prediction (NWP) models has been significant with precipitation being one of the output parameters. However, the representation of the precipitation generated by such models is problematic, particularly at the sub-daily scale. Kidd et al. (2013) compared the output of the ECMWF (European Centre for Medium-range Weather Forecasts) operational model and precipitation products from the Tropical Rainfall Measuring Mission (TRMM), concluding that the ECMWF model poorly represented the diurnal cycle, simulating rainfall too early compared to the satellite products. In addition, the model did not appear to simulate mountain-slope breezes well or adequately capture many of the characteristics of mesoscale* convective systems. What is still lacking is an adequate representation of sub-grid-scale processes in parameterization schemes. The proper representation of sub-daily precipitation in models is critical for hydrological modeling and flow forecasting and non-hydrostatic limited area models are starting to address this critical aspect.

There is therefore a need for precipitation measurements that are available across the whole of the Earth's surface: satellite observations have the potential to fulfill this need.

15.2. Satellite observation of rainfall

Observational capabilities for the constant monitoring of precipitation from space have substantially improved over the past few decades and now contribute to the investigation of climate processes. Measuring precipitation from space has been a key application of remote sensing since the early days of satellite meteorology in the 1960s. The key issue is to match the observations made of the cloud tops or precipitation particles in the atmosphere with precipitation rates at the ground. In this sense, the problem is similar to that of the ground-based radar and is still not fully solved, although many advances have been made in time. A comprehensive presentation of the state of the art of satellite precipitation measurements has recently been published, describing the status of observations and satellite programs, retrieval techniques, algorithms and sensors and validation programs worldwide, as well as the observed characteristics of precipitation processes, and applications that have recently developed from the wealth of satellite precipitation products (Levizzani et al. 2020a, 2020b).

There are continuing needs and challenges to better understand and predict precipitation changes in weather, hydrology and the climate system. While all are interconnected, they differ in their need for the frequency of observations, their

*See Glossary.

timeliness, their resolution and their uncertainty. As such, it is best to examine these needs separately with the notion that they may be combined when specific missions are designed. Crucially, the development of techniques is often driven by the need to address user requirements, considering their spatial and temporal scales and sampling, as well as the latency of the product. However, there is often a gulf between the expectations and requirements of the user community and the technical feasibility of generating such precipitation products. Ultimately, many schemes need to seek a compromise between providing a reasonable estimate of the surface precipitation to meet user requirements with the set of satellite observations available.

15.2.1. *Visible/Infrared*

Visible/Infrared (VIS/IR) retrieval schemes benefit from the frequent and consistent temporal sampling of observations that are available from geostationary (GEO)-based sensors, together with spatial resolutions (approximately 4 km) that are generally commensurate with many user applications. While VIS imagery has the best resolution, potential products are limited to daytime use only due to the daylight. Consequently, these retrieval schemes have focused primarily on IR imagery. IR-based techniques rely upon the notion that clouds with low cloud-top temperatures are taller/deeper and will therefore precipitate more. However, the relationship between cloud top temperature and the precipitation near to the Earth's surface is somewhat indirect and affected by variations in regional regimes and of the life cycle of the clouds themselves. Multi-spectral infrared channels may be used to improve the retrieval of cloud top characteristics, such as particle size and phase, but the indirectness of cloud-top properties to surface precipitation persists. The benchmark IR-only technique is the GOES Precipitation Index (GPI) which assigns 3 mm h^{-1} to any cloud top temperature below 235 K. The GPI was developed to work at monthly/2.5° resolution scales, although it works surprisingly well at a range of finer temporal/spatial scales. A number of schemes have used the GPI as a basis, generating adjusted GPI schemes to account for regional and temporal variations in the cloud-top temperature to precipitation relationships. More recently, retrieval schemes have exploited neural networks to better identify and retrieve precipitation. The Precipitation Estimation from Remotely Sensed Information using Artificial Neural Networks – Cloud Classification System (PERSIANN-CCS) technique uses cloud texture information (e.g. spatial variability, minimum cloud top temperature, etc.) to better define the surface precipitation. However, the delineation of precipitation in IR products remains problematic since cold thin clouds may be confused with cold deep clouds leading to "false" precipitation, while warm shallow clouds which are precipitating may be missed. Overall, IR-based products perform better in tropical regimes where the cloud-top temperature to surface precipitation relationships are more robust.

15.2.2. *Passive microwave*

Observations made by passive microwave (PMW) sensors are sensitive to liquid water and ice in the atmosphere. Low-frequency channels (up to about 40 GHz) are generally more directly sensitive to the liquid hydrometeors*, while the higher-frequency channels (> 40 GHz) are more sensitive to solid (ice) particles, although this is dependent upon the concentrations of liquid/ice and the particle sizes. Over the oceans, the increased emission from the liquid water droplets at the lower frequencies leads to increasing brightness temperatures (Tb) which can be used to retrieve precipitation. However, increasing ice scattering[1] is present at 37 GHz in high intensity events which can reduce the Tbs, thus leading to potentially ambiguous retrievals. The high emissivity* of land surfaces reduces the utilization of low-frequency channels for the detection of precipitation and the high-frequency channels must be used. Scattering by precipitation-sized ice particles results in a reduction in the brightness temperatures. Thresholding approaches of the Tb depression at the various PMW frequencies have been developed and have been shown to be simple to implement and suitable for operational environments. However, not all clouds have ice present and the ice in the upper levels of a cloud may not be a good representation of precipitation at the surface. Ideally, techniques use a combination of PMW frequencies to retrieve precipitation over different surface types.

Many precipitation retrieval techniques using these basic radiometric properties have achieved a degree of success. However, providing an unambiguous retrieval of precipitation over different surfaces is often difficult and compounded by non-unique radiometric signatures to hydrometeor profile/surface rainfall relationships. Computing the hydrometeor-radiative transfer relationships provides not only a basic understanding of the underlying relationships, but also allows for the investigation of using these relationships to develop retrieval techniques. In particular, not only can the surface precipitation be retrieved, but also the vertical profile of the hydrometeors through the atmosphere. An example of such a physically based precipitation retrieval scheme is the Goddard Profiling (GPROF) scheme, which is designed to provide estimates of surface precipitation and vertical profiles constrained by model information (2-m temperature and total precipitable water) and ancillary data sets

---

*See Glossary.

1. Scattering is the deviation of radiation or particles from a straight trajectory in all directions caused by a non-uniformity in the medium. In clouds it is due to the presence of hydrometeors and can be simple or multiple.

(surface types). GPROF is designed to work across a range of observations from PMW sensors to ensure consistent precipitation estimates. Meanwhile the Precipitation Retrieval and Profiling Scheme (PRPS) was developed for cross-track sounding instruments (such as the Microwave Humidity Sounder, MHS** and the *Sondeur Atmosphérique du Profil d'Humidité Intertropicale par Radiométrie*, SAPHIR**) using the Global Precipitation Measurement (GPM) mission Dual-frequency Precipitation Radar (DPR**) measurements as the calibrator (Box 15.1 and Figure 15.1).

The Global Precipitation Measurement (GPM) mission is an international satellite mission designed to advance our knowledge and understanding of global precipitation (Hou et al. 2014) and follows the very successful Tropical Rainfall Measuring Mission (TRMM; Kummerow et al. (2000)). Launched on February 28, 2014, the GPM Core Observatory (CO) is a collaboration between NASA and the Japan Aerospace Exploration Agency (JAXA) and joins a constellation of international partner satellites with precipitation-capable sensors (Skofronick-Jackson et al. 2017). The GPM-CO is in a 65° inclination, 407 km altitude orbit designed to cross the orbital tracks of the constellation allowing for the cross-calibration of all the sensors and to enhance the temporal sampling of observations from passive microwave (PMW) sensors. The GPM-CO carries a Dual-frequency Precipitation Radar (DPR) and the GPM Microwave Imager (GMI**) to provide a broad swath mapping capability. The DPR consists of Ku- (13.6 GHz) and Ka- (35.5 GHz) band radars to measure the three-dimensional structure of precipitation with a minimum detectable rain rate of about 0.2 mm/hr, while the GMI provides PMW observations across 13 channels from 10 GHz through to 183.31 GHz. Importantly, the greater sensitivity of the DPR and the high-frequency channels of the GMI ensure greater sensitivity to light precipitation and snowfall than previous sensors. Through the combination of the GPM-CO and GPM constellation precipitation retrievals, aided by infrared (IR) observations, techniques such as Integrated Multi-satellitE Retrievals for GPM (IMERG) and Global Satellite Mapping of Precipitation (GSMaP) are able to generate precipitation estimates at the 10-km, 30-minute scale.

GPM data and precipitation products are available through the NASA Precipitation Processing System (PPS) which captures, processes, archives and distributes data from the GPM-CO and constellation members. The brightness temperature (Tb) data from the constellation sensors are calibrated against the GMI sensor (to ensure consistency) and used by the GPROF to generate instantaneous precipitation products within about 3 hours of the satellite overpasses. Mapped precipitation products are available like the IMERG scheme at 0.1° × 0.1°, 30-minute resolution within 4 hours. A ground validation (GV) component is included as an integral part of the US GPM program, aimed at verifying and validating the GPM precipitation products. Specific GPM-GV field campaigns are designed to provide greater insight into the processes associated with precipitation and have provided valuable data across a number of different climate regimes.

Box 15.1. *The Global Precipitation Measurement (GPM) mission*

**See Appendix 3.

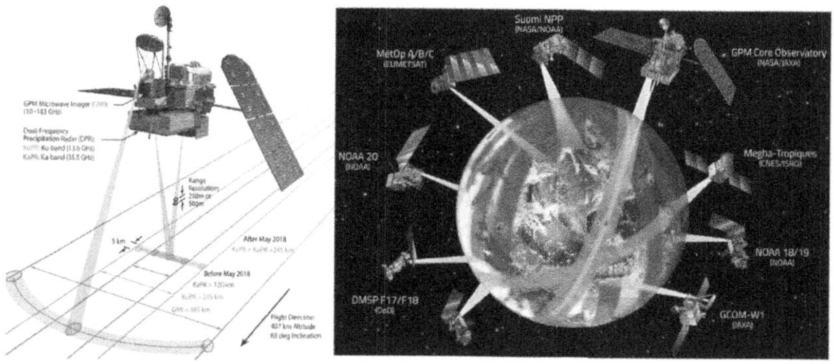

Figure 15.1. *The Global Precipitation Measurement (GPM) mission Core Observatory (left) and the GPM constellation of international partner missions (right) (source: NASA-GSFC)*

EUMETSAT's Satellite Application Facility on Support to Operational Hydrology and Water Management (H-SAF[2]) generates a suite of precipitation products. A Cloud Dynamics and Radiation Database (CDRD) was developed by using a regional/mesoscale model, within a cloud resolving model (CRM), to produce a large set of precipitation simulations and their associated meteorological–microphysical vertical profile. Similarly, a large database of cloud-resolving model simulations over Europe and Africa is used by the Passive microwave Neural network Precipitation Retrieval (PNPR). Based on a single neural network, it is designed to retrieve surface precipitation from cross-track sounding instruments and recently adapted to conical scanning sensors.

Further development of the radiative transfer models also allowed the generation of 1-D variational approaches. The Microwave integrated Retrieval System (MiRS) is an iterative physical inversion system to extract profiles of temperature, moisture, liquid cloud and hydrometeors, together with the surface emissivity and skin temperature. Such schemes perform well over the ocean where the surface background is relatively homogeneous. The Unified Microwave Ocean Retrieval Algorithm (UMORA) retrieves sea surface temperature, surface wind speed, columnar water vapor, columnar cloud water and surface rain rate, while the Hamburg Ocean Atmosphere Parameters and Fluxes from Satellite Data (HOAPS-3) provides turbulent heat fluxes, evaporation, precipitation, freshwater flux and related atmospheric variables over the global ice-free ocean. As with all schemes, careful inter-sensor calibration is required to ensure a homogeneous time series.

2. https://hsaf.meteoam.it.

Despite the advantage of the basis of the PMW techniques being more direct, the resolutions of PMW sensors are generally much poorer than those of IR sensors. The most recent PMW instruments have resolutions of about 5 km at best at the high-frequency channels (approximately 89 GHz), while at lower frequency channels resolutions are typically about 20 km or coarser. Furthermore, PMW instruments are only available on low-Earth-orbiting (LEO) satellites due to technical limitations on the size of the antenna and the footprint of the radiometer. Studies are on their way to overcome these technical limitations although the resolution at GEO orbit will be necessarily poorer. However, at present, the LEO-only PMW constellation severely restricts the temporal sampling possible by any single satellite except at high latitudes and for low inclination orbits. A number of satellites and sensors are needed to adequately resolve the variability of precipitating systems.

15.2.3. *Radar*

The launch of the TRMM mission (Figure 15.2) in 1997 opened a new era in the measurement of precipitation from space through the inclusion of the TRMM Precipitation Radar (PR**), to be followed by the GPM DPR from 2014 onwards, providing vertical profiles of hydrometeors down to about 1 km above the Earth's surface. While the PR consisted of a single Ku-band radar, the DPR sensor has both Ku- and Ka-band radars providing cross-track swaths of 245 km and 125 km (also 245 km after May 21, 2018), respectively, with a spatial resolution of approximately 5.2 km, and minimum detectable precipitation intensity of about 0.5 mm h^{-1}. The processing of the radar backscatter from the hydrometeors is constrained by several factors, such as stratiform/convective fraction and presence/absence of a bright-band, as well as accounting for path attenuation through the use of the surface return. However, the retrieval capabilities from such radar systems depend on the Ku-Ka-band sensitivities to precipitation.

The radar sensors provide the most direct measure of precipitation. However, the utility of their products is limited by long revisit times (typically days), and north–south extent of the data (36°N–36°S for the PR and 66°N–66°S for the DPR). Despite this, such radar data is invaluable for the generation of databases used by PMW retrieval schemes (such as GPROF and PRPS) and for their calibration/validation. In addition, the capability of these sensors to generate detailed three-dimensional information allows for greater insight into and understanding of precipitation systems (Figure 15.3). Furthermore, the longevity of the PR, and now the DPR, permits climatological distributions of precipitation rates to be derived (Liu and Zipser 2015).

**See Appendix 3.

Figure 15.2. *Artist's impression of the Tropical Rainfall Measuring Mission (TRMM) in orbit between 1997 and 2015 (source: NASA-GSFC)*

15.2.4. *Merged products*

No single satellite precipitation product is capable of completely capturing the true surface precipitation, not least because different users have different requirements. Combining observations is seen as a way of gathering information on precipitation that can be combined into a single, improved product, such as through the use of multiple observations to provide a better understanding between the observations and the surface precipitation (e.g. cloud delineation or constraining retrievals). A good example is the use of active (PR) and passive (TRMM Microwave Imager, TMI**) observations gathered by TRMM to reduce the retrieval ambiguities by minimizing the differences in the physical modeling of precipitation. A similar scheme forms the basis of the GPM COmbined Radar-Radiometer algorithm (CORRA) that uses observations from both the DPR and GPM Microwave Imager (GMI).

**See Appendix 3.

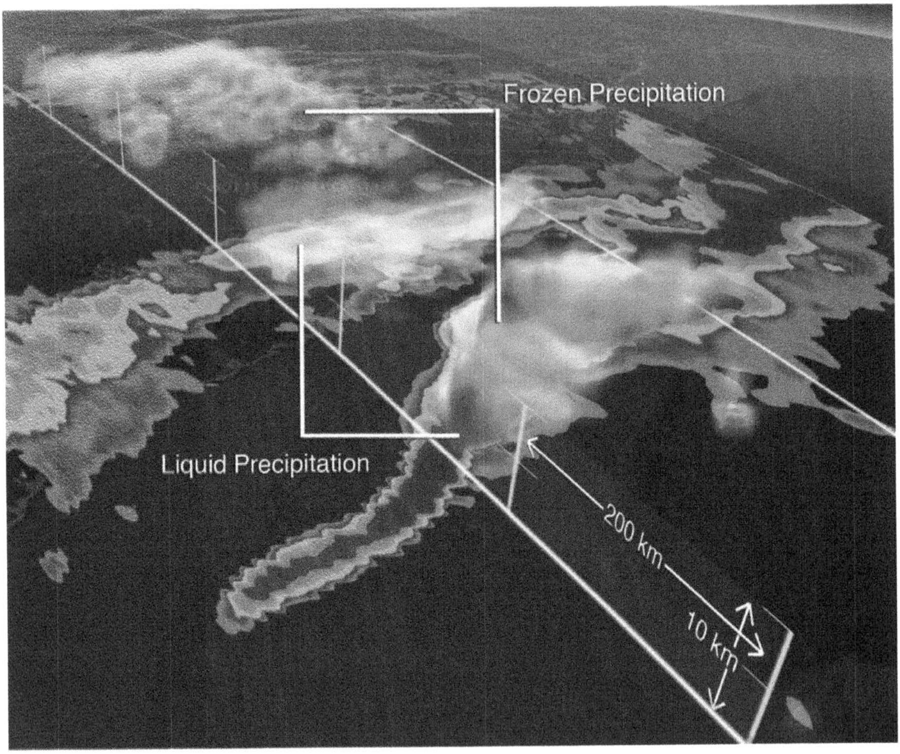

Figure 15.3. *One of the first storms observed by the NASA/JAXA GPM Core Observatory on March 17, 2014, in the eastern United States revealed the full range of precipitation, from rain to snow (source: NASA-GSFC)*

Several empirical schemes have been developed to exploit the directness of the PMW retrievals with the better temporal and spatial sampling of IR observations. Recent examples of techniques that modify the IR thresholds and rain rates include the current PERSIANN suite of schemes. Kidd et al. (2003) developed a combined PMW-IR algorithm that used regional/spatial histograms of PMW derived precipitation intensities to set the IR cloud top temperature thresholds for different rain intensities. The use of neural networks was explored to extract precipitation information from combined IR and PMW observations, and its use is nowadays expanding because of the increase in data availability which creates problems in dealing with large amounts of data. Similarly, the Self-Calibrating Multivariate Precipitation Retrieval (SCaMPR) technique updates their IR-to-rainfall relationships based on PMW observations.

The calibration of the IR observations by the PMW retrievals has shown a good degree of promise. However, such an approach has two main drawbacks. First, the spatial and temporal scale over which the calibration is undertaken is crucial for capturing the variations in the relationship yet providing sufficient information for a robust calibration. Second, the distribution of the precipitation is determined by the cloud as observed by the IR, and therefore may still not relate to the distribution of the precipitation at the surface. Since precipitation estimates derived from PMW observations are demonstrably better than those from the IR, particularly for instantaneous retrievals, it is logical to use these PMW estimates and augment them with information from the IR observations. Schemes were therefore devised which generate motion vectors from the frequent/regular IR data to advect the PMW precipitation estimates. These "merged" schemes allow precipitation estimates to be produced at the temporal sampling of the IR data, but with the PMW precipitation estimates.

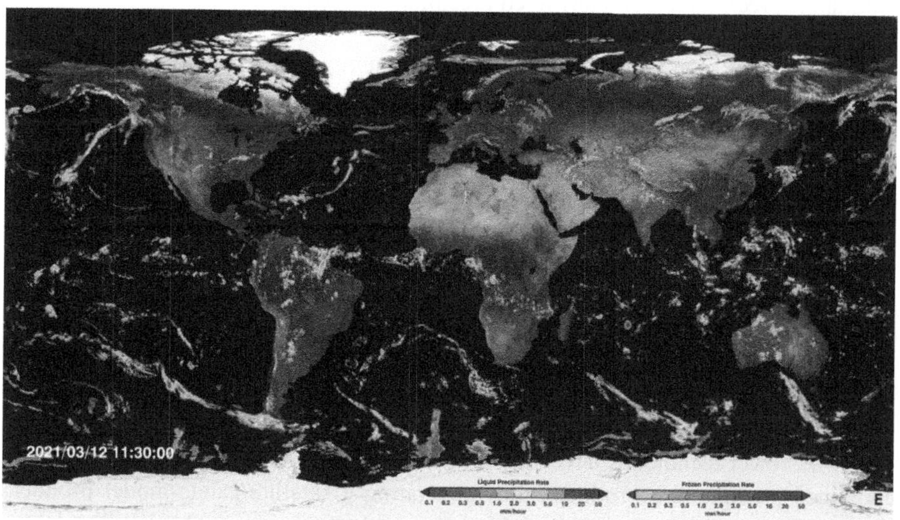

Figure 15.4. *Global IMERG precipitation products on March 12, 2021 11:30 UTC. Note the liquid and solid components, which characterize global satellite retrievals for the first time ever thanks to the GPM sensors (source: NASA-GSFC)*

There are currently three main operational merged products available providing surface precipitation information at scales of about 10 km/30 minutes or better. These are the JAXA Global Satellite Mapping of Precipitation (GSMaP), the NOAA CPC MORPHing Technique (CMORPH) and the NASA-Integrated Multi-satellitE Retrievals for GPM (IMERG) (see Figure 15.4). Although these schemes differ, they are based around three similar algorithm stages: (i) individual PMW measurements are generated/acquired to estimate the surface precipitation; (ii) motion vectors are derived from IR imagery or models to move the precipitation between the successive PMW estimates; and (iii) if necessary, IR-derived estimates fill in gaps if no PMW estimate is available. Subtle differences arise between these schemes as a consequence of options taken at each of the three stages. Despite such schemes having the potential to provide a continuous record of precipitation over time, the complexities in the generation of these products should not be underestimated, primarily due to the reliance upon the original PMW retrievals, the accuracy of which varies between the sources of the data provided by the individual instruments. In addition, the mechanisms used to move/morph the precipitation between individual PMW retrievals and how one PMW retrieval is mapped (in time, space and magnitude) onto the next PMW retrieval can lead to significant variations. Furthermore, these schemes do not necessarily cope well with situations such as orographic precipitation or short-lived precipitation events.

To meet the requirements of the user community, many satellite precipitation products incorporate conventional surface data (e.g. gauges) to some extent: this is often necessary for final "research-quality" to correct for any systematic bias in the satellite estimates. Indeed, the inclusion of surface data sets in long-term precipitation records is essential to ensure consistency. Such an example is the TRMM Multisatellite Precipitation Analysis (TMPA). The TMPA combines precipitation estimates from multiple satellites and gauge analyses (if available) to generate a precipitation product at 0.25° × 0.25° every 3 hours over 50°N–50°S. The near real-time product is calibrated using TRMM precipitation products, while gauges are used for later versions. The TMPA is now superseded by the IMERG product that provides global precipitation products at 0.1° × 0.1° every 30 minutes.

Some newer techniques have started to incorporate model information into their combined precipitation products, thus generating a satellite-gauge-model product. For example, the Multi-Source Weighted-Ensemble Precipitation (MSWEP) scheme generates a combined 3-hourly product at 0.1° resolution using the combined products themselves. However, all such combination schemes are only as good as their components and the combined product may underestimate or miss altogether some precipitation extremes.

15.3. Observation of solid precipitation

There is great interest in hydrological investigations with global snowfall rate products to help close the atmospheric inputs to the water cycle (Field and Heymsfield 2015; Rodell et al. 2015) and to study snowpack patterns and growth.

Snowfall rate retrievals are focused on falling snow as opposed to snow depth and snow extent on the ground that are retrieved using other sensors. Solid precipitation, whose most important components are snow and hail, is extremely difficult to measure both from space (Levizzani et al. 2011) and from the ground (Rasmussen et al. 2012). At the surface, snowfall basically occurs in the cold climate regions and highlands, whereas over the ocean in the tropics frozen hydrometeors are observed in the upper troposphere associated with stratiform precipitation. These facts identify snowfall as an intrinsically three-dimensional process that needs to be correctly identified and measured over the entire globe.

For PMW retrievals, snowfall estimates can be affected by changes in surface emissivity, especially over snow-covered surfaces. The land surface variable emissivity hinders snowfall detection compared to oceanic backgrounds that have lower and more uniform emissivities, especially when compared to direct radar measurements. Indeed, seasonal studies of GMI snowfall retrievals have indicated a dependence on snow-cover characteristics. However, the GMI high frequencies (166 and 183 GHz) have been shown to be particularly useful for falling snow estimates.

The best strategy is thus to compare active and passive estimates, as recently done by several authors using GMI and DPR on GPM also focusing on the snowfall product differences arising from classification, instrumentation, sampling and algorithm differences to provide a more informative falling snow comparison. The CloudSat Cloud Profiling Radar (CPR**) has proven to be an excellent sensor to detect snow-bearing clouds, given its sensitivity to cloud hydrometeors. Extensive studies on the climatology of snowfall from shallow cumuliform clouds have been first conducted and more recently extended to all types of snowfall. CloudSat is an excellent example of how cloud and precipitation radars can guide PMW sensors in snowfall retrievals thanks to their ability to detect snowfall. However, it must be noted that a much better understanding of the scattering properties of hydrometeors is necessary to improve the information from passive and active MW observations since the natural variety of hydrometeor properties also leads to the complexity of how they interact with electromagnetic radiation. The challenge of calculating hydrometeor scattering properties is not only the scattering method itself but also starts with an accurate characterization of the hydrometeors themselves.

PMW measurements from satellites have been used to identify the signature of hail in intense thunderstorms. The scattering signal of hailstones is typically observed as a strong depression of upwelling brightness temperatures from the cloud to the satellite and is instrumental for their detection, including information on the hailstone cross-section.

**See Appendix 3.

The future of measuring solid (and also liquid) precipitation from space is likely in the developing field of new triple-frequency radars exploiting the advantages of using Ka-, Ku- and W-bands. The triple-frequency concept will hopefully contribute to removing current limitations of precipitation retrievals through improvement in the observations of rain and ice microphysics.

15.4. Precipitation and the Earth water cycle

Drawing future climate scenarios together with weather monitoring and forecasting is a complex task which varies greatly depending on the application area. The World Meteorological Organization (WMO) Observing Systems Capability Analysis and Review (OSCAR[3]) is the reference source for the requirements that are expressed in terms of six criteria: uncertainty, horizontal resolution, vertical resolution, observing cycle, timeliness and stability. For each criterion, three values are indicated: threshold (minimum value to be met for the data to be useful), breakthrough (intermediate level representing a significant improvement for the targeted application) and goal (requirement beyond which no further improvements are necessary). The OSCAR confidence level for liquid precipitation is between firm (result of impact studies on actual data used in actual applications) and reasonable (not firm, but based on a strong heritage of experiments with similar data). In the case of solid precipitation intensity at the surface, the confidence level is only tentative (positive impact expected but not yet validated due to lack of suitable data). It is thus clear that the improvement of future precipitation missions with regard to observing cycle, resolution, timeliness and uncertainty level are deemed necessary. Because requirements focus on parameters rather than on individual missions, the overarching goals in the case of precipitation tend to focus both on high quality individual products as well as merged products. The International Precipitation Working Group (IPWG; Levizzani et al. (2018)) is helping WMO, the Coordination Group for Meteorological Satellites (CGMS) and the Committee on Earth Observation Satellites (CEOS) to meet these goals in the near future with the clear notion of making available products for specific user needs from hydrometeorology to climate. Note that robustness of the constellation of passive and active sensors in orbit is a key issue for ensuring the necessary continuity of products (Huffman et al. 2016).

Hydrological predictions try to meet the fundamental societal need for water, which remains insecure over large portions of the Earth. Needs vary from monitoring and predicting water availability in water-scarce areas (especially developing countries) to ensuring efficient use of water where it is often used

3. https://www.wmo-sat.info/oscar/observingrequirements.

without adequate planning (most of the industrialized countries). Accurate observations are thus a key element of hydrologic predictions (Lettenmaier 2017). The last US Decadal Survey (2017) ranges from recommendations to extend ongoing measurements, to new endeavors in detecting the phase (rain or snow) of precipitation, to measuring snow water equivalent and evapotranspiration*, to new fields for application of remotely sensed data, such as water quality, groundwater recharge, effects of urbanization, water-modulated biogeochemical cycling and prediction of hazard chains. Improvements in our observing capabilities from space relate not only to precipitation (rainfall and snowfall), but also snow water equivalent (still quite elusive especially in mountain areas in spite of our ability to estimate snow cover), evapotranspiration, snow and ice melt and recharge and withdrawal of groundwater. Advances are planned in the areas of precipitation, river discharge[4], evapotranspiration, total storage estimates, soil moisture and surface water storage (lakes, wetlands and reservoirs), while the US Decadal Survey (2017) further highlights the need for snow water equivalent measurements. It is thus very relevant to conceive new precipitation missions along a future refinement of other present key hydrology-related missions such as the Gravity Recovery and Climate Experiment (GRACE) and the Soil Moisture and Ocean Salinity (SMOS) and Soil Moisture Active Passive (SMAP), or even the possible Snow Water Equivalent mission. This will allow for a much more integrated view of precipitation in the Earth System, as well as monitoring of extreme precipitation in the context of its causes and consequences.

An improved translation from radiometric observations from space to physical properties is a prerequisite for a complete quantification of the hydrologic cycle components and their interactions. Such "translation grammar" varies considerably from case to case, depending on the frequency of observations, timeliness, latency, spatial resolution and uncertainty, which can scarcely be addressed by a single mission; rather, they require several missions and/or constellations of sensors conceived for different applications. Thus, a complete understanding of precipitation structure, its changes and links to the other water cycle components is still burdened by relevant science challenges.

The first challenge concerns the design of the future satellite precipitation observing system as an essential component of a consistent climate observing system conceived for addressing rising sea level, droughts, floods, extreme events, food security and freshwater availability in a changing climate. Weatherhead et al. (2018) propose giving priority to the Grand Challenges of the World Climate

*See Glossary.

4. River discharge: the volumetric flow rate of water (m^3/s) that is transported through a given cross-sectional area of the river.

Research Program (WCRP): (i) melting ice and global consequences; (ii) clouds, circulation and climate sensitivity; (iii) carbon feedback in the climate system; (iv) understanding and predicting weather and climate extremes; (v) water for the food baskets of the world; (vi) regional sea level change and coastal impacts; and (vii) near-term climate prediction. Monitoring the Earth through satellite missions that are not part of an integrated global system is no longer possible, since interconnected climate processes require observations designed to contribute to a data system that enables better-quality projections of future scenarios. Precipitation is relevant to most of the seven challenges, and its global observation requires a coordinated worldwide effort. In particular, precipitation observations should help to model development focused around processes that most affect storm tracks, tropical rain belts and climate sensitivity (Tapiador et al. 2017). Such an approach will help in reducing model biases and uncertainties, while at the same time advancing the utility of global modeling. Answers to questions such as the future of the permafrost layers, or the dynamics of the terrestrial and ocean carbon sinks, depend on the magnitude of warming and the distribution of precipitation. Validation has become an integral part of model development due to the availability of global gridded products that support the evaluation of the next generation of climate models.

Trends and variability (seasonal, decadal and longer time) are extremely difficult to estimate but are highly important for the verification of climate model projections. This is especially true in areas of the globe where the variations are extreme and influence millions of lives. An example of these studies is illustrated in Figure 15.5.

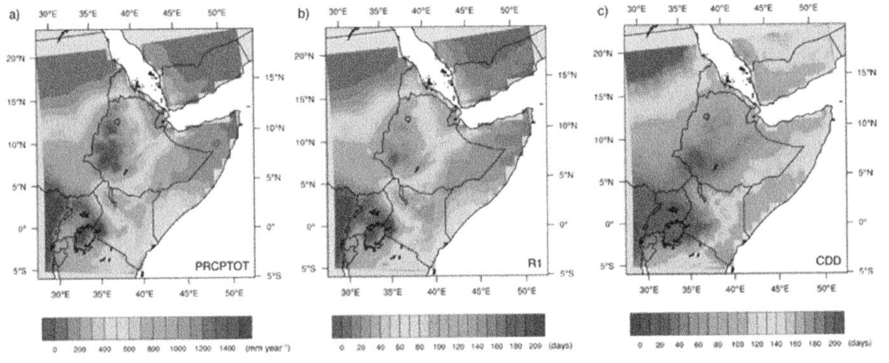

Figure 15.5. *Maps of the annual climatology of precipitation indexes for the period 1983–2015 over East Africa from an ensemble of three satellite precipitation products: (a) PRCPTOT (total precipitation from days with rain rate > 1 mm); (b) R1 (number of precipitating days with rain rate > 1 mm); (c) CDD (number of consecutive dry days, i.e. with precipitation < 1 mm) (source: Cattani et al. (2018), courtesy of MDPI)*

In summary, it has been demonstrated by several studies in the last decade that all components of the water cycle need to be observed and data sets produced to help the modeling effort of the water and energy cycle. The Global Climate Observing System (GCOS) regularly assesses the status of global climate observations of the atmosphere, land and ocean, and maintains definitions of the Essential, Climate Variables (ECVs[5]) required to observe the Earth's changing climate. Precipitation is essential, together with soil moisture, river discharge, evapotranspiration, the water table[6] and other variables; all these datasets together contribute to the open challenge of closing the water cycle (Dorigo et al. 2021).

15.5. References

Cattani, E., Merino, A., Guijarro, J.A., Levizzani, V. (2018). East Africa rainfall trends and variability 1983–2015 using three long-term satellite products. *Remote Sens.*, 10, 931.

Dorigo, W., Dietrich, S., Aires, F., Brocca, L., Carter, S., Cretaux. J.-F., Dunkerley, D., Enomoto, H., Forsberg, R., Güntner, G. et al. (2021). Closing the water cycle from observations across scales: Where do we stand? *Bull. Amer. Meteor. Soc.*, 102(10), E1897–E1935.

Field, P.R. and Heymsfield, A.J. (2015). Importance of snow to global precipitation. *Geophys. Res. Lett.*, 42, 9512–9520.

Hou, A.Y., Kakar, R.K., Neeck, S.A., Azarbarzin, A., Kummerow, C.D., Kojima, M., Oki, R., Nakamura, K., Iguchi, T. (2014). The global precipitation measurement mission. *Bull. Amer. Meteor. Soc.*, 95, 701–722.

Kidd, C., Dawkins, E., Huffman, G.J. (2013). Comparison of precipitation derived from the ECMWF operational forecast model and satellite precipitation datasets. *J. Hydrometeor.*, 14, 1463–1482.

Kidd, C., Huffman, G.J., Becker, A., Skofronick-Jackson, G., Kirschbaum, D., Joe, P., Muller, C. (2017). So, how much of the Earth's surface is covered by rain gauges? *Bull. Amer. Meteor. Soc.*, 98, 69–78.

Kidd, C., Tan, J., Kirstetter, P.-E., Petersen, W.A. (2018). Validation of the version 05 level 2 precipitation products from the GPM core observatory and constellation satellite sensors. *Quart. J. Roy. Meteor. Soc.*, 144(S1), 313–328.

Kirschbaum, D.B., Huffman, G.J., Adler, R.F., Braun, S., Garrett, K., Jones, E., McNally, A., Skofronick-Jackson, G., Stocker. E., Wu, H. et al. (2017). NASA's remotely-sensed precipitation: A reservoir for applications users. *Bull. Amer. Meteor. Soc.*, 98, 1169–1194.

5. https://gcos.wmo.int/en/essential-climate-variables.

6. Water table: the level below which the ground is saturated, very often equal to the phreatic water level.

Kucera, P.A., Ebert, E.E., Turk, F.J., Levizzani, V., Kirschbaum, D., Tapiador, F.J., Loew, A., Borsche, M. (2013). Precipitation from space: Advancing earth system science. *Bull. Amer. Meteor. Soc.*, 94, 365–375.

Kummerow, C., Simpson, J., Thiele, O., Barnes, W., Chang, A.T.C., Stocker, E., Adler, R.F., Hou, A., Kakar, R., Wentz, F. et al. (2000). The status of the tropical rainfall measuring mission (TRMM) after two years in orbit. *J. Appl. Meteor.*, 39, 1965–1982.

Lettenmaier, D.P. (2017). Observational breakthroughs lead the way to improved hydrological predictions. *Water Resour. Res.*, 53, 2591–2597.

Levizzani, V. and Cattani, E. (2019). Satellite remote sensing of precipitation and the terrestrial water cycle in a changing climate. *Remote Sens.*, 11, 2301.

Levizzani, V., Laviola, S., Cattani, E. (2011). Detection and measurement of snowfall from space. *Remote Sens.*, 3(1), 145–166.

Levizzani, V., Kidd, C., Aonashi, K., Bennartz, R., Ferraro, R.R., Huffman, G.J., Roca, R., Turk, F.J., Wang, N.-Y. (2018). The activities of the international precipitation working group. *Quart. J. Roy. Meteor. Soc.*, 144(S1), 3–15.

Levizzani, V., Kidd, C., Kirschbaum, D.B., Kummerow, C.D., Nakamura, K., Turk, F.J. (eds) (2020a). *Satellite Precipitation Measurement 1*. Springer Nature, Cham.

Levizzani, V., Kidd, C., Kirschbaum, D.B., Kummerow, C.D., Nakamura, K., Turk, F.J. (eds) (2020b). *Satellite Precipitation Measurement 2*. Springer Nature, Cham.

Liu, C. and Zipser, E.J. (2015). The global distribution of largest, deepest, and most intense precipitation systems. *Geophys. Res. Lett.*, 42, 3591–3595.

Michaelides, S., Levizzani, V., Anagnostou, E.N., Bauer, P., Kasparis, T., Lane, J.E. (2009). Precipitation: Measurement, remote sensing, climatology and modeling. *Atmos. Res.*, 94, 512–533.

Rasmussen, R.M., Baker, B., Kochendorfer, J., Meyers, T., Landolt, S., Fischer, A.P., Black, J., Theriault, J.M., Kucera, P.A., Gochis, D. et al. (2012). How well are we measuring snow? *Bull. Amer. Meteor. Soc.*, 93, 811–829.

Rodell, M., Beaudoing, H.K., L'Ecuyer, T.S., Olson, W.S., Famiglietti, J.S., Houser, P.R., Adler, R., Bosilovich, M.G., Clayson, C.A., Chambers, D. et al. (2015). The observed state of the water cycle in the early twenty-first century. *J. Clim.*, 28, 8289–8318.

Skofronick-Jackson, G.M., Petersen, W.A., Berg, W., Kidd, C., Stocker, E.F., Kirschbaum, D.B., Kakar, R., Braun, S.A., Huffman, G.J., Iguchi, T. et al. (2017). The global precipitation measurement (GPM) mission for science and society. *Bull. Amer. Meteor. Soc.*, 98, 1679–1695.

Tapiador, F.J., Navarro, A., Levizzani, V., García-Ortega, E., Huffman, G.J., Kidd, C., Kucera, P.A., Kummerow, C.D., Masunaga, H., Petersen, W.A. et al. (2017). Global precipitation measurements for validating climate models. *Atmos. Res.*, 197, 1–20.

Trenberth, K.E., Dai, A., Rasmussen, R.M., Parsons, D.B. (2003). The changing character of precipitation. *Bull. Amer. Meteor. Soc.*, 84, 1205–1213.

US Decadal Survey (ed.) (2017). *Thriving on Our Changing Planet: A Decadal Strategy for Earth Observation from Space*. The National Academies Press, Washington, DC.

Weatherhead, E.C., Wielicki, B.A., Ramaswamy, V., Abbott, M., Ackerman, T.P., Atlas, R., Brasseur, G., Bruhwiler, L., Busalacchi, A.J., Butler, J.H. et al. (2018). Designing the climate observing system of the future. *Earth's Future*, 6, 80–102.

Appendices

Appendix 1

Atmospheric Profiles

Claude CAMY-PEYRET
IPSL, Paris, France

This appendix on atmospheric profiles is intended to supplement information that could not be fully developed in several chapters of the book.

A1.1. AFGL profiles

A useful source of profiles that has been widely used for modeling radiative transfer in the Earth's atmosphere was developed by Anderson et al. (1986) to serve as a basis for code comparisons or simulations across the spectrum from microwave to UV through the infrared and visible. Simplified "typical" atmospheres were selected based on the format of the reference atmosphere known as the U.S. Standard Atmosphere, 1976. This reference covered only "average" thermodynamic profiles, focusing instead on the structure and composition of the upper atmosphere. For simulations in the so-called middle atmosphere, information on minor constituent profiles and seasonal and latitude variations was missing.

It is this type of data that has been published under the title "AFGL atmospheric constituent profiles (0–120 km)" by Anderson et al. (1986) where AFGL stands for Air Force Geophysical Laboratory. There are six AFGL profiles with an index i that identifies them with i = [1;6]. They correspond to the profiles in Table A1.1. New profile databases have been proposed as described below.

Satellites for Atmospheric Sciences 2,
coordinated by Thierry PHULPIN, Didier RENAUT, Hervé ROQUET and Claude CAMY-PEYRET.

AFGL profile index	AFGL profile name	Information on conditions
1	Tropical	15N, annual average, day
2	Mid-latitude summer	45N, July, day
3	Mid-latitude winter	45N, January, day
4	Sub-arctic summer	60N, July, day
5	Sub-arctic winter	60N, January, day
6	U.S. Standard Atmosphere, 1976	45° 32' 33", average, day

Table A1.1. *AFGL profiles*

The AFGL profiles are given on a grid of 50 levels in geometric altitude from 0 to 120 km with a discretization of 1 km between 0 and 25 km, 2.5 km between 25 and 50 km and 5 km above 50 km up to 120 km. The classification of the first five profiles is associated with different seasonal and latitudinal conditions. The last profile ($i = 6$) is an annual zonal average applicable for mid-latitudes. The temperature profiles (kinetic temperature) as a function of the geometric altitude between 0 and 120 km are shown in Figure A1.1.

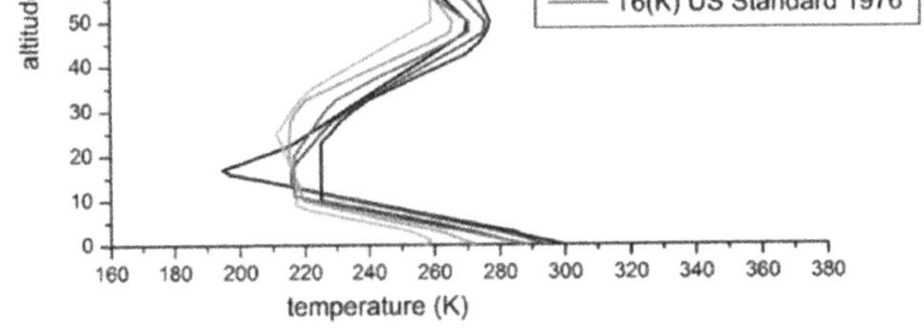

Figure A1.1. *Temperature variation as a function of altitude for the six AFGL profiles. For a color version of this figure, see www.iste.co.uk/phulpin/satellites2.zip*

We identify the different characteristic levels that correspond to the significant changes (value and/or sign) of the vertical thermal gradient from bottom to top:

– the tropopause with a marked minimum temperature located at 17 km for tropical atmosphere, a tropopause that varies with the season for other atmospheres with an altitude of 9 km for the atmosphere in the vicinity of the Arctic Circle in winter of the northern hemisphere;

– the stratopause with a maximum of approximately 50 km corresponding to maximum heating by photodissociation of ozone;

– the minimum temperature at the mesopause approximately 80 km;

– the increase of the temperature above this last altitude due to the photodissociation of molecular oxygen and the gradual transition to the thermosphere whose kinetic temperature is greatly modulated during the solar activity cycle.

Pressure variations as a function of altitude are exponential: the pressure decreases rapidly over eight decades between 0 and 120 km, with small modulations related to variations in the vertical temperature gradient as seen in Figure A1.2.

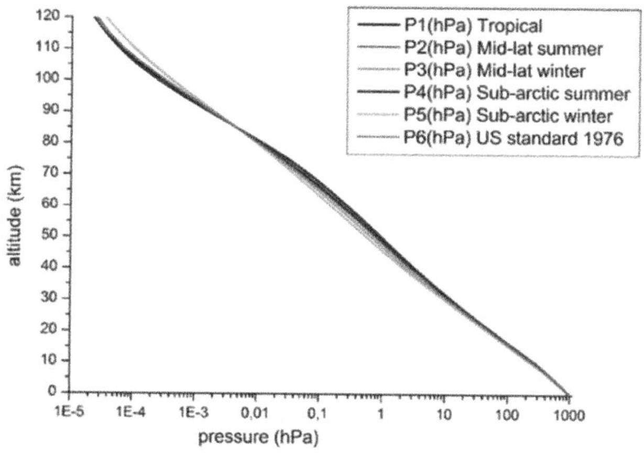

Figure A1.2. *Pressure variation as a function of altitude for the six AFGL profiles. For a color version of this figure, see www.iste.co.uk/phulpin/satellites2.zip*

In the following, the AFGL concentration profiles of H_2O, CH_4 and NO_2 are discussed (with associated comments) as examples of the variation with an altitude of three minor constituents typical of the Earth's atmospheric composition. Concentrations of the chemical species are expressed in volume mixing ratio (ppmv or parts per million by volume) which are noted here xX(z) with X = [H_2O, CH_4, NO_2]. The difference between the mixing ratio (without further specification) and mixing ratio with respect to dry air is important for precision measurements (in situ in general) when taking into account (or not) the presence of water vapor.

As seen in Figure A1.3 for H_2O in the vicinity of the surface, increasing values of humidity are observed when going from the sub-arctic winter profile (the driest) to the sub-arctic summer, mid-lat winter, mid-lat summer and finally to the tropical (the wettest). The cold trap (minimum of the H_2O) in the vicinity of the tropopause is characteristic and varies according to the altitude of this thermodynamic “barrier”. The increase of the H_2O profile up to the stratopause comes from the oxidation and photodissociation of CH_4 which produces H_2O. At higher altitudes, the photodissociation of H_2O by UV radiation (less and less filtered by ozone) is explaining in turn the decrease of water vapor (the simplified AFGL model no longer differentiates between the six profiles above 80 km).

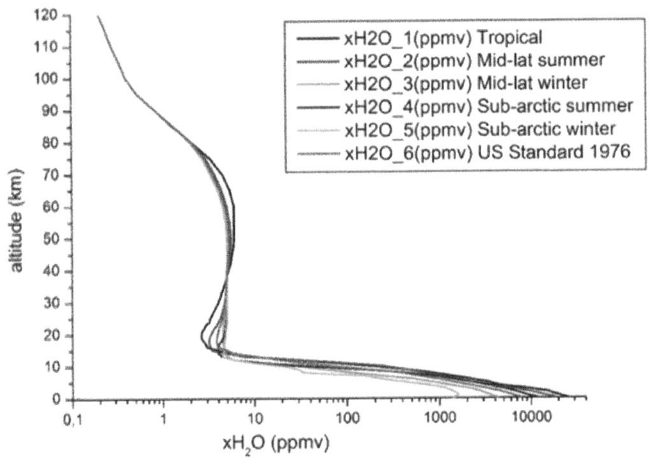

Figure A1.3. *Variation of the H_2O mixing ratio as a function of altitude for the six AFGL profiles. For a color version of this figure, see www.iste.co.uk/phulpin/satellites2.zip*

Compared to the original AFGL profiles (which date back to the 1980s), the CH_4 mixing ratio profiles of Figure A1.4 have been multiplied by a factor to reproduce the increasing trend of the average concentration of methane (a greenhouse gas) at the surface with a value close to 1.86 ppmv in 2016. Again, the AFGL model for CH_4 is a simplified version of reality: the six profiles are identical above 60 km, and the U.S. Standard profile is quite different from the other five. But the vertical distribution of methane is realistic: a constant vertical profile of CH_4, a species quite well mixed in the troposphere (the vertical resolution of the AFGL 1D model is

1 km in this altitude domain); above the tropopause, the CH_4 destruction processes by radical OH, by $O(^1D)$ and by direct photodissociation rapidly decrease the mixing ratio of CH_4.

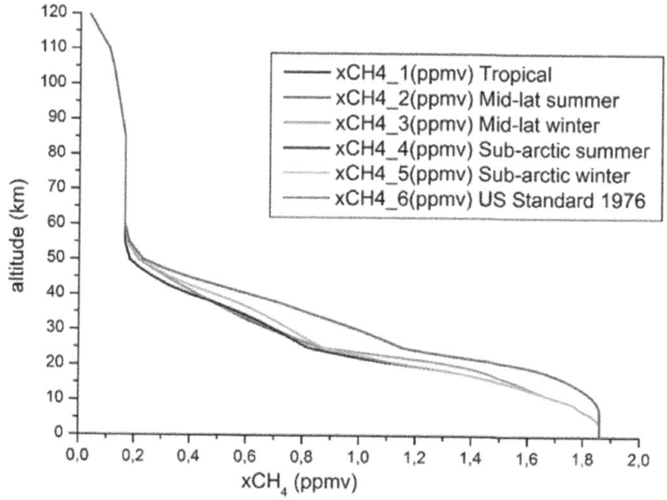

Figure A1.4. *Variation of the CH_4 mixing ratio as a function of altitude for the six AFGL profiles. For a color version of this figure, see www.iste.co.uk/phulpin/satellites2.zip*

For nitrogen dioxide NO_2, a reactive species, the AFGL profile is not provided for each of the i = [1;6] profiles. Implicitly, it corresponds to a standard atmosphere with a very simplified stationary 1D model (for noon). The vertical profile nevertheless shows a very low NO_2 concentration in the unpolluted troposphere (which corresponds here to a background level of 2.3×10^{-5} ppmv or 0.023 ppbv), then a rise above the tropopause is associated with the conversion of the source of nitrogen oxides in the stratosphere, that is, nitrous oxide (N_2O). As a result of its photodissociation or reaction with $O(^1D)$, this source species is transformed into NO_2 and NO (its partner in photochemical equilibrium in the presence of ozone), both species being gathered as $NO_x = NO + NO_2$.

The photodissociation of NO_2 above the mesopause leads to its gradual transformation into NO which becomes the most abundant species during the day in the NO_x partition. Time-dependent 3D photochemical and dynamic models (via the solar zenith angle) are needed to go beyond the very "simplified" AFGL profile for NO_2 shown in Figure A1.5.

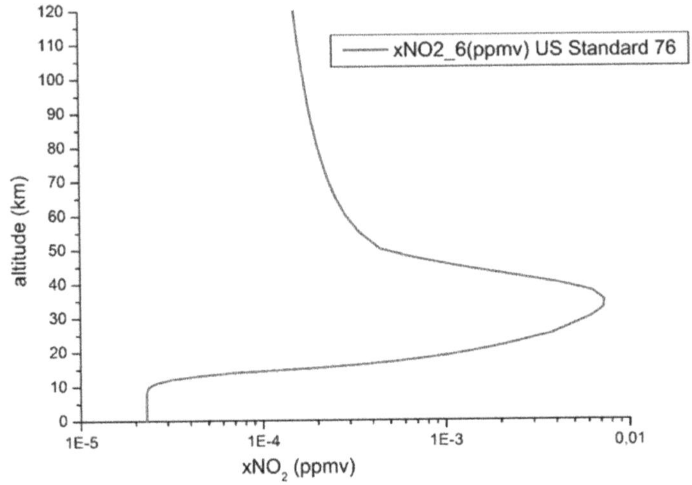

Figure A1.5. *Variation of the NO_2 mixing ratio as a function of altitude for the AFGL profile. For a color version of this figure, see www.iste.co.uk/phulpin/satellites1.zip*

A1.2. Other profile databases

To enlarge the profile base and better adhere to geophysical variability, we can use the TIGR database[1] (Thermodynamic Initial Guess Retrieval), Chevallier et al. (1998), which includes 2,311 profiles and where the most extreme situations are represented.

Other profile databases have been introduced for more specific objectives such as the SAFREE database for simulations of sea surface temperature restitution (François et al. 2002) or the GAPRI database for simulations of surface temperature restitution on land surfaces ((Mattar et al. 2015).

A1.3. References

Anderson, G.P., Clough, S.A., Kneizys, F.X., Chetwynd, J.H., Shettle, E.P. (1986). AFGL atmospheric constituent profiles (0-120 km). AFGL-TR-86-0110. Environmental Research Papers.

1. https://ara.lmd. polytechnique.fr/index.php?page=tigr.

Chevallier, F., Chéruy, F., Scott, N.A., Chédin. A. (1998). A neural network approach for a fast and accurate computation of a longwave radiative budget. *J. Appl. Meteorol.*, 37, 1385–1397.

François, C., Brisson, A., Le Borgne, L., Marsouin, A. (2002). Definition of a radiosounding database for sea surface brightness temperature simulations. Application to sea surface temperature retrieval algorithm determination. *Remote Sens. Environ.*, 81, 309–326. doi:10.1016/S0034-4257(02)00008-1.

Mattar, C., Durán-Alarcón, C., Jiménez-Muñoz, J.C., Santamaría-Artigas, A., Olivera-Guerra, L., Sobrino, J.A. (2015). Global atmospheric profiles from reanalysis information (GAPRI): A new database for earth surface temperature retrieval. *Int. J. Remote Sens.*, doi:10.1080/01431161.2015.1054965.

NOAA/NASA/USAF (1976). U.S. Standard Atmosphere. Technical memorandum, U.S. Government Publishing Office, Washington, DC.

Appendix 2

Spectroscopic Parameters for Radiative Transfer

Claude CAMY-PEYRET
IPSL, Paris, France

This appendix on spectroscopic parameters is intended to supplement information that could not be fully developed in several chapters of the book.

A2.1. Spectroscopic parameters and line profiles

When trying to model in detail the absorption or emission of molecules present in the atmosphere[1], spectroscopic parameters must be available to compute the corresponding spectrum of the molecule in the gas phase. For this reason, these parameters are based on the long-standing work in the field of experimental and theoretical molecular spectroscopy, which has made it possible to produce very useful spectroscopic databases as input to direct radiative transfer calculation models. We will first determine these properties specific to each molecular species which depend on the temperature, the pressure of the medium and the abundance of the absorbing molecule, and depending on the type of molecule concerned, we will use either a line-by-line description of the radiative properties of the molecule, or a description via absorption cross section for molecules whose spectrum is too dense

1. Except in the upper atmosphere where photodissociation phenomena break molecules into their atomic constituents. Here, we will not cover, for example, atomic oxygen lines in the submillimetric domain.

Satellites for Atmospheric Sciences 2,
coordinated by Thierry PHULPIN, Didier RENAUT, Hervé ROQUET and Claude CAMY-PEYRET.

or whose line-by-line modeling proves impracticable (and empirical cross sections, i.e. measured experimentally, will be used in this situation).

A2.1.1. *Line-by-line parameters*

In the terrestrial atmosphere that concerns us here (air for which N_2 and O_2 are the main constituents), in order to describe a molecular transition of species A in absorption " ← ' between a lower level identified by the index ' and an upper level identified by the index ", spectroscopic databases for the atmosphere use the following parameters (the spectral variable used in atmospheric spectroscopy is usually the wavenumber $\tilde{\nu}$ expressed in cm^{-1}):

– the position of the line $\tilde{\nu}_0$ (at zero pressure) in cm^{-1};

– the intensity of the line $S(T_{ref})$ at a reference temperature $T_{ref} = 296$ K that is expressed in $cm^{-1}/(molecule\ cm^{-2})$;

– energy E" of the lower level of transition in cm^{-1};

– the broadening coefficient $\gamma_{A\text{-}air}(T_{ref})$ in cm^{-1} atm^{-1};

– the exponent n_{air} of temperature dependence $\gamma_{A\text{-}air}(T) = \gamma_{A\text{-}air}(T_{ref})\ (T_{ref}/T)^{nair}$;

– the self-broadening coefficient of the line $\gamma_{A\text{-}A}(T) = \gamma_{A\text{-}A}(T_{ref})\ (T_{ref}/T)^{nA}$. This latter coefficient is important to consider for A = H_2O since water vapor is the most abundant minor constituent, at least in the lower atmosphere where collisions H_2O–H_2O play a significant role;

– the shifting coefficient of the line resulting from collisions between the radiatively active molecule and air molecules noted $\delta_{A\text{-}air}(T_{ref})$ in cm^{-1} atm^{-1}. In general, this parameter is not always very well documented, as it has only been measured in the laboratory for a limited number of transitions. Otherwise, we take either a value by default, or a value resulting from theoretical calculations (which have their limitations);

– the rotational and vibrational (and electronic in some cases, such as O_2) quantum numbers of the upper and lower levels of the transition;

– nuclear statistical weights (related to the spin of equivalent nuclei where available) needed to calculate partition functions (see below) or to calculate emission spectra under conditions of non-local thermodynamic equilibrium (non-LTE);

– quality indicators for these different parameters;

– information about the origin of these parameters.

In general, the parameters $\tilde{\nu}_0$ and E" result from theoretical models (vibration–rotation Hamiltonian) and are the most precise considering the numerous works on the analysis of laboratory spectra recorded with high-performance instruments (Fourier transform, laser spectrometry, frequency comb technique).

The line intensity parameter $S(T_{ref})$ is more difficult to determine experimentally with precision (we reach 0.5% in the best cases), and therefore theoretical models (vibration–rotation transition moment operator) are used for this purpose. This is all the more necessary as the lines to be measured are numerous and often superimposed on each other, which makes it quite difficult to determine intensities. For example, the Ozone Atlas of Flaud et al. (1990) (calculated using the theoretical tools mentioned above but based on the best experimental data of the time) contains about 93,000 lines of O_3 from the microwave (0.3 cm^{-1}) to the mid-infrared (3,200 cm^{-1}) region.

A2.1.2. *Absorption cross sections*

When it is not possible to generate them reliably, the parameters of individual lines (in the case of a very dense spectrum) or when models are not available, laboratory spectra are recorded under very precise conditions of purity, temperature and pressure of the mixture (pure gas plus collision partner, in general, N_2 or synthetic dry air is used as a buffer gas in order to approach atmospheric conditions). The intensity transmitted through a cell of length L (in cm) containing the absorbing molecule A diluted in the buffer gas in homogeneous conditions of pressure P (in hPa) and temperature T (in K) is written as (Beer–Lambert law):

$$I(\tilde{\nu}) = I_0(\tilde{\nu}) \exp[- L \times c_A \times \sigma_A(\tilde{\nu}; T, P)] = I_0(\tilde{\nu}) \exp[-\tau(\tilde{\nu})]$$

In this equation, c_A is the concentration of species A (in molecule/cm^3) and the absorption cross section $\sigma_A(\tilde{\nu}; T, P)$ is expressed in cm^2/molecule. We can check the homogeneity of the optical depth $\tau(\tilde{\nu})$, that is, the argument of the exponential since [cm] × [molecule/cm^3] × [cm^2/molecule] is indeed a dimensionless quantity. The experimental determination of $\sigma_A(\tilde{\nu}; T, P)$ must be done (by adapting the resolution, gas concentration and path length conditions) to the spectral region of interest and selecting a set of (T,P) pairs that best cover the conditions encountered in the atmosphere. The corresponding tables (three dimensions) called look-up tables (LUTs) allow for the interpolation according to the spectral variable $\tilde{\nu}$ and the thermodynamic variables T and P of the medium.

For an isolated line under homogeneous conditions of pressure and temperature, we can make the link between the intensity of the line, the profile of the line and the cross section by writing as follows:

$$\tau(\tilde{\nu}; T, P) = L \times c_A \times S(T) \times \varphi_A(\tilde{\nu} - \tilde{\nu}_0; T, P) = L \times c_A \times \sigma_A(\tilde{\nu}; T, P)$$

So that in the vicinity of the isolated line of position $\tilde{\nu}_0$, we have:

$$\sigma_A(\tilde{\nu}; T, P) = S(T) \times \varphi_A(\tilde{\nu} - \tilde{\nu}_0; T, P)$$

The function $\varphi_A(\Delta\tilde{\nu}; T, P)$ (expressed in $1/cm^{-1}$) describes the normalized profile of the line such as:

$$\int_{-\infty}^{\infty} \phi_A(\Delta\tilde{\nu}; T, P) d\Delta\tilde{\nu} = 1$$

Dimensions are checked again since the product of the intensity of the line by the profile is expressed as:

$$[cm^{-1}/(molecule\ cm^{-2})] \times [1/cm^{-1}] = [cm^2/molecule]$$

The line profile depends on the pressure and temperature conditions as described below

A2.1.3. *Line profiles*

In the upper part of the atmosphere where the effect of collisions between the radiatively active molecular species A (which absorbs or emits the radiation) and other molecules is reduced, the profile of lines is determined by the molecular mass m_A and the thermal agitation, that is, the kinetic temperature T of the medium. It is therefore of the Doppler type:

$$\varphi_D(\Delta\tilde{\nu}) = (1/\gamma_D)\,(\ln(2)/\pi)^{1/2} \exp[-\ln(2)\,(\Delta\tilde{\nu}/\gamma_D)^2]$$

In this equation, $\Delta\tilde{\nu} = \tilde{\nu} - \tilde{\nu}_0$ is the deviation from the center of the line and we find $\varphi_D(\pm\gamma_D) = 1/2\ \varphi_D(0)$. The parameter γ_D is the half-width half-maximum Doppler (HWHM) of the line and is expressed as: $\gamma_D = \tilde{\nu}_0\,[2\ln(2)\,k_B\,T/(m_A\,c^2)]^{1/2}$ or $\gamma_D \approx 3.58 \times 10^{-7}\,\tilde{\nu}_0\,(T/M_A)^{1/2}$ when T is expressed in K and the molar mass M_A is in g/mol. The constant k_B is the usual Boltzmann constant.

For the lower layers of the atmosphere, when collisions that contribute to the widening of the lines dominate, it is a Lorentz profile:

$$\varphi_L(\Delta\tilde{\nu}) = (1/\pi)\frac{\gamma_L}{\Delta\tilde{\nu}^2+\gamma_L{}^2}$$

In this equation, $\Delta\tilde{\nu} = \tilde{\nu} - (\tilde{\nu}_0 + \delta_L)$ is the deviation from the center of the line due to the shift δ_L under the effect of collisions and we find $\varphi_L(\pm\gamma_L) = 1/2\ \varphi_L(0)$. The parameter γ_L is the half-width half-maximum Lorentz (HWHM) of the line and is expressed as: $\gamma_L = (P_{air}/P_0)\ \gamma_{A\text{-}air}(T_{ref})\ (T_{ref}/T)^{nair}$ where $P_0 = 1$ atm $=1013.25$ hPa.

In the intermediate pressure domain (assuming the absence of correlation between the effects of widening by collisions and the effect of thermal agitation), we must use the Voigt profile which is written as the convolution product of the two profiles:

$$\varphi_V(\Delta\tilde{\nu}) = \varphi_D(\Delta\tilde{\nu}) \otimes \varphi_L(\Delta\tilde{\nu})$$

The Voigt profile is widely used in radiative transfer codes but for some molecules and under certain pressure conditions we may need (depending on the spectral resolution of satellite instruments) the use of more complicated line profiles that are discussed in Hartmann et al. (2008).

When looking at a particular species that has a dense spectrum of lines identified by the index $k=[1,n_A]$, we can make the link between the line-by-line description and the description in terms of cross section by writing the contribution (of the radiatively active species A) to the optical depth for a homogeneous medium at temperature T and total pressure P (for a concentration c_A of the diluted absorbing gas in the N_2 buffer or in air) in the form of:

$$\tau_A(\tilde{\nu};T,P) = L \times c_A \times \sum_{k=1}^{k=n_A} S_k(T)\phi_k(\tilde{\nu} - \tilde{\nu}_k;T,P) = L \times c_A \times \sigma_A(\tilde{\nu};T,P)$$

$$\text{or } \sigma_A(\tilde{\nu};T,P) = \sum_{k=1}^{k=n_A} S_k(T)\phi_k(\tilde{\nu} - \tilde{\nu}_k;T,P)$$

The parameters $\gamma_{A\text{-}air}(T_{ref})$, n_{air} and $\delta_{A\text{-}air}(T_{ref})$ appear in the expression of the profile of the individual lines ϕ_k depending on whether we consider a Doppler, Lorentz or Voigt line profile.

Apart from the temperature dependence already introduced of the broadening coefficients (often measured experimentally), if we know the assignment of the line and therefore the lower level E" of the transition, the dependence of the line intensity as a function of temperature can be rigorously computed at the Boltzmann equilibrium for a medium at the kinetic temperature T by the formula:

$$S(T)/S(T_{ref}) = [Z(T_{ref})/Z(T)] \exp[-c_2 E'' (1/T-1/T_{ref})] \, rap(T,T_{ref})$$

The Z(T) function is the total partition function of the absorbing molecule. It is sometimes denoted TIPS for total internal partition sum, as can be found in Laraia et al. (2011).

The coefficient $rap(T,T_{ref}) = [1-\exp(-c_2 \tilde{\nu}_0 /T)]/[1-\exp(-c_2 \tilde{\nu}_0 /T_{ref})]\}$ translates the temperature dependence of the so-called induced emission term, and this term becomes very close to the unit when $c_2 \tilde{\nu}_0 >> T$ (i.e. the visible domain). But this term must be taken into account in the far-infrared or submillimeter range when we have $c_2 \tilde{\nu}_0 << T$. The constant c_2 used here (in connection with the use of wavenumber as the spectral variable) is called the "second radiation constant" and is equal to 1.438776877 K/cm^{-1}.

A2.1.4. *Continua and collision-induced spectra*

Numerous laboratory works and experimental measurements under well-documented atmospheric conditions have shown the limits of line modeling with a simple Voigt profile. In the case of water vapor, in particular, it is observed that the Voigt profile does not correctly reproduce the line profile far in the wings. A practical method which makes it possible, on the one hand, to limit the number of points of the profile to be calculated (gain in computer time) and which can use, on the other hand, atmospheric measurements and empirical modeling is to preserve the Voigt profile up to a certain distance from the center of the lines (often ± 25 cm^{-1}) and to attribute the absorption of the lines outside this interval to what is called the continuum (Clough et al. 2005). This continuum (which also exists for species other than H_2O) is modeled as a cross section by an empirical function of temperature, total pressure of the medium (and partial pressure of vapor which is in significant abundance in the lower layers of the troposphere). The latest version of the most used continuum is MT_CKD v3.5[2].

Other effects related to molecular collisions also require models that go beyond the line-by-line framework such as line-mixing effects or collision-induced absorption (CIA) spectra. All of the concepts and many useful details on the effect of collisions on molecular spectra are provided in Hartmann et al. (2005).

2. See: https://github.com/AER-RC/MT_CKD.

A2.1.5. *Line-by-line modeling of optical depth*

Since the atmosphere is a medium whose temperature and pressure conditions vary, it will always be necessary for radiative transfer calculations in this inhomogeneous medium to discretize the problem by carrying out what is called atmospheric layering (layers in which we will assume homogeneous conditions of temperature, pressure and concentrations). We must use an atmospheric layering adapted to the viewing conditions (nadir, limb, occultation, etc.) which will minimize numerical errors related to discretization. This operation is usually performed automatically by radiative transfer codes to find the right compromise between accuracy and calculation time. In any case, for a homogeneous layer (crossed over a length L by the atmospheric path) at temperature T, at pressure P, and containing n_e absorbing species $j = [1,n_e]$ of concentrations c_j having a spectrum of lines $k = [1,n_j]$, we can write (by generalizing to several species the formula given for species A):

$$\tau(\tilde{\nu};T,P) = \mathrm{L} \times \sum_{j=1}^{j=n_e} c_j \sum_{k=1}^{k=n_j} S_{k,j}(T)\phi_{k,j}(\tilde{\nu} - \tilde{\nu}_{k,j};T,P)$$

In this formula, the double sum relates to the species j and to the lines *k* of the species *j*, hence the use of the double index *k,j* for the intensities, positions, and profiles of lines. We see that when we have to consider a large number of absorbing species (generally of the order of a dozen), for each species we must take into account a large number of lines (in general of the order of a hundred) and that, for the spectral domain considered, we must calculate the optical depth for a large number of spectral samples $\tilde{\nu}$ (of the order of a thousand), the calculation of the optical depth for a single layer can be expensive in computing time, especially since the calculation of $\varphi_{k,j}$ profiles in the Voigt case (or for more complicated profiles) is itself the limiting factor. Once the contribution for a homogeneous layer (in which the path is L_i) has been calculated, we must combine the contributions of the different segments $i = [1,n_i]$ to the total atmospheric path by considering the optical depth $\tau_i(\tilde{\nu};T_i,P_i)$ for the values of T_i, P_i and $c_{j,i}$ specific to each homogeneous layer. To be complete and assuming that the diffusion phenomena are neglected (absorption and infrared emission without aerosol), the discretized equation of radiative transfer at local thermodynamic equilibrium for a ray propagating from layer 1 to layer n_j is written as:

$$R_i(\tilde{\nu}) = R_{i-1}(\tilde{\nu}) \exp[-\tau_i(\tilde{\nu};T_i,P_i)] + \{1- \exp[-\tau_i(\tilde{\nu};T_i,P_i)]\}\ B(\tilde{\nu},T_i)$$

In this equation that gives the luminance (or radiance) at the exit of the layer *i*, the first term represents the transmission of radiation from the layer *i* – 1 transmitted after absorption in layer i and the second term represents the proper emission of layer i. For the first layer and depending on the viewing geometry, the term $R_0(\tilde{\nu})$ represents either the surface term $\varepsilon(\tilde{\nu})\ B(\tilde{\nu}, T_{surf})$ for the nadir view of a surface of emissivity $\varepsilon(\tilde{\nu})$ and temperature T_{surf}, or the incident spectrum at the top of the atmosphere in the case of an external source (sun, moon, star).

Once the spectrum at the output of the atmospheric path has been calculated, it will remain to convolve the luminance $R(\tilde{\nu})$ calculated on the so-called "fine resolution" grid by the spectral response function of the instrument (we use the term ILS for instrument line shape or ISRF for spectral response function) which leads us to write the theoretical spectrum at the instrument output in the form:

$$R_{inst}(\tilde{\nu}) = \int_{-\infty}^{\infty} ILS(\tilde{\nu}' - \tilde{\nu}; \tilde{\nu}) R(\tilde{\nu}') d\tilde{\nu}$$

An instrumental function in the form of an integral (or kernel) has been introduced here to reflect the fact that some spectrometers may have a device function that depends on the spectral variable (off-axis detectors in the case of an interferometer or detector matrix in the case of a grating spectrometer).

A2.2. Spectroscopic databases

Several open-access spectroscopic databases (subject to identification) provide line-by-line spectroscopic parameters and absorption cross sections to be used as input to radiative transfer codes. Among the best known (and for the most recent versions of these databases) are HITRAN[3] (Gordon et al. 2016) and GEISA[4] (Jacquinet-Husson et al. 2016). These databases are multi-usage in the sense that they cover the entire electromagnetic spectrum from radio-frequencies to ultraviolet, through microwaves, submillimeter waves, far-infrared, all infrared, near-infrared and visible. Specialized databases are sometimes specific to each radiative transfer code.

A2.3. Atmospheric radiative transfer codes

A very large number of radiative transfer models (RTMs) have been developed for atmospheric applications and are more or less widely used. Dedicated codes have often been developed specifically for Earth observation and atmospheric sounding instruments and satellite missions. Wikipedia provides an interesting list of these

3. See: https://hitran.org.
4. See: https://www.sciencedirect.com/science/article/abs/pii/S0022285216301011?via%3Dihub.

codes in a fairly complete table[5]. Among the widely used codes, we can mention only the following:

– 4A/OP (Scott and Chedin 1981) is often used by CNES for IASI and IASI-NG[6];

– LBLRTM (Clough et al. 2005) is a successor to FASCODE and widely used by the North American community[7];

– RTTOV (Saunders et al. 1999) is widely accepted as an efficient tool by the ECMWF user community[8].

A2.4. References

Clough, S.A., Shephard, M.W., Mlawer, E.J., Delamere, J.S., Iacono, M.J., Cady-Pereira, K., Boukabara, S., Brown, P.D. (2005). Atmospheric radiative transfer modeling: A summary of the AER codes. *J. Quant. Spectrosc. Radiat. Transfer*, 91, 233–244. doi:10.1016/j.jqsrt.2004.05.058.

Flaud, J.-M., Camy-Peyret, C., Rinsland, C.P., Smith, M.A.H., Malathy Devi, V. (1990). *Atlas of Ozone Spectral Parameters from Microwave to Medium Infrared*. Academic Press, Cambridge.

Gordon, I.E., Rothman, L.S., Hill, C., Kochanov, R.V., Tan, Y., Bernath, P.F., Birk, M., Boudon, V., Campargue, A., Chance K.V. et al. (2017). The HITRAN2016 molecular spectroscopic database. *J. Quant. Spectrosc. Radiat. Transfer*, 203, 3–69. doi:/10.1016/j.jqsrt.2017.06.038.

Hartmann, J.-M., Boulet, C., Robert, D. (2008). *Collisional Effects on Molecular Spectra: Laboratory Experiments and Model Consequences for Applications*. Elsevier, Amsterdam.

Jacquinet-Husson, N., Armante, R., Scott, N.A., Chédin, A., Crépeau, L., Boutammine, C., Bouhdaoui, A., Crevoisier, C., Capelle, V., Boonne, C. et al. (2016). The 2015 edition of the GEISA spectroscopic database. *J. Mol. Spectrosc.*, 327(31–72). doi:10.1016/j.jms.2016.06.007.

Laraia, A.L., Gamache, R.R., Lamouroux, J., Gordon, I.E., Rothman, L.S. (2011). Total internal partition sums to support planetary remote sensing. *Icarus*, 215(391–400). doi:10.1016/j.icarus.2011.06.004.

---

5. See: https://en.wikipedia.org/wiki/Atmospheric_radiative_transfer_codes.
6. See: https://ara.lmd.polytechnique.fr.
7. See: http://rtweb.aer.com/lblrtm.html.
8. See: https://nwp-saf.eumetsat.int/site/software/rttov/download.

Saunders, R.W., Matricardi, M., Brunel, P. (1999). An improved fast radiative transfer model for assimilation of satellite radiance observations. *Quarterly J. Royal Meteorological Soc.*, 125, 1407–1425. doi:10.1256/smsqj.55614.

Scott, N.A. and Chedin, A. (1981). A fast line-by-line method for atmospheric absorption computations: The automatized atmospheric absorption atlas. *J. Appl. Meteorol.*, 20, 802–812. doi:10.1175/1520-0450.

Appendix 3

List of Sensors and Satellites Cited in the Book

Here, we list the names of the instruments cited in the book, as well as their platforms (or in some cases associated missions), and the chapters in which they are mentioned. The names of these instruments are generally acronyms that are developed here. The characteristics of these instruments and satellites are given accurately on the WMO OSCAR site[1].

Instrument	Acronym	Satellite	Chapter(s) (1) or (2) denotes volume number
3MI	Multi-viewing Multi-channel Multi-polarization Imager	MetOp-SG-A	**(1)** 6, 9
AATSR	Advanced Along-Track Scanning Radiometer	ENVISAT	**(2)** 3, 4, 6, 9, 10
ABI	Advanced Baseline Imager	GOES-16 to -18	**(1)** 1, 2 **(2)** 4, 6
ACE-FTS	Atmospheric Chemistry Experiment – Fourier Transform Spectrometer	SCISAT	**(2)** 11, 13
AGRI	Advanced Geostationary Radiation Imager	FY-4A and B	**(2)** 4, 6

1. https://space.oscar.wmo.int/.

Satellites for Atmospheric Sciences 2,
coordinated by Thierry PHULPIN, Didier RENAUT, Hervé ROQUET and Claude CAMY-PEYRET.

AHI	Advanced Himawari Imager	Himawari-8 and -9	**(1)** 1 **(2)** 4
AIRS	Atmospheric Infrared Sounder	Aqua	**(1)** 3, 5 **(2)** 1, 4, 7, 9, 13, 14
AMI	Active Microwave Imager	ERS-1 and -2	**(1)** 4, 5 **(2)** 3, 4, 6
AMSR	Advanced Scanning Microwave Radiometer	ADEOS-2	**(2)** 4, 14
AMSR-E	Advanced Scanning Radiometer for EOS	Aqua	**(1)** 3, 9 **(2)** 3, 4, 6, 14
AMSR2	Advanced Scanning Microwave Radiometer – 2	GCOM-W	**(2)** 3, 4, 6
AMSU (-A/-B)	Advanced Microwave Sounding Unit (-A/-B)	NOAA-15 to -19, MetOp-A to -C, Aqua	**(1)** 1, 2, 3, 4, 5, 9 **(2)** 1, 3, 4, 6
APT	Automatic Picture Transmission	ESSA-2, -4, -6, and -8, Nimbus-1 and -2, TIROS-8, NOAA-1	**(1)** 3
ASAR	Advanced Synthetic Aperture Radar	ENVISAT	**(1)** 4 **(2)** 6
ASCAT	Advanced SCATterometer	MetOp-A to C	**(1)** 4, 5 **(2)** 3, 4, 6
ASCENDS	Active Sensing of CO2 Emissions over Nights, Days, and Seasons		**(2)** 10
ASTER	Advanced Spaceborne Thermal Emission and Reflection radiometer	Terra	**(1)** 3
ATLID	ATmospheric LIDar	EarthCARE	**(2)** 8
ATMS	Advanced Technology Microwave Sounder	SNPP, NOAA-20	**(1)** 1, 2 **(2)** 4, 6, 9, 12
ATOVS	Advanced TIROS Operational Vertical Sounder: instrument suite consisting of HIRS and AMSU (-A/-B) or MHS instruments	NOAA-15 to -19, MetOp-A and B	**(1)** 2, 5
ATSR-2	Along-Track Scanning Radiometer – 2	ERS-2	**(1)** 4 **(2)** 8, 9, 10

AVCS	Advanced Vidicon Camera System	ESSA-3, ESSA-5, ESSA-7, ESSAA-9, Nimbus-1 and -2, NOAA-1	**(1)** 3
AVHRR	Advanced Very High Resolution Radiometer	TIROS-N, NOAA-6 to -19, MetOp-A to -C	**(1)** 1, 2, 4, 5, 7, 8 **(2)** 9, 10, 14
BUV	Backscatter Ultraviolet Spectrometer	Nimbus-4	**(1)** 3, 4
CALIOP	Cloud-Aerosol LIdar with Orthogonal Polarization	CALIPSO	**(1)** 9 **(2)** 9, 10, 14
CERES	Clouds and Earth's Radiant Energy System	TRMM, Terra, Aqua, SNPP, NOAA-20	**(1)** 2, 3 **(2)** 14
CLAES			**(2)** 11
CPR	Cloud Profiling Radar for CloudSat	CloudSat	**(2)** 6, 15
CO2M	Copernicus CO2 Monitoring	Sentinel 7	**(2)** 13
CrIS	Cross-track Infrared Sounder	SNPP, NOAA-20	**(1)** 1, 2, 4, 7, 9 **(2)** 1, 4, 7, 9, 10, 14
CZCS	Coastal Zone Color Scanner	Nimbus-7	**(1)** 1, 3
DDMI	Delay Doppler Mapping Instrument	CYGNSS	**(2)** 3, 6
DORIS	Doppler Orbitography and Radiopositioning Integrated by Satellite	Cryosat, Cryosat-2, HY-2A to D, JASON-1 to -3, SARAL, Sentinel-3A and B, SPOT-2 to -5, TOPEX-Poseidon, Sentinel-6A	**(1)** 3, 4
DPR	Dual-frequency Precipitation Radar	GPM Core Observatory	**(2)** 6, 15
EPIC	Earth Polychromatic Imaging Camera	DSCOVR	**(2)** 4, 9, 13, 14
ERB	Earth Radiation Budget	Nimbus-6 and -7	**(1)** 2, 3 **(2)** 14

ERBE	Earth Radiation Budget Experiment	NOAA-9 and -10	**(1)** 2, 3 **(2)** 14
ESMR	Electrically Scanning Microwave Radiometer	Nimbus-5 and -6	**(1)** 1, 3
ETM+	Enhanced Thematic Mapper+	Landsat-7	**(1)** 3
FCI	Flexible Combined Imager	MTG-I	**(1)** 1, 5
FLORIS	FLuORescence Imaging Spectrometer	FLEX	**(2)** 3
GEOCARB			
GEMS	Geostationary Environment Monitoring Spectrometer	GEO-KOMPSAT-2B	**(1)** 4, 6 **(2)** 8
GERB	Geostationary Earth Radiation Budget	Meteosat-8 to -11	**(1)** 4, 5 **(2)** 14
GIIRS	Geostationary Interferometric Infrared Sounder	FY-4A and B	**(1)** 2 **(2)** 1, 4
GLM	Geostationary Lightning Imager	GOES-16 to -18	**(1)** 1, 2
GMI	GPM Microwave Imager	GPM Core Observatory	**(2)** 3, 4, 14, 15
GOME	Global Ozone Monitoring Experiment	ERS-2	**(1)** 4, 9 **(2)** 7, 8, 9, 14
GOME-2	Global Ozone Monitoring Experiment – 2	MetOp-A to -C	**(1)** 4, 5, 9 **(2)** 7, 8, 9, 14
GOMOS	Global Ozone Monitoring by Occultation of Stars	ENVISAT	**(1)** 4 **(2)** 11
GPS/MET	Global Positioning System/Meteorology	Orbview-1	**(2)** 14
GRAS	GNSS Receiver for Atmospheric Sounding	MetOp-A to -C	**(1)** 4, 5
HALOE	HALogen Occultation Experiment	UARS	**(2)** 8, 11
HES			**(1)** 2
HIRS(/2, 3, and 4)	High Resolution Infrared Sounder	Nimbus-6, TIROS-N, NOAA-6 to -19, MetOp-A and -B	**(1)** 1, 2, 3, 4, 5, 9 **(2)** 4, 3, 14
HIS	Hyperspectral Imaging Sensor		**(2)** 4, 14

HRIR	High Resolution Infrared Radiometer	Nimbus-1 to -3	**(1)** 3
HSB	Humidity Sounder for Brazil	Aqua	**(1)** 3
HIRA	Chinese Infrared Sounder	FY-3D-3E	**(2)** 11
HIRDLS	HIgh Resolution Dynamics Limb Sounder	Aura	**(1)** 3
(H)SCAT	SCATterometer	HY-2A, B, and D	**(2)** 3, 4, 14
IASI	Infrared Atmospheric Sounding Interferometer	MetOp-A to -C	**(1)** 2, 4, 5, 6, 7, 9, 10 **(2)** 1, 4, 7, 8, 9, 10, 11, 13, 14
IASI-NG	Infrared Atmospheric Sounding Interferometer – New Generation	MetOp-SG-A	**(1)** 4, 5, 6, 9 **(2)** 1, 13
ICI	Ice Cloud Imager	MetOp-SG-B	**(2)** 3
IDCS	Image Dissector Camera System	Nimbus-3 and -4	**(1)** 3
IKPA	Russian Advanced Infrared ATmopsheric Sounder		**(2)** 1
IMG	Interferometric Monitor for Greenhouse	ADEOS	**(2)** 8
InSAR	Interferometric SAR		**(1)** 9
IPDA lidar	Integrated Path Differential-Absorption lidar	MERLIN	**(2)** 10
IRIS	Infra-Red Interferometer Spectrometer	Nimbus-3 and -4	**(1)** 1, 3 **(2)** 4
IRS	InfraRed Sounder	MTG-S	**(1)** 4, 5, 6, 9 **(2)** 1
ITPR	Infrared Temperature Profile Radiometer	Nimbus-5	**(1)** 3 **(2)** 4
JAMI	Japanese Advanced Meteorological Imager	Himawari-6	**(1)** 1 **(2)** 4
JMR	JASON Microwave Radiometer	JASON-1	**(1)** 3
LI	Lightning Imager	MTG-I	**(1)** 4, 5
LIMS	Limb Infrared Monitor of the Stratosphere	Nimbus-7	**(1)** 3 **(2)** 4
LITE	Lidar In-space Technology Experiment		**(2)** 14

LRA			**(1)** 3
LRIR	Limb Radiance Inversion Radiometer	Nimbus-6	**(1)** 1, 3 **(2)** 4
MERIS	MEdium Resolution Imaging Spectrometer	ENVISAT	**(1)** 4, 9
Merlin		CNES/DLR	**(2)** 10, 13
METimage	METeorological imager	MetOp-SG-A	**(1)** 5
MHS	Microwave Humidity Sounder	NOAA-18 and -19, MetOp-A to -C	**(1)** 2, 4, 5, 7, 9 **(2)** 1, 4, 6, 15
MI	Meteorological Imager	COMS	**(2)** 4
Microcarb	Micro satellite Carbone	CNES	**(2)** 10, 13
MIPAS	Michelson Interferometer for Passive Atmospheric Sounding	ENVISAT	**(1)** 4 **(2)** 11
MIRAS	Microwave Imaging Radiometer using Aperture Synthesis	SMOS	**(2)** 3, 6
MISR	Multi-angle Imaging Spectro-Radiometer	Terra	**(1)** 3 **(2)** 8, 9, 14
MLS	Microwave Limb Sounder	Aura	**(2)** 3, 8, 11
MODIS	MODerate resolution Imaging Spectroradiometer	Terra, Aqua	**(1)** 2, 3, 9 **(2)** 3, 8, 9, 10, 14
MOPITT	Measurements Of Pollution In The Troposphere	Terra	**(1)** 3, 4
MRIR	Medium Resolution Infrared Radiometer	Nimbus-2 and -3, TIROS-2 to -7	**(1)** 3
MSSCC	Multi-color Spin Scan Cloud Camera	ATS-3	**(1)** 1, 2
MSU	Microwave Sounding Unit	TIROS-N, NOAA-6 to -14	**(1)** 1, 2 **(2)** 4, 6
MTVZA-GY	Imaging/Sounding Microwave Radiometer – improved	Meteor-M N1, Meteor-M N2(-1 to -6)	**(2)** 4, 6
MUSE	Monitor of Ultraviolet Solar Energy	Nimbus-3 and -4	**(1)** 3
MVIRI	Meteosat Visible Infrared Imager	Meteosat-1 to -7	**(1)** 1, 4, 5, 6 **(2)** 14

MVISR	Multichannel Visible Infrared Scanning Radiometer	FY-1A to D	**(1)** 1
MWHS (-1/-2)	MicroWave Humidity Sounder (-1/-2)	FY-3A to E	**(2)** 4
MWR	MicroWave Radiometer	ENVISAT	**(1)** 4
MWRI	MicroWave Radiation Imager	FY-3A to D	**(2)** 4, 6
NEMS	Nimbus-E Microwave Sounder	Nimbus-5	**(1)** 3
NSCAT	NASA SCATterometer	ADEOS	**(2)** 3
OCO	Orbiting Carbon Observatory	OCO-2	**(1)** 9 **(2)** 9, 10, 13
OLS	Operational Linescan System	DMSP-F01 to 19	**(1)** 2
OMI	Ozone Monitoring Instrument	Aura	**(1)** 3, 4 **(2)** 7, 8, 9
OMPS-limb	Ozone Mapping and Profiler Suite	SNPP	**(2)** 9, 10, 14
OMPS-nadir	Ozone Mapping and Profiler Suite	SNPP, NOAA-20	**(1)** 1, 2 **(2)** 9, 10
OSCAT	Oceansat SCATterometer	OceanSat-2, ScatSat-1	**(2)** 3, 4, 6, 10
PARASOL	Polarization and Anisotropy of Reflectances at the top of the Atmosphere coupled with an Observation Satellite carrying a Lidar	CALIPSO	**(2)** 8
PMR	Pressure Modulator Radiometer	Nimbus-6	**(1)** 3 **(2)** 4
POLDER	Polarization and Directionality of Earth Reflectance	Adeos-I and Adeos-II, PARASOL	**(1)** 6, 9 **(2)** 8, 14
Poseidon (-2, -3, and -4)	Radar altimeter	TOPEX-Poseidon, JASON-1 to -3, Sentinel-6A and B	**(1)** 3
PR	Precipitation Radar	TRMM	**(2)** 6, 15
RA	Radar Altimeter	ERS-1 and -2	**(1)** 4
RapidScat	Rapid Scatterometer on the ISS	ISS	**(2)** 4

RFSCAT	Rotating Fan-beam SCATterometer	CFOSAT	**(2)** 3, 4
ROSA	Radio Occultation Sounder for Atmosphere	Megha-Tropiques	**(1)** 8
SAM II	Stratospheric Aerosol Measurement – II	Nimbus-7	**(1)** 3
SAGE	Stratospheric Aerosol and Gas Experiment		**(2)** 8, 9, 10
SAGE-II SAGE-II		ERBS	**(2)** 9, 10
SAGE-III/ISS SAGE-III		ISS	**(2)** 9, 10
SAMS	Stratospheric And Mesospheric Sounder	Nimbus-7	**(1)** 3 **(2)** 4
SAPHIR	Atmospheric Sounder of the Intertropical Moisture Profile by Radiometry	Megha-Tropiques	**(1)** 6, 8 **(2)** 4, 6, 9, 14, 15
SAR	Synthetic Aperture Radar	RadarSat-1 and -2	**(2)** 6
SASS	SeaSat-A Scatterometer System	SeaSat	**(2)** 3
SBUV	Solar Backscatter UltraViolet	Nimbus-7, NOAA-9 to -11, NOAA-13 and -14, NOAA-16 to -19	**(1)** 2, 3 **(2)** 8
SCAMS	SCAnning Microwave Spectrometer	Nimbus-6	**(1)** 3 **(2)** 4
ScaRaB	Scanner for Radiation Budget	Meteor-3-6, Resurs-O1-4, Megha-Tropiques	**(1)** 6 **(2)** 14
SCIAMACHY	Scanning Imaging Absorption spectrometer for Atmospheric CHartographY	ENVISAT	**(1)** 4, 5 **(2)** 4, 7, 8, 9, 10, 11, 13
SCMR	Surface Composition Mapping Radiometer	Nimbus-5	**(1)** 3
SCR	Selective Chopper Radiometer	Nimbus-4 and -5	**(1)** 3 **(2)** 4
SeaWIFS	Sea-viewing WIde Field-of-view Sensor	OrbView-2	**(1)** 3

SeaWinds	SeaWinds	ADEOS-2 QuikSCAT	**(1)** 3 **(2)** 4, 6
Sentinel-4	–	MTG-S	**(1)** 1, 4, 5, 6
Sentinel-5	–	MetOp-SG-A	**(1)** 4, 5
SEVIRI	Spin-stabilized Enhanced Visible and InfraRed Imager	Meteosat-8 to -11	**(1)** 1, 4, 5, 9 **(2)** 1, 3, 4, 9
SIFTI	Static Infrared Fourier Transform Interferometer	TRAQ (remained in draft form)	**(1)** 6
SIRAL	SAR Interferometer Radar ALtimeter	CryoSat, CryoSat-2	**(2)** 3
SIRS	Satellite Infra-Red Spectrometer	Nimbus-3 and -4	**(1)** 3 **(2)** 4
SLSTR	Sea and Land Surface Temperature Radiometer	Sentinel-3A and B	**(2)** 3, 8, 9, 10
SMAP	Soil Moisture Active-Passive	SMAP	**(2)** 3, 6
SMMR	Scanning Multichannel Microwave Radiometer	Nimbus-7, SeaSat	**(1)** 1, 3 **(2)** 3, 6
SSCC	Spin Scan Cloud Camera	ATS-1	**(1)** 1, 2
SSM (/T/T-2)	Special Sensor Microwave (Temperature/Humidity)	DMSP-F04, DMSP-F07 to 15	**(2)** 4
SSM/I	Special Sensor Microwave – Imager	DMSP-F08, DMSP-F10 to 15	**(2)** 3, 6, 14
SSMIS	Special Sensor Microwave Imager/Sounder	DMSP-F16 to 19	**(1)** 9 **(2)** 3, 4, 6
SSU	Stratospheric Sounding Unit	TIROS-N, NOAA-6 to -9, NOAA-11 and -14	**(1)** 1, 2 **(2)** 4
S-VISSR	Stretched Visible and Infrared Spin Scan Radiometer	FY-2A to H	**(1)** 1
TANSO-FTS	Thermal and Near Infrared Sensor for Carbon Observation – Fourier Transform Spectrometer	GOSAT 1, 2, 3	**(2)** 13

TEMPO			(1) 4 (2) 8
TES	Tropospheric Emission Spectrometer	Aura	(1) 3, 4 (2) 4, 8, 10, 13
THIR	Temperature-Humidity Infrared Radiometer	Nimbus-4 to -7	(1) 3
TMI	TRMM Microwave Imager	TRMM	(2) 14, 15
TOMS	Total Ozone Mapping Spectrometer	TOMS Earth Probe	(1) 1, 3 (2) 8, 9
TOVS	TIROS Operational Vertical Sounder: instrument suite consisting of HIRS, SSU and MSU	TIROS-N, NOAA-6 to -14	(1) 1, 2 (2) 4, 14
TROPOMI	TROPOspheric Monitoring Instrument	Sentinel-5 Precursor	(1) 4, 9 (2) 7, 9, 10, 13
TRSR	Turbo Rogue Space Receiver	GFO, ICESat, ICESat-2, JASON-1 to -3, Ørsted	(1) 3
UVNS	Ultraviolet Visible Near Infrared Spectrometer	MetOp SG-A 1, 2, 3	(2) 13
VAS	Vertical Atmospheric Sounder	GOES-4 to -7	(1) 1, 2
VHRR	Very High Resolution Radiometer	NOAA-2 to -5	(1) 1
VIIRS	Visible Infrared Imaging Radiometer Suite	SNPP, NOAA-20	(1) 1, 2 (2) 9, 10
VISSR	Visible Infrared Spin-Scan Radiometer	Himawari-1 to -4	(1) 1, 2 (2) 4
VTPR	Vertical Temperature Profiling Radiometer	NOAA-2 to -5	(1) 1 (2) 4
WindSat	WindSat	Coriolis	(2) 3, 6

List of satellites or missions in the field of the atmosphere

Satellite	Acronym	Agency
A-Train	Afternoon train of satellites	NASA
ADEOS	Advanced Earth Observing Satellite	JAXA

ADL-AEOLUS	Satellite mission for atmospheric wind	ESA
ALOS	Advanced Earth observation satellite	JAXA
AQUA	Afternoon train of satellites	NASA
ARKIKA-M1	Satellites to monitor Earth's atmosphere and surface in the Arctic	ROSCoSMOS
ATS (-1, -3, and -6)	Application Technology Satellite	NASA
BIOMASS	Seventh ESA Earth Explorer Mission (P-band radar)	ESA
CALIPSO	Cloud-Aerosol Lidar and Infrared Pathfinder Satellite Observation	NASA/CNES
CBERS	China–Brazil Earth Resources Satellite program	INPE
CFOSAT	Chinese-French Ocean Satellite	CMA/CNES
CLARREO	Climate Absolute Radiance and Refractivity Observatory	NASA
COSMIC	Constellation Observing System for Meteorology, Ionosphere and Climate	NOAA
CryoSat	Cryosphere Satellite	ESA
Earthcare	Earth observation satellite for polar sheet study	ESA
ENVISAT	Environmental Satellite	ESA
ERBS	Earth Radiation Budget Satellite	NASA
ERS	Earth Resources Satellite	ESA
FORMOSAT	Earth observation satellite	Taiwan
FORUM	Far-infrared Outgoing Radiation Understanding and Monitoring mission	ESA
FY Feng-Yun	CMA satellites	CMA
GCOM-W	Global Change Observation	JAXA
GeoCARB	Geostationary CARBon cycle observatory	NASA
GHGsat	GreenHouse Gas satellite	
GOES	Geostationary Operational Environmental Satellites	NOAA
GOSAT	Greenhouse gases Observing SATellite	JAXA
GOES	Geostationary Orbit Environmental Satellite-(East/West)	NOAA
GPM	Global Precipitation Mission	NASA
GRACE	Gravity Recovery And Climate Experiment	NASA/DLR

HY-XX Hai Yang XX	Chinese ocean satellite missions	CNSA
ICESat	Ice, Clouds and land Elevation Satellite	NASA
ISS	International Space Station	
JPSS	Joint Polar Satellite System	NOAA/NASA/DOD
LANDSAT	Earth observation satellites for high resolution imagery	NASA
MEGHA-TROPIQUES	Low inclination orbiting satellite to study tropical atmosphere	CNES/ISRO
MERLIN	Methane Remote sensing lidar mission	CNES/DLR
METEOSAT	Geostationary satellites for operational meteorology	EUMETSAT
METOP	Polar orbiting satellites for operational meteorology	EUMETSAT
METOP-SG	Polar orbiting satellites for operational meteorology (Second Generation)	EUMETSAT
Microcarb	Micro satellite for carbon dioxide monitoring	CNES
MSG	Meteosat Second Generation	EUMETSAT
MTG	Meteosat Third Generation	EUMETSAT
MTP	Meteosat Transition Programme	EUMETSAT
MTSAT	Multi-functional Transport Satellite	JMA
NIMBUS		NASA
NOAA-1 to -20		NOAA
NPOESS	National Polar-orbiting Operational Environmental Satellite System	NOAA
NPP	National Polar-orbiting Partnership	NOAA
OCO	Orbiting Carbon Observatory	NASA
PLEIADES	PLE	CNES
POES	Polar Orbiting Environmental Satellite	NOAA
QSAT	Water vapor ("q") from Satellite	
ResursSat		ISRO
S5P	Sentinel 5 Precursor	ESA
S3	Sentinel 3	ESA
S4	Sentinel 4	ESA
S5	Sentinel 5	ESA
SCARBO	Space Carbon Observatory	
SPOT		CNES

Suomi-NPP		NOAA
TERRA		NASA
TIROS	Television InfraRed Observation Satellite	NOAA
TRMM	Tropical Rainfall Measuring Mission	NASA
UARS	Upper Atmosphere Research Satellite	NASA

Appendix 4

Microwave Sensors and Infrared Hyperspectral Sounders

The following tables present a non-exhaustive list of past, current or future instruments[1] in the microwave domain or hyperspectral sounders in the infrared domain.

A4.1. Non-exhaustive list of passive microwave instruments

The type 1 instrument is for across track scanning instruments and type 2 is for conical scanning instruments.

Acronym	Instrument	Agency	Type	Satellite	Use of data	
AMR	Advanced Microwave Radiometer	NASA	1	JASON-2 JASON-3	2008	2021
AMSU-A	Advanced Microwave Sounding Unit – A	NOAA	1	Aqua MetOp-A–19	1998	2027
AMSU-B	Advanced Microwave Sounding Unit – B	NASA	1	NOAA-15–17	1998	2021

1. An overview of satellites and instruments is available at: https://space.oscar. wmo.int/.

Satellites for Atmospheric Sciences 2,
coordinated by Thierry PHULPIN, Didier RENAUT, Hervé ROQUET and Claude CAMY-PEYRET.

Aquarius	Aquarius	NASA	1	SAC-D	2011	2015
ATMS	Advanced Technology Microwave Sounder	NASA	1	JPSS-2–4 NOAA-20 SNPP	2011	2039
HSB	Humidity Sounder for Brazil	INPE	1	Aqua	2002	2023
JMR	JASON Microwave Radiometer	NASA	1	JASON-1	2001	2013
MHS	Microwave Humidity Sounding	EUMETSAT	1	MetOp-A–C NOAA-18–19	2005	2027
MIRAS	Microwave Imaging Radiometer using Aperture Synthesis	ESA	1	SMOS	2009	2021
MSR	Microwave Scanning Radiometer	JAXA	1	MOS-1 MOS-1B	1987	1996
MSU	Microwave Sounding Unit	NOAA	1	NOAA-6–14	1978	2007
MW radiometer	MicroWave radiometer	NASA	1	SWOT	2022	2026
MWHS-1	MicroWave Humidity Sounder – 1	CMA	1	FY-3A FY-3B	2008	2021
MWHS-2	MicroWave Humidity Sounder – 2	CMA	1	FY-3C–I	2013	2030
MWR	MicroWave Radiometer	ESA	1	Sentinel-3A–3D	2016	2032
MWR	MicroWave Radiometer	CONAE	1	SAC-D	2011	2015
MWTS-1	MicroWave Temperature Sounder – 1	CMA	1	FY-3A FY-3B	2008	2021
MWTS-2	MicroWave Temperature Sounder – 2	CMA	1	FY-3C FY-3D	2013	2022

SAPHIR	*Sondeur Atmospherique du Profil d'Humidite Intertropicale par Radiometrie*	CNES	1	Megha-Tropiques	2011	2021
AMSR2	Advanced Microwave Scanning Radiometer – 2	JAXA	2	GCOM-W	2012	2022
AMSR-E	Advanced Microwave Scanning Radiometer for EOS	JAXA	2	Aqua	2002	2023
GMI	GPM Microwave Imager	NASA	2	GPM Core Observatory	2014	2026

A4.2. Non-exhaustive list of active microwave instruments

Acronym	Instrument	Agency	Type	Satellite	Use of data	
AMI-SCAT	Active Microwave Instrument – SCATterometer	ESA	Radar scatterometer	ERS-1 ERS-2	1991	2011
ASCAT	Advanced SCATterometer	ESA	Radar scatterometer	MetOp-A MetOp-B MetOp-C	2006	2027
NSCAT	NASA SCATterometer	NASA	Radar scatterometer	ADEOS	1996	1997
SCA	SCATterometer	ESA	Radar scatterometer	MetOp-SG-B1 MetOp-SG-B2 MetOp-SG-B3	2025	2045
OSCAT	OceanSat SCATterometer	ISRO	Radar scatterometer	OceanSat-2 OceanSat-3 OceanSat-3A ScatSat-1	2009	2028
SeaWinds	SeaWinds	NASA	Radar scatterometer	ADEOS-2 QuikSCAT	1999	2009
WindRAD	Wind RADar	CMA	Radar scatterometer	FY-3E FY-3I	2021	2030

ALT	Radar ALTimeter	NASA	Radar altimeter	SeaSat	1978	1978
ALT	Radar ALTimeter	NSOAS	Radar altimeter	HY-2C HY-2F HY-2G	2020	2030
AltiKa	Ka-band Altimeter	CNES	Radar altimeter	SARAL	2013	2021
Altimeter	Altimeter	NASA	Radar altimeter	SWOT	2022	2026
KaRIN	Ka-band Radar INterferometer	NASA	Radar altimeter	SWOT	2022	2026
SIRAL	SAR Interferometer Radar ALtimeter	ESA	Radar altimeter	CryoSat Cryosat-2	2005	2024
SRAL	Synthetic aperture Radar ALtimeter	ESA	Radar altimeter	Sentinel-3A Sentinel-3B Sentinel-3C Sentinel-3D	2016	2032
SWIM	Surface Waves Investigation and Monitoring instrument	CNES	Radar altimeter	CFOSAT CFOSAT follow-on	2018	2027
Poseidon-4	Poseidon-4	ESA	Radar altimeter	Sentinel-6A Sentinel-6B	2020	2032
CPR	Cloud Profiling Radar for Earth-CARE	JAXA	Cloud and precipitation radar	EarthCARE	2023	2026
DPR	Dual-frequency Precipitation Radar	JAXA	Cloud and precipitation radar	GPM Core Observatory	2014	2026
PR	Precipitation Radar	JAXA	Cloud and precipitation radar	TRMM	1997	2015
Rainradar	Rainradar	CMA	Cloud and precipitation radar	FY-3G FY-3J	2023	2030
RainCube	Radar In a CubeSat	NASA	Cloud and precipitation radar	ISS RainCube	2018	2020

A4.3. Non-exhaustive list of hyperspectral sounders

Name	Satellite	Agency	Pixel size in nadir	Spectral range in cm^{-1}	Spectral resolution in cm^{-1}	Radiometric noise NeDT @ 280 K
AIRS	Aqua	NASA	14 km	650–1,136 1,216–1,613 2,170–2,674	0.5 1.0 2.0	0.35 K 0.20 K 0.14 K
IASI	MetOp-A MetOp-B MetOp-C	CNES EUMETSAT	12 km	645–2,670	0.25 unmodified	0.2–0.5 (2) K
CrIS	SNPP NOAA-20 JPSS-2 JPSS-3 JPSS-4	NASA	14 km	650–1,095 1,210–1,750 2,155–2,550	0.625 1.25 2.5	0.24–0.39 K 0.19–0.28 K 0.18–0.21 K
IKFS	Meteor-M N2 Meteor-M N2-2 Meteor-M N2-3,4,5,6	Roscosmos	35 km	665–2,000	0.5	0.5 K
HIRAS	FY-3E FY-3F FY-3H FY-3I	CMA	16 km	650–1,136 1,210–1,750 2,155–2,550	0.625 1.25 2.5	0.2–0.8 K @ 250K 0.2–0.3 K @ 250 K 0.3–0.5 K @ 250 K
GIIRS	FY-4A FY-4B FY-4C, D, E, F, G	CMA	16 km 12 km	680–1,130 1,650–2,250	0.625 0.625	0.5 mW/m2/sr/cm-1 0.1 $mW/m^2/sr/cm^{-1}$
IRS	MTG-S-1 MTG-S-2	EUMETSAT	4 km	700–1,210 1,600–2,175	0.625	
IASI-NG	MetOp-SG-A1 MetOp-SG-A2 MetOp-SG-A3	CNES EUMETSAT	12 km	645–2,760	0.125	0.13–0.65 (1.43) K

Glossary

This glossary presents the main categories of instruments on board satellites and used in meteorology and climate science. The definitions are taken from the glossary established by Didier Renaut and published in *La Météorologie* in May 2017[1]. This glossary also offers definitions for some generic terms used to characterize satellite measurements.

Active sensor: a space instrument that irradiates a scene with a man-made electromagnetic radiation and measures part of the radiation reflected, backscattered or re-emitted by this same scene. Lidars and radars are examples of active sensors. Thanks to the emission of radiation in the form of pulses, the active sensor can provide a range-resolved measurement in the medium probed or an accurate measurement of the altitude of the surface.

Calibration: an operation to convert the magnitude of the output of a measuring instrument (e.g. the level of mercury in a thermometer or the detected backscatter power of a meteorological radar) in the magnitude of the physical input variable (e.g. the temperature or radar reflectivity) actuating that instrument.

Climate data record: a time series of measurements of sufficient length, consistency and continuity to determine climate variability and climate change.

1. https://lameteorologie.fr/issues/2017/97.

Satellites for Atmospheric Sciences 2,
coordinated by Thierry PHULPIN, Didier RENAUT, Hervé ROQUET and Claude CAMY-PEYRET.

Cloud fraction (or nebulosity): the Cloud cover or cloudiness defines the cloud-covered fraction of a geographic area. It is expressed as a percentage of the total area (between 0 and 100) or as a real number (between 0.0 and 1.0).

Column content (of a gas): number per unit area of molecular species or aerosol particles in a given atmospheric path. The column can be slant (SCD = slant column density) or vertical (VCD = vertical column density). A column can be partial (between two atmospheric levels) or total (between the surface and the top of the atmosphere or TOA).

The unit is molecule cm^{-2} or mol m^{-2}. For O_3 and SO_2, columns can also be expressed in Dobson unit (DU). 1 DU is equivalent to $2.68678011 \times 10^{20}$ molecule m^{-2}. The constant comes from the Loschmidt constant $2.68678011 \times 10^{25}$ molecule m^{-3} (exact number in SI), which is the number density of an ideal gas in standard conditions of temperature $T_0 = 273.15$ K and pressure $P_0 = 1013.25$ hPa.

The factor $10^{-5} = 10^{-3} \times 10^{-2}$ corresponds to an atmospheric path of one milli-atmosphere-cm. A standard vertical total column of ozone is 300 DU.

Constellation: in the spatial context, a constellation is a set of artificial satellites performing identical functions and distributed in such a way as to ensure their common mission at all times. A constellation can be made up of numerous orbiting satellites in low orbit or in medium orbit to carry out, for example, telecommunications, satellite navigation missions or observation. Observation of the atmosphere for meteorological purposes relies, among other things, on a constellation of operational geostationary satellites and on a constellation of polar-orbiting Sun-synchronous satellites.

Convective system: the cloudy zone of the atmosphere in which convective movements take place bringing warm and moist air from the surface to the top of the atmosphere. A mesoscale convective system (MCS) is a set of thunderstorms gathering together into clusters or lines over time.

Data assimilation: a mathematical method for defining the initial conditions of numerical prediction models by achieving an optimal combination (in the statistical sense) between observations and a previous short-range forecast.

Emissivity: the emissivity of the surface of a material is its effectiveness in emitting energy as thermal radiation.

ENSO (El Niño Southern Oscillation): the periodic change in the ocean–atmosphere system in the tropical Pacific that impacts weather patterns worldwide. This happens every 3–7 years (5 years on average) and usually lasts 6–18 months.

Essential climate variable (ECV): a physical, chemical or biological variable or a group of linked variables that contributes critically to the characterization of the Earth's climate.

Evapotranspiration: the sum of water evaporation and transpiration from a surface area (typically vegetated) into the atmosphere.

Footprint: the region of the ground observed (or illuminated if the instrument is active) instantaneously by a space instrument. The footprint is determined by the instantaneous field of view of the receiver for a passive instrument or by the divergence of the emission beam for an active instrument.

Geostationary Earth orbit (GEO): a circular orbit around the Earth characterized by zero orbital inclination (i.e. an orbit in the equatorial plane), and an orbital period equal to the Earth's rotational period. The sub-satellite point of a geostationary orbit always remains the same point on the equator. The geostationary orbit around the Earth is at an altitude of 35,786 km above the Earth's geoid; we commonly speak of satellites at 36,000 km.

GNSS radio occultation: a semi-passive space measurement technique based on the refraction by the atmosphere of radio waves emitted by satellites in high orbit in the GNSS system (Global Navigation Satellite System), which includes the American GPS (Global Positioning System), Russian Glonass and European Galileo, which represents in all more than 70 satellites. The receiver, placed on a low-orbiting satellite, detects and measures the radio waves from a GNSS satellite after refraction by their passage tangentially through the Earth's atmosphere. From the measurement of the refraction characterized by the angle of curvature followed by the wave, vertically resolved information is deduced about the temperature, humidity, and pressure of the atmospheric regions crossed, even for cloudy regions.

Ground track: known as a suborbital track, it is the vertical projection of the satellite's orbit onto the surface of the Earth.

Horizontal/vertical spatial resolution: as the measurements provided by a satellite instrument are generally very indirect, it is necessary to distinguish between the spatial resolution of the instrument and the spatial resolution of the measurement. The spatial (horizontal, vertical) resolution of the instrument is the smallest distance (horizontal distance, vertical distance) separating two neighboring targets that the instrument is able to distinguish. The spatial resolution of the measurement is the smallest distance (horizontal distance, vertical distance) separating two points in space for which the final measurements are calculated. For the same instrument, the spatial resolution of the measurement depends on the parameter to be restored. It can be affected by different spatial averaging operations.

Hydrometeor: any type of water or ice particles that have formed in the atmosphere or at the Earth's surface as a result of condensation or solidification. It is the form in which atmospheric water is condensed, suspended, falling or deposited on other objects: atmospheric hydrometeors are therefore cloud droplets, rain, drizzle, snow and hail. It can also refer to water deposits on the ground (dew or ice).

Hyperspectral: a spatial sensor is said to be hyperspectral if it makes its observations in many very narrow spectral bands. It restores the radiation spectrum (emitted, absorbed, scattered) with excellent spectral resolution.

Imaging radiometer: equipped with a scanning system, the radiometer becomes an imager. If its spatial resolution remains modest, it then covers a very large territory (dimensions of around a thousand kilometers). The scanning can be mechanical (in the case of a scanner), with a sequential reading of the image, or electronic thanks to matrices or strips of detectors (in the case of the push broom scanner), with a simultaneous reading of part of the image. In addition, the instrument has a number of spectral channels.

Imaging spectrometer: equipped with a scanning system, the spectrometer becomes a spectro-imager. High- or medium-spatial-resolution Earth observation instruments with multispectral analysis capabilities belong to this category, such as the Tropomi on Sentinel-5P or the OLCI instrument on Sentinel-3.

Intercalibration (of satellite instruments): involves systematically comparing the measurements acquired by these instruments with those acquired by reference instruments chosen for their accuracy, stability and durability and, if possible, traceable to the International System (SI) of units. The intercalibration of instruments is necessary to ensure the interoperability of their measurements and to allow for the analysis of long-term series of observations for the study of the climate.

Interoperability: the interoperability of observation systems characterizes their ability to deliver observations that can be combined in a coherent way even though they come from different sources. This implies that the sampling protocols are known and compatible, that the measurements are traceable to common references and that the data is processed and formatted according to common or compatible standards.

Kalman filter: a sequential data assimilation method for estimating an optimal state of the environment under study and its error covariance matrix, as well as their propagation over time using a linear model.

Leaf area index: the ratio between the surface area of a leaf and the surface area of the ground it covers. This index thus expresses the vertical density of the plant cover in m^2/m^2.

LEO, MEO, GEO (satellite categories): satellites in circular or near-circular orbit are classified into three categories according to their altitude, traditionally denoted LEO, MEO and GEO (respectively, for Low, Medium and Geostationary Earth Orbit):

– LEO: between 400 and 1,400 km (nowadays, it is instead between 600 and 900 km), with a period of revolution around 100 minutes.
– MEO: between 20,000 and 23,000 km, for navigation satellites (GPS-type), with a period around 12 hours.

– GEO: the altitude, nearly 36,000 km, is accurately determined so that the period of revolution of the satellite is equal to the period of rotation of the Earth, one sidereal day (a little less than 24 hours).

More than 99% of satellites in circular or near-circular orbit fall into one of these three categories. There are many gaps with regard to satellite altitudes. For example, we will not find a satellite with an altitude between 8,000 and 18,000 km.

Lidar: an active instrument using the emission and backscattering of light pulses emitted by a laser to locate and analyze atmospheric targets or the Earth's surface. There are various types of lidar: simple backscatter lidar, Doppler lidar, differential absorption lidar (DIAL), etc. Operating in the optical range, the lidar does not pass through the clouds once they have become slightly thick, unlike the radar (Chapter 13 of Volume 2).

Lidar altimeter: a lidar instrument (active) used for measuring the altitude of the Earth's surface located vertically below the satellite. A lidar altimeter provides topographic measurements of land masses, oceans and polar ice caps. Like all lidars, this altimeter is blind in the presence of thick clouds.

Low Earth orbit (LEO): a family of orbits around the Earth reaching an altitude of up to 2,000 km. Among these are the **polar orbit**, characterized by an orbital inclination close to 90°, allowing the two geographical poles to fly over at each revolution, and the **Sun-synchronous orbit**, whose altitude and inclination are such that the angle between the orbital plane and the direction of the Sun remains almost constant. A satellite placed in such an orbit passes over a given point on the Earth's surface at the same local solar time. As opposed to geostationary satellites, meteorologists often use the term **non-geostationary satellite** for a satellite in low orbit.

Mesoscale: the mesoscale corresponds to the intermediate spatial scale between the planetary scale (known as synoptic), which is that of lows and anticyclones on a continent (> 2,000 km) and the micro-scale which characterizes individual clouds (< a few km). Its horizontal extension extends between a few tens of kilometers up to a few hundred kilometers.

Metadata: information that characterizes data, for example, time and location of a satellite image.

Mixing ratio (of a gas): concentration of gas divided by the concentration of the air at the same place. It is a unitless number, expressed in parts per thousand, per million (ppm), or per billion (ppb).

Multidecadal Atlantic oscillation: periodic variation of the surface temperature of the North Atlantic with characteristic times of several decades, in relation to disturbances in atmospheric and oceanic circulations at the scale of the North Atlantic basin and beyond.

Numerical weather prediction model: computer-based software set of mathematical equations representing an approximation of the behavior of the atmosphere to obtain a simulation of its future states.

Operational: refers to a system or service that meets high performance and availability criteria to provide data almost continuously to critical applications. Distinguishes itself from R&D missions or services intended for technological or scientific demonstration where the exploitation of data can be exploratory and is not subject to real-time constraints.

Outgoing longwave radiation (OLR): radiation (or amount of radiation: flux) emitted into space in the infrared wavelengths at the top of the atmosphere.

Passive sensor: a space instrument that measures part of the electromagnetic radiation naturally reflected, backscattered, or emitted by a scene (reflected solar radiation, thermal infrared radiation emitted, etc.).

Phase 0: the initial phase of a space project during which the mission is expressed in terms of needs, the environment and its constraints are assessed and possible concepts are identified.

Phase 1: the second phase of a space project where the needs are refined and validated. The different possible concepts are explored, and the critical elements are identified and evaluated in order to estimate the associated risks.

Pixel: the pixel is the smallest homogeneous element of an image. Pixels are usually rectangular, sometimes square. We can distinguish the pixels of the detector matrix (of size fixed by construction) from the pixels of the restored image which may or may not be geo-rectified (taking into account the sphericity of the Earth).

Polarimeter: an instrument for measuring different elements of the Stokes vector of the incident radiation. We talk about spectro-polarimeter when it also does spectral analysis (PARASOL, 3MI).

Radar: an active space instrument using the emission and backscatter of microwave pulses to locate and analyze atmospheric targets or the Earth's surface. We often classify the frequency of radars in several bands (the limits of the bands depend somewhat on the sources).

Band L	400 MHz–1.5 GHz	20–75 cm
Band S	1.5–3.9 GHz	7.7–20 cm
Band C	3.9–6.2 GHz	4.8–7.7 cm
Band X	6.2–10.9 GHz	2.7–4.8 cm
Band K	10.9–36 GHz	0.83–2.7 cm
Band Ku	11–14 GHz	2.1–2.7 cm
Band Ka	20–30 GHz	1.0–1.5 cm
Millimetric	40–300 GHz	1–7.5 mm

Radar altimeter: a radar (active) instrument used primarily for measuring the altitude of the Earth's surface located vertically below the satellite. It provides topographic measurements of land and, above all, very precise measurements (within a few centimeters) of the topography of the oceans.

(Radar) Reflectivity: the amount of transmitted power returned to the radar receiver after hitting hydrometeors, compared to a reference power density at a distance of 1 m from the radar antenna.

Radio interferences: disturbances of the natural electromagnetic signals emitted by the Earth's surface and the atmosphere in the microwave range by anthropogenic emissions resulting from means of telecommunications such as mobile telephony.

Radiometer: a passive space instrument whose predominant function is the measurement of an electromagnetic flux. Its spatial resolution is generally modest (from a few hundred meters to a few tens of kilometers). Observation instruments with high spatial resolution (metric) such as

HRG on Spot-5 or the instrument of the Pléiades satellites are instead called "high-spatial-resolution imagers" or "cameras".

Revisit time (or period): the time interval separating two successive sightings of the instrument at the same point in space. This depends both on the orbit of the satellite and on the scanning and swath of the instrument. The inverse quantity, called the revisit frequency or rate, is also often used.

Roughness length: a horizontal scale characterizing surface heterogeneities. It influences the intensity of turbulent transfers between the surfaces and the atmosphere as well as the surface emissivity in the microwave range.

(Satellite) Data level:

– level-0 products: telemetry as received from the instrument, after assembly with supplemental information needed for further analysis;

– level-1 products: measurements associated with a temporal and geospatial reference system, possibly with some corrections applied and/or calibration from processing of level-0;

– level-2 products: retrievals in geophysical space, relating to the environment, obtained from level-1;

– level-3 products: spatially and/or temporally complete retrieval data, resampled from level-2;

– level-4 products: indicators, products (as fluxes) or analyses, derived from level-3.

(Satellite) Platform: the subsystem of a satellite that provides services, i.e. all the generic functions necessary for its activity in orbit. The other subsystem is the payload which groups the instruments of the satellite.

Scan, field of view: remote sensing or meteorological satellites have instruments on board that observe the ground and most often take "images" in the visible, infrared or microwave domain. They are taken by the *scanning* of the instrument. In the microwave domain, the two main types of scanning are orthogonal scanning and conical scanning. In the infrared (longwave, shortwave, near-infrared) and VIS/UV, the images are acquired either by a pointing mirror to direct the LoS at the proper position on ground or by integrating (in a short time) the collected radiation along the swath using a 2D detector matrix to acquire a line of the image:

– orthogonal scanning: the instrument takes the images with a scan perpendicular to the ground track, generally symmetrically with respect to the central sight, known as the *nadir*. Seen from the satellite, the angle between aiming at nadir and aiming at a point P is called the *half-aperture angle* f. The extreme aiming angle (right or left), the angle of *maximum half-aperture* or f_0. The field of view is equal to $2 f_0$;

– conical scanning: the scan of the instrument describes a cone whose central axis has a greater or lesser inclination with respect to the nadir. The rotational movement of the instrument describes a full revolution or fraction of a revolution. The width of the sweep's ground track, the *swath*, determines the field of view.

Scatterometer: a radar (active) instrument used to measure the backscatter coefficient of surfaces with very high precision. **Wind scatterometers** are the most common; with an appropriate choice of radar frequency and multi-directional (scanning) observation, they can measure wind speed and direction just above oceans with a wide swath.

Sounder: a space-based instrument capable of range-resolved measurements along the line of sight. The term "sounder" will be reserved here for passive instruments, even if active instruments such as lidar or radar are also sounders. A distinction is made between limb sounders that target a source of radiation (sun, star, Earth's atmosphere, etc.) tangentially to the Earth, which provides accurate measurements in the upper atmosphere, and vertical sounders, which target the Earth's surface at nadir and are more sensitive to the middle and lower atmosphere. Sounders are in principle spectrometers with very high spectral resolution in order to be able to analyze the absorption and emission lines of atmospheric constituents. It is from this very fine spectral analysis that the capability for altitude resolved measurements is achieved. Sounders are most often equipped with an "imager" function in order to produce maps.

Spectrometer: a passive space instrument whose predominant function is the measurement and analysis of part of the electromagnetic spectrum. There are several types of spectrometers. The most widely used are grating spectrometers or interferometers of different types: Michelson, Mertz, etc.

Specular reflection (sun glint): the property of a surface to redirect an incident radiation beam into a reflected beam at the same angle on the other side of the normal to the surface (like a mirror).

Sub-satellite point (SSP): the sub-satellite point is, at a given instant, the point on the Earth's surface where the straight line joining the centers of mass of the satellite and the Earth passes. It should be noted that this point generally differs from the vertical of the satellite (nadir), through which the straight line perpendicular to the surface of the Earth passes, passing through the satellite and whose continuation defines the track of the satellite.

Synoptic scale: there is a usual classification of atmospheric phenomena according to their spatio-temporal scale. The synoptic scale characterizes phenomena whose size is about a thousand kilometers and with a duration of a few days. For example, this is the typical scale for monitoring weather disturbances.

Synthetic aperture radar (SAR): an imaging (scanning) radar instrument capable of providing high-resolution images of the Earth's surface, day and night and through clouds. The computer processing of successive radar signals coming from the same target makes it possible to obtain a resolution equivalent to that of a very large "synthetic" antenna.

Time resolution: as with spatial resolution, a distinction must be made between the time resolution of the instrument and the time resolution of the measurement. The time resolution of the instrument is identical to the revisit period. The temporal resolution of the measurement is the smallest time interval separating two successive final measurements for the same point in space. For the same instrument, the temporal resolution of the measurement depends on the parameter to be retrieved. It can be affected by different time-averaging operations.

Uncertainty: the standard deviation of a sufficiently large number of measurements of the same quantity by the same instrument or method. Hence, the random part of the instrument reading or output; it represents the limit of measurement precision. The uncertainty of an instrument is caused by the unpredictable effects upon its performance of such factors as friction, backlash, electronic, photonic or detector noise (Chapter 12 of Volume 2).

Virtual constellation: in the context of the CEOS, a "virtual constellation" is a coordinated set of space-based and/or ground-based facilities contributed by different partners, which focus on the observation of a particular parameter or group of parameters of the Earth System.

List of Authors

Thomas AUGUST
European Organisation for the
Exploitation of Meteorological
Satellites (EUMETSAT)
Darmstadt
Germany

Bill BELL
Copernicus Climate Change Service
European Centre for Medium Range
Weather Forecasts
Shinfield Park
Reading
UK

Régis BORDE
European Organisation for the
Exploitation of Meteorological
Satellites (EUMETSAT)
Darmstadt
Germany

Dominique BOUNIOL
CNRM
Université de Toulouse
Météo-France
CNRS
France

Hélène BROGNIEZ
LATMOS
Université de Versailles
St Quentin-en-Yvelines
Guyancourt
France

Claude CAMY-PEYRET
IPSL
Paris
France

Cyril CREVOISIER
Laboratoire de Météorologie Dynamique
IPSL
Palaiseau
France

Juan CUESTA
LISA – UMR7583
Créteil
France

Gaëlle DUFOUR
LISA – UMR7583
Créteil
France

John EYRE
Met Office
Fitzroy Road
Exeter
UK

Christopher KIDD
Earth System Science
Interdisciplinary Center
University of Maryland
and
NASA Goddard Space Flight Center
Greenbelt
USA

Vincenzo LEVIZZANI
National Research Council of Italy
Institute of Atmospheric Sciences
and Climate (CNRISAC)
Bologna
Italy

Jean-François MAHFOUF
CNRM
Université de Toulouse
Météo-France
CNRS
France

Thibaut MONTMERLE
Météo-France
Operations Department for Prediction
Nowcasting Division
Toulouse
France

Jean PAILLEUX
Météo et Climat
France

Thierry PHULPIN
Retired from CNES and Météo-France
Toulouse
France

Laurence PICON
LMD
Sorbonne Université
Paris
France

Paul POLI
European Organisation for the
Exploitation of Meteorological
Satellites (EUMETSAT)
Darmstadt
Germany

Didier RENAUT
Retired
Saint-Martin-de-Boscherville
France

Hervé ROQUET
Direction de l'Enseignement Supérieur
et de la Recherche de Météo-France
Saint-Mandé
France

Frank ROUX
Laboratoire d'Aerologie
Université de Toulouse
CNRS
France

Sarah SAFIEDDINE
LATMOS/IPSL
Sorbonne Université
UVSQ
CNRS
Paris
France

Jörg SCHULZ
European Organisation for the Exploitation of Meteorological Satellites (EUMETSAT)
Darmstadt
Germany

Pasquale SELLITTO
LISA – UMR7583
Créteil
France

Jean-Noël THÉPAUT
Copernicus Climate Change Service
European Centre for Medium Range Weather Forecasts
Shinfield Park
Reading
UK

Camille VIATTE
LATMOS/IPSL
Sorbonne Université
UVSQ
CNRS
Paris
France

Index

T, U

V, W

Summary of Volume 1

Chapter 5. The Role of EUMETSAT (Europe)
François MONTAGNER

Chapter 6. The Role of the National Center for Space Studies (CNES, France)
Carole DENIEL and Pierre TABARY

Printed and bound by CPI Group (UK) Ltd, Croydon, CR0 4YY

13/09/2026

14961068-0008